# Industrial Automation and Control System Security Principles:

## Protecting the Critical Infrastructure

## Second Edition

# Industrial Automation and Control System Security Principles:
## Protecting the Critical Infrastructure
### Second Edition

By Ronald L. Krutz, PhD, PE

ISA™

**Notice**

The information presented in this publication is for the general education of the reader. Because neither the author nor the publisher has any control over the use of the information by the reader, both the author and the publisher disclaim any and all liability of any kind arising out of such use. The reader is expected to exercise sound professional judgment in using any of the information presented in a particular application.

Additionally, neither the author nor the publisher has investigated or considered the effect of any patents on the ability of the reader to use any of the information in a particular application. The reader is responsible for reviewing any possible patents that may affect any particular use of the information presented.

Any references to commercial products in the work are cited as examples only. Neither the author nor the publisher endorses any referenced commercial product. Any trademarks or tradenames referenced belong to the respective owner of the mark or name. Neither the author nor the publisher makes any representation regarding the availability of any referenced commercial product at any time. The manufacturer's instructions on the use of any commercial product must be followed at all times, even if in conflict with the information in this publication.



Printed in the United States of America.
10 9 8 7 6 5 4 3 2

ISBN: 978-1-941546-82-6



ISA
67 T. W. Alexander Drive
P.O. Box 12277
Research Triangle Park, NC 27709

**Library of Congress Cataloging-in-Publication Data in process**

# Dedication

*My loving family…*

*The greatest blessing a man can have……*

## Acknowledgment

I want to thank my wife, Hilda, for her encouragement and support during the writing of this book.

I also want to thank ISA editors Chloe Tuck, Liegh Elrod, and Susan Colwell for their valuable insights and review of the text.

~ *RLK*

# Contents

# About the Author

## Ronald L. Krutz, PhD, PE, CISSP, ISSEP

Dr. Krutz is Chief Scientist for Security Risk Solutions, Inc. He has more than 30 years of experience in industrial automation and control systems, distributed computing systems, computer architectures, information assurance methodologies, and information security training. He has been a Senior Information Security Consultant at Lockheed Martin, BAE Systems, and REALTECH Systems Corporation, an Associate Director of the Carnegie Mellon Research Institute (CMRI), and a faculty member in the Carnegie Mellon University Department of Electrical and Computer Engineering. Dr. Krutz founded the CMRI Cyber Security Center and was founder and director of the CMRI Computer, Automation and Robotics Group. He was also a lead instructor for $(ISC)^2$ Inc. in its Certified Information Systems Security Professionals (CISSP) training seminars

He coauthored the CISSP Prep Guide for John Wiley and Sons and is coauthor of the *Wiley Advanced CISSP Prep Guide*; the *CISSP Prep Guide, Gold Edition*; the *Security + Certification Guide*; the *CISM Prep Guide*; the *CISSP Prep Guide, Second Edition: Mastering CISSP and ISSEP (Information Systems Security Engineering Professional)*; the *Network Security Bible*; the *CISSP and CAP (Certification and Accreditation Professional) Prep Guide, Platinum Edition: Mastering CISSP and CAP*; the *Certified Ethical Hacker (CEH) Prep Guide*; *Cloud Computing Security*; and *Web Commerce Security*. He is also the author of *Securing SCADA Systems* and of three textbooks in the areas of microcomputer system design, computer interfacing, and computer architecture. Dr. Krutz has seven patents in the area of digital systems and has published more than 30 technical papers.

Dr. Krutz is also a Senior Fellow of the International Cyber Center of George Mason University.

Dr. Krutz holds BS, MS, and PhD degrees in Electrical and Computer Engineering, is a Registered Professional Engineer in Pennsylvania, and is a Senior Life Member of the IEEE.

# Foreword

I have known Ron (Dr. Krutz) for quite some time, and he has, for many years, done a great job of teaching people about information protection. His background in electrical engineering is particularly important for the subject matter because cybersecurity is not just about bits and bytes. It is about the interaction of digital content and mechanisms with the physical world. So let's talk about that for a minute.

All of the standards, policies, frameworks, and everything else are meaningless without the fundamental understanding that industrial automation and control systems are about automating processes that were once, or in some cases could never be, done by people. The most advanced technologies of this sort build each resulting thing or control each associated system as a unique activity never to be done again and never done before. Take for example 3D printing of custom parts for prototypes. Each iteration of the resulting device is unique and has to meet various engineering constraints. Once the prototype is done, the same design specs, or very similar ones, have to be manufactured, in some cases at high speed in high volume out of a different material. If there is a flaw, the result could be enormous losses. Integrated circuit fabrication lines are costing billions of dollars to build and errors of a few nanometers could make all of the produced chips fail. A failure in a chemical processing plant could, and in history has, killed thousands of people. Control systems run the machines of war and can fail in truly catastrophic ways. We have rovers on Mars and robots that clean floors in homes. We have quad copters and better shoes, and solar panels, and self-parking cars. And we have our future.

All of these are tied up in, built by, operated by, and dependent on control systems. And depending on the control system and requirements for its proper operation, delays of a few microseconds, location errors of nanometers, less than one degree of temperature change, or any number of other such things can cause not only their failure, but the failure of the things that depend on them being right.

All of this "getting things right" is historically designed, engineered, and implemented with the assumption that "the other guy" is doing the right thing. It uses the assumption that the inputs are what they should be, the outputs are used as they should be, and the thing we are responsible for can do its job in the environment specified. But then come the bad actors, and this assumption is no longer true. Here is a design problem that nobody can really solve. Lift anything of any size and shape 1 inch without breaking it. It is under-specified, so I will always be able to find something too heavy or ill shaped or imbalanced or made of the wrong material that causes it to fail. In some sense, that is the cybersecurity problem.

The obvious solution to this problem is to simply provide additional specifications. Go ahead... I'll wait... forever! In truth, cybersecurity is not a science and there is not adequate scientific understanding behind it to even create a theory of what a complete specification would be. We have lists of things that can go wrong and ways to address them, but no such list can be shown to be comprehensive, and any attempt to create comprehensive coverage either results in so generic a model that it has almost no practical use or an even longer list with an idea of what the coverage is. As evidence of this, at a conference of security educators some years back, we were each to bring a sample homework assignment for a graduate class. When I told everyone mine, they all said it was way too hard and that nobody could ever solve it. My assignment was to securely add two integers, always producing the correct result.

My response to their complaint was, if we cannot add two integers together correctly, how can we reasonably hope to do anything else in any of their assignments? This was nothing novel to them. They knew they were trying to roll water up a hill to the moon with their hands. But making it so stark was problematic because it revealed the fundamental challenge we face. This was in the 1990s. It is now 20 years later and the challenge is harder because more complex things that interact more and in more complex ways are tied together with higher and higher consequences and more motivated and capable attackers.

Ah! I may have failed to mention that. All of the normal design assumptions are based on benevolence by all actors. Random things happen, of course, and that is why it takes all of this design to deal with the realities of the environment. But the safety systems in a plant will certainly fail if a worker or designer or implementer or inspector or all of them together collude to disable or bypass the safety feature. The resulting injury would constitute a cybersecurity failure. The chief information security officer might get fired for failing to address the problem of collusion between eight different workers to cause a plant failure if the consequences and publicity were extreme enough. But understand clearly—the budget for all of this security is viewed as an undesired, after-the-fact, add-on to the operation of the company and directly takes away from the bottom line. So the CISO who fails to address this need likely has no budget for the proposed improvements that would have mitigated it to a limited extent.

I would write my own book on this line, but I think my point is made. Ron has taken on the Herculean task of trying to explain all of this in a single book. And here it is…

Fred Cohen

# Preface

## Industrial Automation and Control System Security: A Component of a Nation's Critical Infrastructure

As defined by ANSI/ISA-62443-1-1 (99.01.01)-2007, industrial automation and control systems (IACSs) include (but are not limited to) distributed control systems (DCSs), programmable logic controllers (PLCs), remote terminal units (RTUs), intelligent electronic devices, supervisory control and data acquisition (SCADA) systems, networked electronic sensing and control, and monitoring and diagnostic systems.

A SCADA system provides the ability to obtain information from remote installations and to send limited control commands to those installations. Industrial control systems (including DCSs, PLCs, and intelligent electronic devices) comprise real-time elements that control critical industrial processes in a wide variety of applications.

Before the advent of local area networking, computer-based industrial automation and control systems were generally isolated from the outside world and used their own proprietary communication protocols. Eventually, as networking technology improved, interconnectivity among plants and other corporate units emerged as a way of obtaining increased knowledge of plant operations and more efficient management of resources.

With the maturation of the Internet and browsers, the TCP/IP protocol and Ethernet LANs found their way into SCADA systems as well as process and manufacturing plant control systems.

In addition, computing platforms, such as PCs running Windows, were adopted for reasons of lower cost and standardization. However, with these advantages came the disadvantages of vulnerabilities and exposure to threats that plague these platforms.

There is also an emerging trend in many organizations toward consolidating some overlapping activities in IACSs and corporate IT systems. This trend is motivated by the cost savings achievable by avoiding the use of disparate platforms, networks, software, and maintenance tools and by an increased capability to run the total organization more efficiently and effectively.

An important issue associated with the merging of these two systems is that, in many cases, both IACSs and corporate IT environments use the same security model. This overlap introduces the possibility of the corporate Internet connection exposing critical operations to additional threats and compromising the real-time, deterministic requirements of plant control systems. The emergence of the Stuxnet worm, aimed specifically at PLCs that transmit and receive real-time control bits, highlights the sophisticated threats that exist today and the critical need for IACS-optimized system security methods. Follow-up malware, such as Havex, Flame, or Flamer that have appeared, portend a trend of future attacks on these critical systems.

This book develops a novel approach to securing industrial automation and control systems by generating applicable, useful, protection principles through the merging and adaptation of the best industrial and governmental standards and practices.

# 1

# Industrial Automation and Control System Fundamental Concepts

*The material in this chapter provides basic coverage of industrial automation and control system components and terminology, including supervisory control and data acquisition (SCADA) systems and distributed control systems (DCSs). However, because the focus of this book is the security of these systems, this chapter is not designed to be a comprehensive tutorial on industrial automation and control systems. The material assumes that the reader is familiar with these systems and communications terminology.*

*Note that the term DCS is also used to refer to digital control systems, but, unless otherwise stated, in this book DCS will denote a distributed control system.*

## Industrial Automation and Control Systems

Over the years, as industrial automation and control systems have evolved into DCSs and SCADA systems, the associated terminology has meant different things to different people. Control engineers, software engineers, plant personnel, and management have attached different meanings to commonly used terms, such as DCSs, industrial control systems (ICSs), supervisory control systems, and SCADA systems.

This situation has been further complicated by the migration from relay logic, to programmable logic controllers (PLCs), to microcomputers, to the use of local area networks (LANS), Windows platforms, standard buses, and so on.

To establish a solid foundation for the material in this book, the definitions from ANSI/ISA-62443-1-1 (99.01.01)-2007 and NIST Special Publication (SP) 800-82[2] will be used.

ANSI/ISA-62443-1-1 (99.01.01)-2007 defines ICSs, DCSs, PLCs, and SCADA systems as belonging to the class of industrial automation and control systems (IACSs). Specifically, ANSI/ISA-62443-1-1 (99.01.01)-2007 states that an industrial automation and control system is:

> A collection of personnel, hardware, and software that can affect or influence the safe, secure, and reliable operation of an industrial process. These systems include, but are not limited to:
>
> a. Industrial control systems, including distributed control systems (DCSs), programmable logic controllers (PLCs), remote terminal units (RTUs), intelligent electronic devices, supervisory control and data acquisition (SCADA), networked electronic sensing and control, and monitoring and diagnostic systems. (In this context, process control systems include basic process control system and safety instrumented system [SIS] functions, whether they are physically separate or integrated.)
>
> b. Associated information systems, such as advanced or multivariable control, online optimizers, dedicated equipment monitors, graphical interfaces, process historians, manufacturing execution systems, and plant information management systems.
>
> c. Associated internal, human, network, or machine interfaces used to provide control, safety, and manufacturing operations functionality to continuous, batch, discrete, and other processes.

NIST SP 800-82 defines an industrial automation and control system as "a general term that encompasses several types of control systems, including supervisory control and data acquisition (SCADA) systems, distributed control systems (DCSs), and other control system configurations, such as skid-mounted programmable logic controllers (PLCs) often found in the industrial sectors and critical infrastructures."

In this text, the ANSI/ISA-62443-1-1 (99.01.01)-2007 terminology and definitions will take precedence over others where there are differences.

## SCADA Systems

ANSI/ISA-62443-1-1 (99.01.01)-2007 defines a SCADA system as "a type of loosely coupled distributed monitoring and control system commonly associated with electric power transmission and distribution systems, oil and gas pipelines, and water and sewage systems." NIST SP 800-82 describes SCADA systems as:

> Highly distributed systems used to control geographically dispersed assets, often scattered over thousands of square kilometers, where centralized data acquisition and control are critical to system operation. They are used in distribution systems, such as water distribution and wastewater collection systems, oil and natural gas pipelines, electrical power grids, and railway transportation systems. A SCADA control center performs centralized monitoring and control for field sites over long-distance communications networks, including monitoring alarms and processing status data.

ANSI C37.1[3] defines SCADA as a system operating with coded signals over communication channels to control RTU equipment. The supervisory system may be combined with a data acquisition system by using coded signals over communication channels to acquire information about the status of RTU equipment for display or for recording functions.

A SCADA system comprises both hardware and software, and the classical SCADA system model includes the following components:

- **Human-Machine Interface (HMI)** – A program that provides the operator with an easy-to-read graphical and textual display of the SCADA system elements. The HMI displays warnings and alerts and supports the operator in analyzing system performance, spotting trends, and changing controls. Some commonly used HMI software packages are Wonderware, RSView, and iFIX. An HMI is typically a resident in the master control center as well as on the plant floor.
- **Master Terminal Unit (MTU) or SCADA Server at a Master Control Center** – A SCADA element at the control center that provides two-way data communication and control of field devices, such as RTUs, PLCs, programmable automation controllers (PACs), and intelligent electrical devices (IEDs). Two-way communications between the MTU and field devices are usually low bandwidth and can be implemented

through a variety of technologies, including telephone, VHF/UHF radio, spread-spectrum radio, satellite, and microwave.

- **Remote Terminal Unit (RTU)** – In a typical application, the RTU serves as a data concentrator and is an interface between the MTU and field devices, such as PLCs, PACs, or IEDs. The RTU gathers information from the field devices and stores it until interrogated by the MTU. Conversely, the RTU receives commands for the field devices and passes these on to be executed. There are some PLCs or PACs that incorporate the functions of the RTU by communicating with the MTU and providing remote data acquisition and control of field devices, such as actuators and pumps.
- **Programmable Logic Controller (PLC)** – A PLC is defined as "a solid-state control system that has a user-programmable memory for storing instructions for the purpose of implementing specific functions, such as I/O control, logic, timing, counting, three-mode (PID) control, communication, arithmetic, and data and file processing."[3] A PLC uses a standard instruction set, such as IEC 61131-3, to implement control logic functions. IEC 61131-3 is a vendor-independent international standard for PLC programming languages for industrial automation. It includes standards for the following PLC programming languages:
  - Function Block Diagram (FBD)
  - Instruction List (IL)
  - Ladder Diagram (LD)
  - Sequential Function Charts (SFC)
  - Structured Text (ST)
- **Programmable Automation Controller (PAC)** – A PAC is similar to a PLC, but has additional capabilities, such as more robust communications, higher-speed processors, integration with organizational databases, and a common development environment for integration of software and hardware components.
- **Intelligent Electronic Device (IED)** – IED is a general term for a device that can communicate directly with the MTU or through the RTU and provide direct connections for controlling and polling field equipment, such as actuators.

Classical SCADA systems provide a dependable means of collecting information from multiple RTUs. However, in many current applications, SCADA systems are used in production environments and perform calculations and data analysis in real time on a plant floor, with HMI capabilities. Thus, in many instances SCADA and HMI reside on the same system to perform operations at the equipment level in a real-time, data-driven production environment. The discussions in this book are meant to encompass both classical and current SCADA environments.

The SCADA control center serves as the central location for monitoring and analyzing data acquired from the field devices and for sending control commands to these devices. Because of its criticality, in many instances there is a duplicate control center at a backup site connected to the main site by a wide area network (WAN). Additional important capabilities that reside in the control center are archiving historical data, providing operations information to business managers and accounting, and restoring lost data. Data historians, real-time relational databases, and yield accounting systems usually provide these capabilities.

The data historian acquires and stores data and provides for prompt recovery of that data, if required. It differs from a real-time database in that the historian only archives information and provides no outputs to other system devices. It offers additional capabilities, such as compressing data to store large amounts of data more efficiently and organizing the data into save-sets, which are histories of the system for specific time periods.

A real-time relational database supports interactive storage and retrieval of detailed contextual information when production processes are involved and can interface with business applications, other databases, forms, and XML files.

A yield accounting system interacts with the real-time, relational database, processes plant production data, and generates production accounting information. Typical information provided by a yield accounting system includes:

- Inventory discrepancies
- Material balances in different areas of the plant and for the total plant
- Plant material movement

- Total amount of product made by a particular unit
- Total volume of a finished product shipped to a customer over a specified time interval

For planning and scheduling purposes, condensed versions of production data are transmitted from the yield accounting system to the enterprise resource planning (ERP) system for processing.

SCADA systems can be configured in a wide variety of architectures with various components, depending on the application and other plant and corporate requirements. Figure 1-1 is an example of a SCADA system client-server model that incorporates fundamental SCADA elements. The figure is presented to provide a conceptual view of a SCADA system and is not intended to represent all the various SCADA system architectures that are employed in the field.

### Distributed Control Systems

A DCS includes a supervisory controller running on a control server and a number of distributed controllers. The supervisory controller transmits data set points to the remote controllers and acquires data from them. The distributed controllers control process elements by communicating with them over a fieldbus-type network based on the information received from the supervisor.

ANSI/ISA-62443-1-1 (99.01.01)-2007[5] classifies a distributed control system as:

> A type of control system in which the system elements are dispersed but operated in a coupled manner.
>
> **NOTE** Distributed control systems may have shorter coupling time constants than those typically found in SCADA systems.
>
> **NOTE** Distributed control systems are commonly associated with continuous processes, such as electric power generation; oil and gas refining; and chemical, pharmaceutical, and paper manufacture, as well as discrete processes such as automobile and other goods manufacturing, packaging, and warehousing.

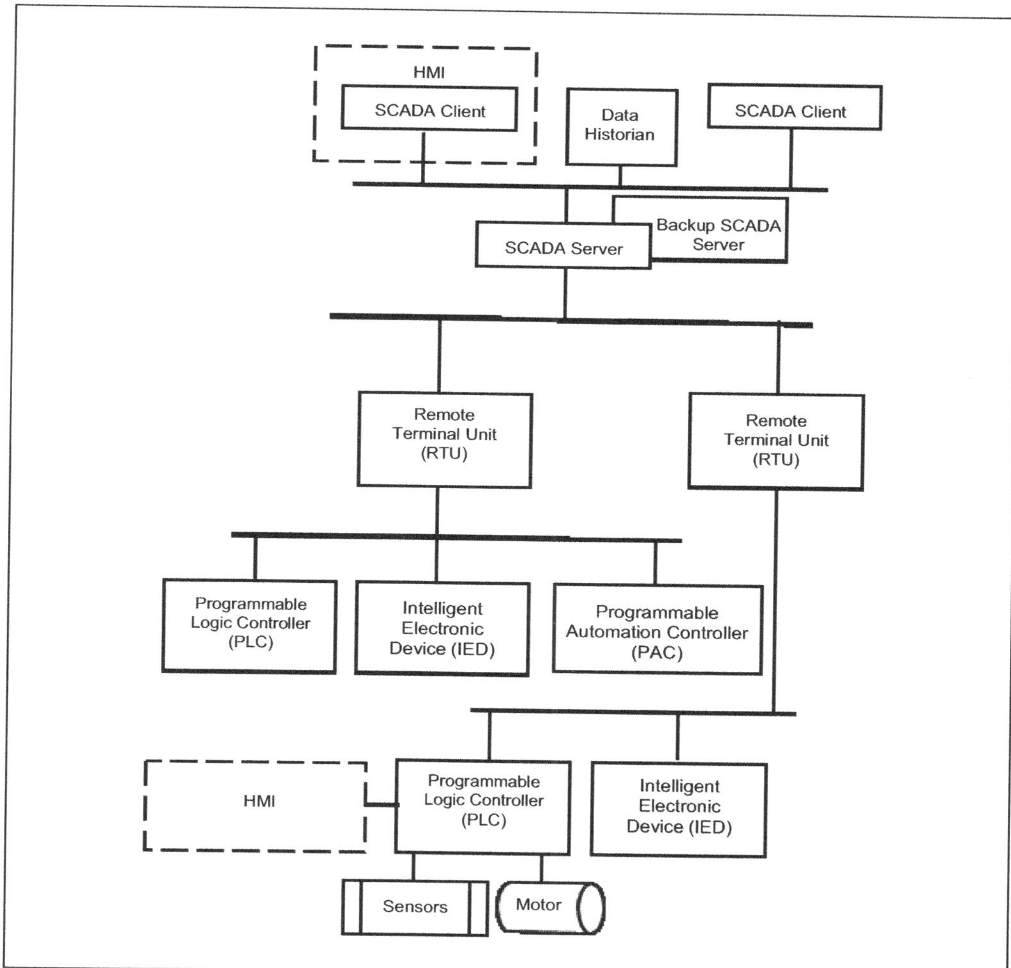


**Figure 1-1. A Client-Server SCADA Architecture Example**

NIST SP 800-82[6] defines a distributed control system as "control achieved by intelligence that is distributed about the process to be controlled, rather than by a centrally located single unit."

In a DCS, subsystems have their own controller elements to manage local processes and to communicate with an operator console for overall supervisory functions. The individual controllers are integrated and communicate through local area networks. Figure 1-2 illustrates a typical DCS. In the figure, the engineering workstation supports distributed system security functions, serves as a development facility, and is used to set alerts and alarm conditions. The operator workstation provides the operator interface and is used by the operator to monitor the DCS, sense alerts and critical situations, and conduct system diagnostics.

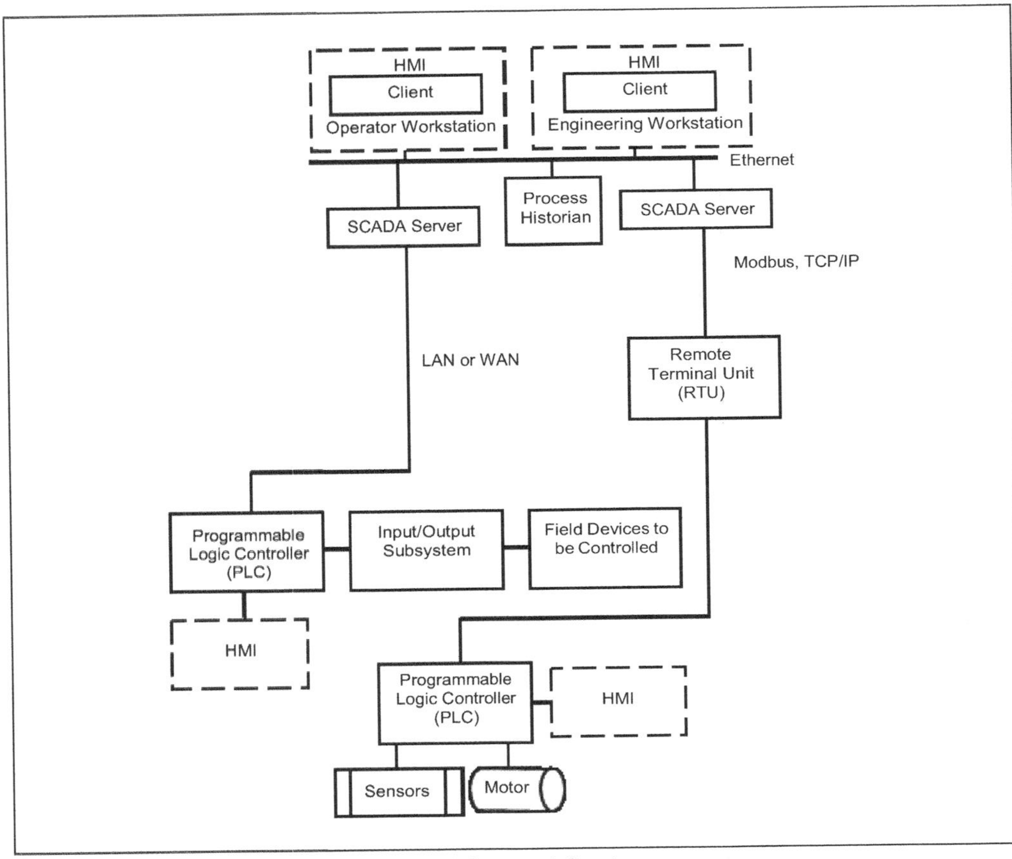


**Figure 1-2. Example of a Distributed Control System**

## Safety Instrumented Systems

Safety instrumented systems (SISs) have been widely used in the process industry to maintain a process in a safe condition during a hazardous situation; for example, if critical set points are exceeded or safe operating conditions are breached. SISs are sometimes referred to as safety shutdown (SSD) systems or emergency shutdown (ESD) systems.

ANSI/ISA-84.00.01-2004, Part 1 (IEC 61511-1 Mod)[7] defines an SIS as an "instrumented system used to implement one or more safety instrumented functions (SIFs). An SIS is composed of any combination of sensor(s), logic solver(s), and final element(s)." The standard also defines an SIF as a "safety function with a specified safety integrity level, which is necessary to achieve functional safety and which can be either a safety instrumented protection function or a safety instrumented control function." Figure 1-3 illustrates the relationship between safety instrumented functions and other process functions.

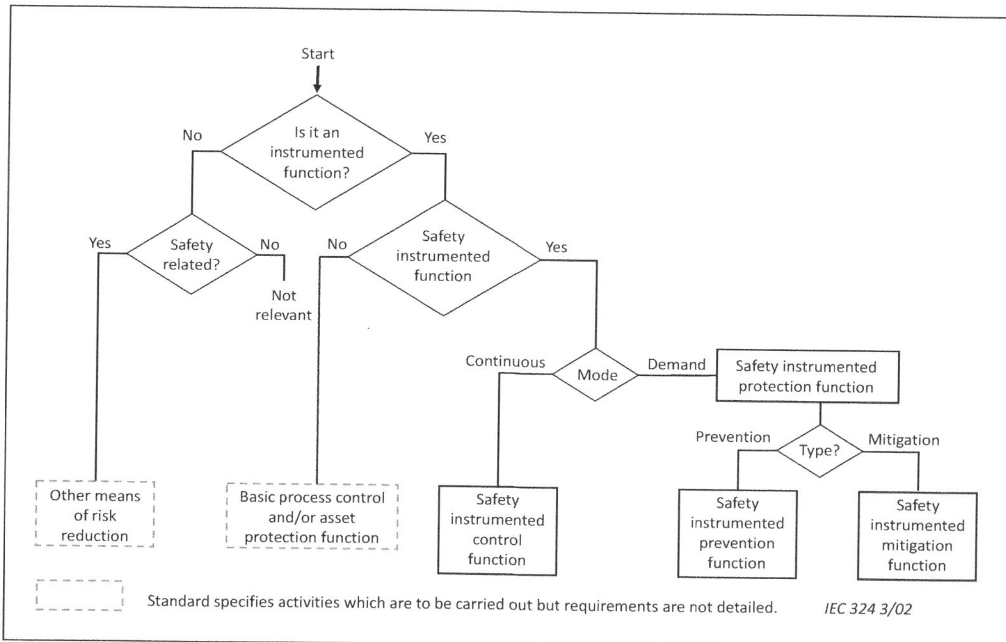


**Figure 1-3. Safety Instrumented Functions versus Other Process Functions**
*Source: ANSI/ISA-84.00.01-2004 Part 1 (IEC 61511-1 Mod)*

An SIS is usually implemented in parallel with conventional control systems to reduce the risks associated with explosive or other dangerous environments. Thus, the proper application of an SIS should be preceded by a formal risk assessment to identify possible threats, vulnerabilities, and likelihoods of occurrence. An excellent guide to risk management is NIST SP 800-39[8] which takes a global, multi-tiered approach to risk management, where system risk is one part of a hierarchy of risk management levels. The document is aimed toward IT applications, but many of the principles are applicable to industrial automation and control systems. The document describes the following three tiers of an organization-wide risk management paradigm:

1. **Tier 1: Organization** – Establishes and implements governance structures consistent with the organizational mission and goals

2. **Tier 2: Mission/Business Processes** – Designs, develops, and implements mission/business processes to support the mission defined in Tier 1

3. **Tier 3: Information Systems** – Integrates risk management activities into the system development life cycle (SDLC) of organizational information systems and addresses the resilience of organizational information systems

NIST SP 800-39 will be discussed in more detail in Chapter 5 of this book. ANSI/ISA-84.00.01-2004 Part 1(IEC 61511-1 Mod) promotes two important concepts associated with an SIS, namely the safety life cycle and safety integrity levels, and defines these terms as follows:

- **Safety Life Cycle** – Necessary activities involved in the implementation of safety instrumented function(s) occurring during a period of time that starts at the concept phase of a project and finishes when all the safety instrumented functions are no longer available for use.

- **Safety Integrity Level (SIL)** – Discrete level (one of four) for specifying the safety integrity requirements of the safety instrumented functions to be allocated to the safety instrumented system. Safety integrity level 4 is the highest level of safety integrity; safety integrity level 1 is the lowest.

Figure 1-4 provides a flow diagram of the SIS life-cycle phases and functional safety assessment steps. The clauses cited in the figure refer to clauses in the ANSI/ISA-84.00.01-2004 document.

## Industrial Automation and Control System Protocol Summary

A protocol defines the rules for entities to use in communicating with each other. In order to better manage communications and compartmentalize the activities required to establish, maintain, use, and close communication links, high-level models are used. In particular, layered architectures are useful in defining the various communication functions in a hierarchical manner. Two widely popular implementations of the layered architecture paradigm are the Open Systems Interconnection (OSI) model and the Transmission Control Protocol/Internet Protocol (TCP/IP) model.

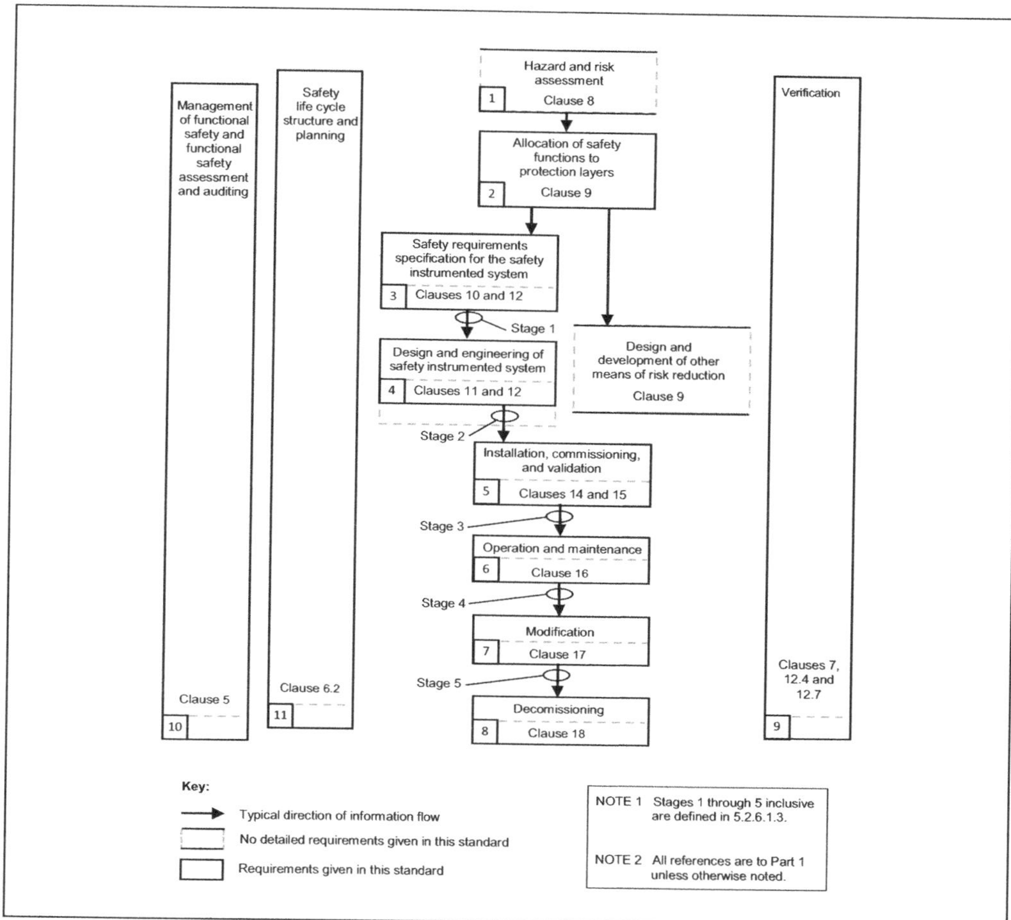


**Figure 1-4. SIS Life Cycle and Functional Safety Elements**
*Source: ANSI/ISA-84.00.01-2004 Part 1 (IEC 61511-1 Mod)*

## The OSI Model

In the early 1980s, the International Organization for Standardization (ISO) developed the OSI reference model. The model functions by encapsulating data and transferring it to an adjacent lower level, where the data is again encapsulated by the lower level, for example by adding a header. Table 1-1 summarizes the OSI model layers and the protocols associated with each layer.

**Table 1-1. OSI Model Layers**

| Layer | Role | Protocols |
|---|---|---|
| 7-Application | Provides for application support and verifies receiver availability. | SMTP (Simple Mail Transport Protocol)<br>FTP (File Transfer Protocol)<br>SNMP (Simple Network Management Protocol) |
| 6-Presentation | Conducts translations and conversions for the application layer, including compression, decryption, and encryption. | MPEG (Motion Pictures Experts Group)<br>HTTP (Hypertext Transfer Protocol)<br>JPEG (Joint Photographic Experts Group) |
| 5-Session | Sets up communication connections, controls the data transmission during a session, and terminates the connection. | NFS (Network File System) which supports file sharing<br>RPC (Remote Procedure Call) |
| 4-Transport | Sets up a logical connection between the transmitter and receiver. The transport layer is connection-oriented and ensures data delivery to the receiver. | TCP (Transmission Control Protocol)<br>UDP (User Datagram Protocol) |
| 3-Network | Performs error checking, addressing, routing, and node traffic control. | IP (Internet Protocol)<br>ICMP (Internet Control Message Protocol) |
| 2-Data Link | Includes the media access control (MAC) sublayer and logical link control sublayer. Performs error-free packet transmission. | PPP (Point-to-Point Protocol)<br>ARP (Address Resolution Protocol) |
| 1-Physical | Converts packets into electrical or optical signals for sending on the transmission media, such as wiring or cables. Defines network electrical and mechanical interfaces to the network transmission media. | EIA-232C (RS-232C)<br>EIA-422-B (RS-422) |

## The TCP/IP Model

In the 1970s, the U.S. Department of Defense developed the TCP/IP, which is the basic protocol of the Internet. It was developed prior to the OSI model and is named from two of its most important protocols: the Transmission Control Protocol (TCP) and the Internet Protocol (IP). The OSI model is a more formalized layered suite of protocols, but it inherited some of its architecture from TCP/IP. Table 1-2 summarizes the four TCP/IP layers and the protocols associated with each layer.

**Table 1-2. TCP/IP Model Layers**

| Layer | Role | Protocols |
|---|---|---|
| 4-Application | Provides the interface to the user. | Telnet<br>FTP<br>SNMP<br>SMTP |
| 3-Host-to-Host or Transport | Provides connection-oriented end-to-end communications, supports error-free packet delivery, and organizes data into packets. | TCP<br>UDP |
| 2-Internet | Performs addressing and routing functions. | IP<br>ARP<br>ICMP<br>PPP |
| 1-Network Access | Combines the operations of the OSI physical and data link layers. | EIA-422-B (RS-422)<br>EIA-232C (RS-232C)<br>IEEE 802.2 Logical Link Control |

With this background of the OSI and TCP/IP models, the salient points of some control system protocols can be reviewed. This summary is not intended to be a comprehensive coverage of all protocols in use but a sampling of some of the popular approaches.

## Object Linking and Embedding for Process Control

NIST 800-82[9] defines OPC (Object Linking and Embedding [OLE] for Process Control) as "a set of open standards developed to promote interoperability between disparate field devices, automation/control, and business systems." The OPC standards specifications are now known as the Data Access Specification. OPC was built on the OLE Component Object Model (COM) and Distributed Component Object Model (DCOM) technologies of Microsoft, which specified a standard set of objects, interfaces, and methods for use in process control and manufacturing automation applications. Due to some vulnerabilities in Microsoft's software, OPC should only be applied between the demilitarized zone (DMZ) and control networks and not used between the corporate network and the DMZ.

The following list summarizes some of the popular OPC specifications:

- **OPC Alarms and Events** – Supports process alarms, messaging, and event notifications on demand
- **OPC Batch** – Provides OPC functionality to batch processes
- **OPC Command** – Provides interfaces that support OPC clients and servers in managing commands sent to devices
- **OPC Data Access** – Provides for the transmission of data from PLCs, IEDs, and DCSs to HMI clients
- **OPC Data eXchange** – Supports interoperability among a variety of vendors and communication and management services among Ethernet fieldbus connections
- **OPC Historical Data Access** – Provides archiving of historical data
- **OPC Security** – Safeguards critical data through access control and integrity protection mechanisms
- **OPC Unified Architecture** – Defines specifications that do not use Microsoft COM in order to enhance cross-platform usage

## OPC Unified Architecture

OPC Unified Architecture (OPC UA) is an OPC Foundation[10] web services technology designed to provide the functions of OPC, but to support interoperability among different platforms and achieve independence from Microsoft COM/DCOM. It can also interface easily with corporate enterprise systems and manufacturing management systems. Some of the features of OPC UA are:

- Integrated security
- Multiplatform compatibility
- Scalable
- Object-oriented
- Ability to identify specific components
- Information modeling, including how data is communicated

## Modbus/TCP Model

Modbus is an OSI Layer 7 serial communication protocol that was developed by Modicon in 1979 for use with data acquisition and control systems. Modbus enables reliable communication among various control devices, including PLCs. Several significant modifications to Modbus have evolved, such as Modbus/TCP, which was developed to be compatible with the TCP/IP model. Additional Modbus variants include Modbus ASCII, Modbus RTU, Modbus RTU/IP, and Modbus over UDP.

## The Distributed Network Protocol

The Distributed Network Protocol (DNP3) is an open SCADA protocol that is used for serial or TCP/IP communications between control devices. It is widely used by utilities, such as water companies and electricity suppliers, for the exchange of data and control instructions between master control stations and remote computers or controllers called *outstations*. Figure 1-5 is an example of a SCADA application, using DNP3 with a data concentrator in a hierarchical mode.

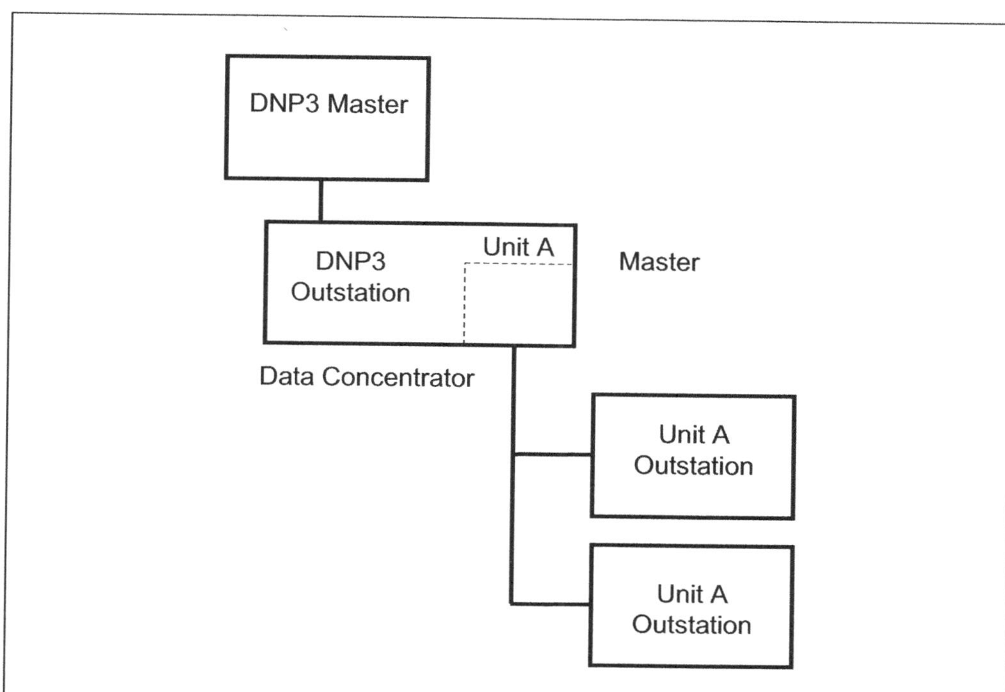


**Figure 1-5. DNP3 Example Usage**

## Utility Communications Architecture Version 2.0/IEC 61850

The Utility Communications Architecture (UCA) Version 2.0 comprises a set of communications protocols for use by electric utilities. In 1999, UCA 2.0 evolved to the layered architecture standard IEC 61850 for substation automation. IEC 61850 enabled IEDs to obtain power grid condition data via an Ethernet process bus and from merge units, which provide interfaces to field devices. Legacy devices utilize an IEC 61850 wrapper around legacy protocols for compatibility. IEC 61850 includes the following features:

- Abstract definitions of services and data
- Definitions for IED communications
- IED communications through substation buses
- Mapping of services to protocols
- Object-oriented approach
- Standardized configuration language (SCL)

Figure 1-6 provides an overview of an IEC 61850-based substation.

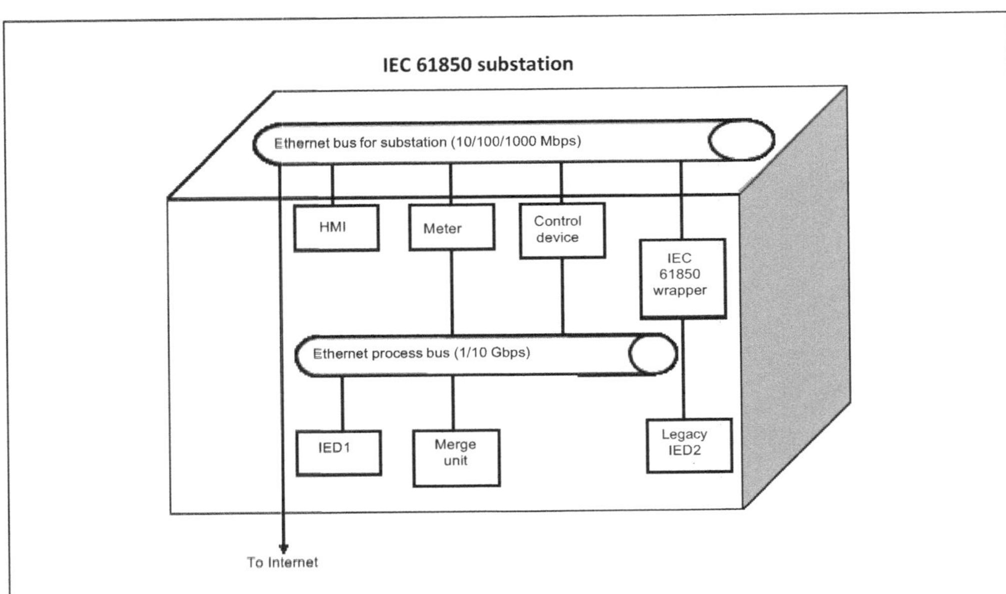


**Figure 1-6. AN IEC 61850 Substation**

## PROFIBUS

Process fieldbus or PROFIBUS is an open standard that defines the electrical, mechanical, and functional aspects of a fieldbus that addresses deterministic data acquisition implementations. It is compatible with European international fieldbus standard EN 50 170. There are three versions of PROFIBUS: PROFIBUS Process Automation (PA), PROFIBUS Decentralized Peripherals (DP), and PROFIBUS Fieldbus Message Specification (FMS). PROFIBUS PA communicates with control devices and data acquisition components through a common serial bus and is specifically designed for use in hazardous and explosive environments through current limiting elements. PROFIBUS DP communicates over a high-speed network with sensors and actuators through a central controller and is suited for factory automation requirements. PROFIBUS FMS supports communications among a large number of high-level applications at moderate transmission rates.

## Controller Area Network

Controller area network (CAN) protocols (also known as ISO Standard 11898-1) were developed in 1986 by Robert Bosch GmbH in Germany for control applications in automobile production lines. The protocol uses an altered form of Ethernet Carrier Sense Multiple Access with Collision Detection (CSMA/CD) to implement a deterministic protocol for SCADA systems. In a deterministic system, an element must respond to events in a predictable manner with a known response time. For example, when controlling a robot arm, its position must be known, and the time it takes to obtain position information must be within certain bounds. Similarly, the time it takes to send commands to the arm and have it respond must be predictable and bounded.

## EtherNet/IP

EtherNet/IP is an application layer protocol designed for industrial automation and control systems applications and is a member of a family of networks that implement the Common Industrial Protocol (CIP™) at its upper layers. EtherNet/IP provides control, configuration, and data exchange capabilities on an Ethernet network. The protocol defines network entities and devices as objects and uses a TCP/IP object for the TCP/IP protocol and an Ethernet link object to set up an EtherNet/IP link. EtherNet/IP uses the OSI model and implements CIP at the session layer and above, and adapts CIP to the specific Ethernet/IP technology at the Transport layer and the lower layers. A typical EtherNet/IP architecture is shown in Figure 1-7.

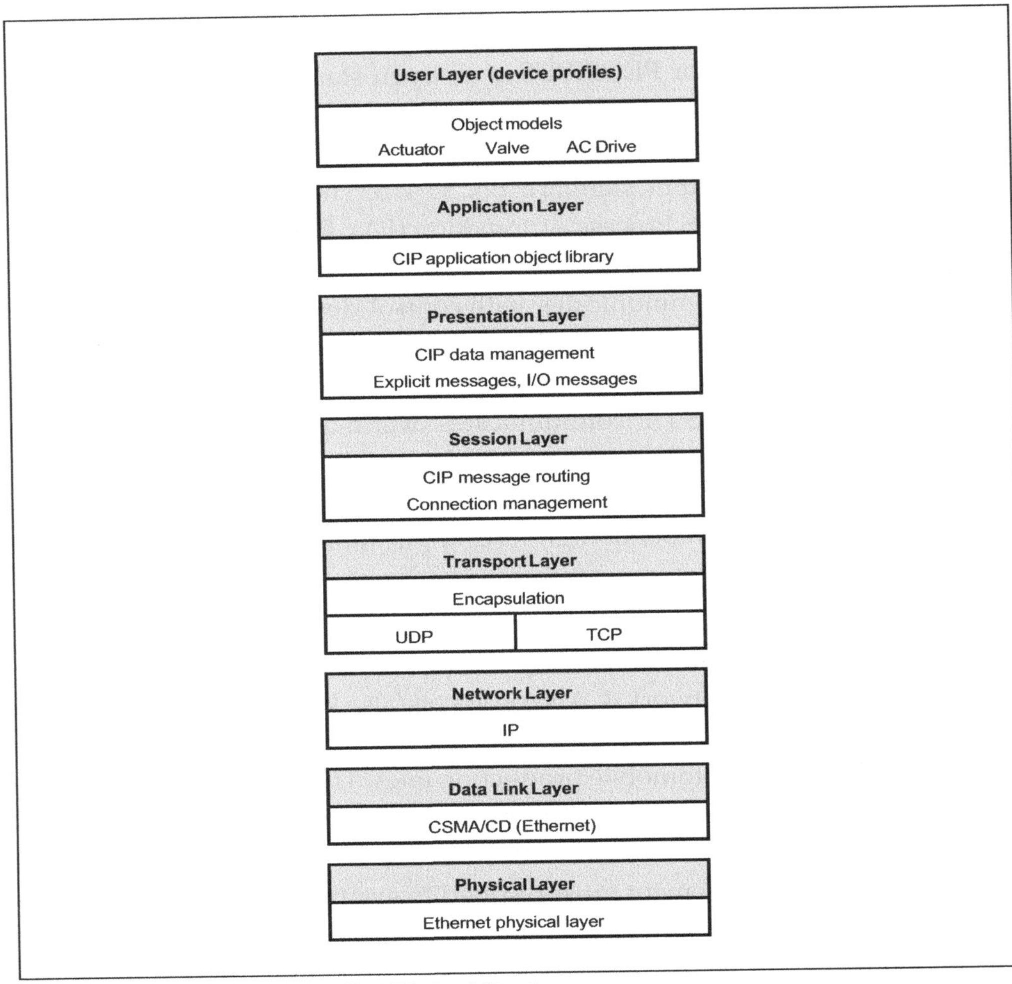


**Figure 1-7. A Typical EtherNet/IP Architecture**

## openSAFETY Protocol

openSAFETY[11] is an application layer protocol sponsored by the Ethernet Powerlink Standardization Group (EPSG),[12] which supports the communication of safety-related data over a variety of industrial networks and ensures data transfer integrity. The performance of openSAFETY data communications is a function of the network and the specific data transfer protocols being used. For example, openSAFETY can be used with Modbus and EtherNet/IP. It is compatible with IEC 61508 and can provide safe implementations up to SIL-3. An important characteristic of openSAFETY is that it employs a variable-length frame with a uniform format that supports safe distribution of values. To meet these requirements, openSAFETY verifies parameters using cyclic redundancy check (CRC) and encapsulates its messages, which independently tunnel through the extant protocols being used in the data acquisition and control network. Figure 1-8 illustrates the relationship of openSAFETY to the other protocols in the OSI layered architecture.

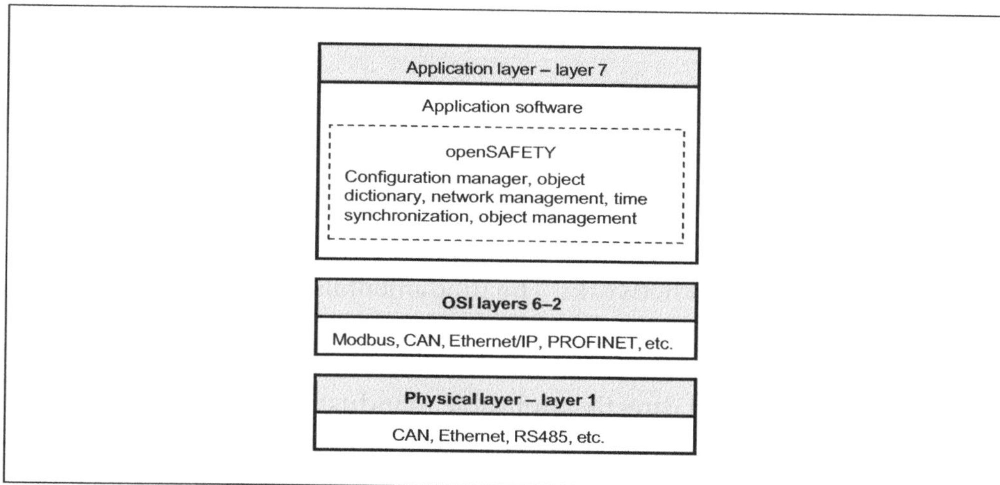


**Figure 1-8. A Typical openSAFETY Architecture**

## Issues in Industrial Automation and Control Systems Security

No one questions the need to protect a nation's critical infrastructure and essential production processes. The available means and methods required to accomplish this defense are varied and have multiple characteristics associated with them. These characteristics include:

- Hardware cost
- Software cost
- Reliability
- Level of assurance
- The effect of security controls on performance
- Maintenance issues
- Safety issues
- Effect on product quality
- Response to emergency situations

- Interfaces to management systems
- Compatibility with 24/7 operations

These issues and others will be explored in the chapters of this book to provide a foundation for securing industrial automation and control systems.

## Summary

Over the past two decades, methodologies and best practices in the field of information system security have been developed, refined, and codified. They comprise a rich compendium of proven and effective measures to secure computer systems and networks. The fundamentals of these developments will be presented in Chapter 2 as a basis for the evolution of security technologies applicable to industrial automation and control systems. A number of these paradigms are not directly applicable to industrial automation and control systems, but many are and many can be employed with modifications tailored to the unique characteristics of these systems. These adapted methods can be effective in securing industrial automation and control systems against threats, such as viruses, Trojan horses, worms, and attacks that result in the unauthorized modification and manipulation of critical data, denial of service and availability, unauthorized modification of audit logs, and general disruption of critical operations.

The information system security approaches embrace a wide variety of disciplines and include:

- Policies, procedures, and standards
- Access control mechanisms
- Physical security
- Identification and authentication technologies, such as biometrics and security tokens
- Audit trails
- Real-time monitoring of system and network information
- Risk management
- Business continuity and disaster recovery planning

- Firewalls to filter message traffic
- Intrusion detection and protection systems
- Public key and secret key cryptography
- Malware detection, isolation, and elimination

## Review Questions for Chapter 1

*Read the questions carefully and select the one BEST answer.*

1.1 According to NIST SP 800-82, the term *industrial control system* encompasses which of the following?

A. SCADA systems

B. Distributed control systems (DCSs)

C. Programmable logic controllers (PLCs)

D. All of the above

1.2 According to ANSI/ISA-62443-1-1 (99.01.01)-2007, "a collection of personnel, hardware, and software that can affect or influence the safe, secure, and reliable operation of an industrial process" is a(n):

A. Industrial automation and control system

B. Distributed control system

C. Discrete control system

D. Programmable logic controller

1.3 A "type of loosely coupled distributed monitoring and control system commonly associated with electric power transmission and distribution systems, oil and gas pipelines, and water and sewage systems" is defined by ANSI/ISA-62443-1-1 (99.01.01)-2007 as which of the following?

A. Distributed control system

B. SCADA system

C. Safety instrumented system (SIS)

D. Human-machine interface (HMI)

1.4 Which one of the following items is NOT a component of a classical SCADA system model?

A. Human-machine interface (HMI)

B. Remote terminal unit (RTU)

C. Enterprise resource planning (ERP)

D. Programmable logic controller (PLC)

1.5 Which of the following is NOT a characteristic of a SCADA master terminal unit (MTU)?

A. Has two-way communication with field devices

B. Usually located at the master control center

C. Has high-bandwidth communication requirements

D. Communicates through telephone, VHF/UHF radio, spread spectrum radio, satellite, and/or microwave

1.6 In a typical SCADA application, what component serves as a data concentrator and is an interface between the MTU and field devices?

A. HMI

B. RTU

C. Data historian

D. Relational database

1.7 IEC standard 61131-3 was designed to make it easier to implement control logic functions. IEC 61131-3 is which of the following?

A. Vendor-independent international standard for PLC programming languages

B. Standard guide to industrial control systems

C. Safety instrumented systems (SIS) standard for the process industry

D. Integrated enterprise-wide risk management standard

1.8 Which of the following is NOT a characteristic of a data historian?

A. Provides for prompt recovery of data

B. Performs data compression

C. Supports interactive storage and retrieval of detailed production information

D. Provides system histories over given time periods

1.9 A yield accounting system used in conjunction with a SCADA system provides which of the following functions?

A. Plant material movement

B. Inventory discrepancies

C. Material balances

D. All of the above

1.10 ANSI/ISA-62443-1-1 (99.01.01)-2007 defines which of the following as "a type of control system in which the system elements are dispersed but operated in a coupled manner"?

A. Distributed control system

B. Discrete control system

C. Dispersed control system

D. Coupled control system

1.11 ANSI/ISA-84.00.01-2004 Part 1(IEC 61511-1 Mod) defines which of the following as an "instrumented system used to implement one or more safety instrumented functions (SIF)"?

A. Safety surety system

B. Failure proof system

C. Safety instrumented system

D. Safety function system

1.12 Necessary activities involved in the implementation of safety instrumented functions occurring during a period of time that starts at the concept phase of a project and finishes when all the safety instrumented functions are no longer available for use is known as which of the following?

A. Safety duration

B. Safety life cycle

C. Safety boundary

D. Safety period

1.13 The safety integrity level (SIL) is a discrete level for specifying the safety integrity requirements of the safety instrumented functions to be allocated to the safety instrumented systems. Which of the following statements is TRUE?

A. Safety integrity level 4 has the lowest level of safety integrity; safety integrity level 1 has the highest.

B. Safety integrity level 4 has the highest level of safety integrity; safety integrity level 1 has the lowest.

C. Safety integrity level 3 has the highest level of safety integrity; safety integrity level 1 has the lowest.

D. Safety integrity level 3 has the lowest level of safety integrity; safety integrity level 1 has the highest.

1.14 The Open Systems Interconnection (OSI) reference model has how many layers?

A. Six

B. Five

C. Seven

D. Four

1.15 Which layer of the OSI model converts packets into electrical or optical signals for sending on the transmission media?

A. Application

B. Presentation

C. Physical

D. Session

1.16 What are the four layers of the TCP/IP model?

A. Application, Host-to-Host, Internet, and Network Access

B. Application, Presentation, Internet, and Network Access

C. Application, Network, Internet, and Network Access

D. Application, Host-to-Host, Internet, and Physical

1.17 NIST SP 800-82 defines which of the following as "a set of open standards developed to promote interoperability between disparate field devices, automation/control, and business systems"?

A. Modbus/TCP

B. OPC

C. DNP3

D. PROFIBUS

1.18 The protocol that supports the communication of safety-related data over a variety of industrial networks and ensures data transfer integrity is known as which of the following?

A. Safety instrumented system (SIS)

B. Safety instrumented function (SIF)

C. openSAFETY

D. ClosedSafety

1.19 Which of the following statements relative to the IEC-layered architecture standard IEC 61850 for substation automation is FALSE?

A. It is a set of protocols for electric utilities.

B. IEC-enabled devices can obtain power grid condition data via an Ethernet process bus.

C. Merge units provide interfaces to field devices.

D. It is not compatible with legacy protocols.

1.20 Which of the following are versions of PROFIBUS?

A. PROFIBUS PA, PROFIBUS CP, PROFIBUS CIP

B. PROFIBUS PA, PROFIBUS DP, PROFIBUS FMS

C. PROFIBUS PA, PROFIBUS CP, PROFIBUS FMS

D. PROFIBUS PA, PROFIBUS CIP, PROFIBUS FMS

## References

1. ANSI/ISA-62443-1-1 (99.01.01)-2007. *Security for Industrial Automation and Control Systems Part 1-1: Terminology, Concepts, and Models.* Research Triangle Park, NC: ISA (International Society of Automation).

2. NIST SP 800-82. *Guide to Industrial Control Systems (ICS) Security. Revision 2*. Washington, DC: NIST (National Institute of Standards and Technology), 2015.

3. IEEE Std. C37.1-1994. *IEEE Standard Definition, Specification, and Analysis of Systems Used for Supervisory Control, Data Acquisition, and Automatic Control*. Piscataway, NJ: IEEE (Institute of Electrical and Electronics Engineers), 1994.

4. *The Automation, Systems, and Instrumentation Dictionary*. 4th ed. Research Triangle Park: ISA (International Society of Automation), 2003.

5. ANSI/ISA-62443-1-1 (99.01.01)-2007. *Security for Industrial Automation and Control Systems Part 1-1: Terminology, Concepts, and Models.* Research Triangle Park, NC: ISA (International Society of Automation).

6. NIST SP 800-82. *Guide to Industrial Control Systems (ICS) Security. Revision 2*. Washington, DC: NIST (National Institute of Standards and Technology), 2015.

7. ANSI/ISA-84.00.01-2004 Part 1 (IEC 61511-1 Mod). *Functional Safety: Safety Instrumented Systems for the Process Industry Sector – Part 1: Framework, Definitions, System, Hardware and Software Requirements.* Research Triangle Park, NC: ISA (International Society of Automation).

8. NIST SP 800-39. *Managing Information Security Risk: Organization, Mission, and Information System View*. Washington, DC: NIST (National Institute of Standards and Technology), 2011.

9. NIST SP 800-82. *Guide to Industrial Control Systems (ICS)*. Revision 2. Washington, DC: NIST (National Institute of Standards and Technology), 2015.

10. OPC Foundation. 2016. http://www.opcfoundation.org/

11. openSAFETY. 2011. http://www.open-safety.com

12. Ethernet Powerlink Standardization Group. 2013. http://www.ethernet-powerlink.org/index.php?id=43

# 2 Information System Security Technology

*Chapter 1 provided a review of some fundamental industrial automation and control system concepts, assuming that the reader has some familiarity with these systems. Because this book is focused on the security of industrial automation and control systems, Chapter 2 will delve into the details of information system security in order to provide a basis for developing effective security approaches for industrial automation and control systems. The field of information system security has evolved over decades and is an abundant source of valuable knowledge and techniques that have been proven over time. By taking into account the unique characteristics of industrial automation and control systems, the technology of information system security can be adapted and tailored to provide effective security for the many types of industrial automation and control applications.*

## Information System Security Fundamentals

Information system security is defined as comprising the three basic elements of confidentiality, integrity, and availability, or C-I-A. These three concepts form the C-I-A triad, as shown in Figure 2-1.

Additional complementary elements that support information system security are identification, authentication, authorization, accountability, auditing, and nonrepudiation. These important topics are summarized in the following sections.

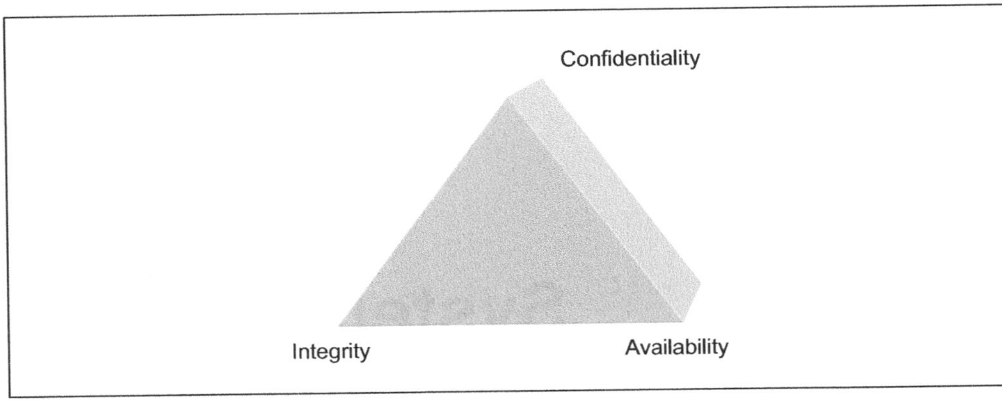


**Figure 2-1. The C-I-A Triad**

## Confidentiality

Confidentiality refers to protecting documents, messages, computer files, and so on from being disclosed, either intentionally or unintentionally, in an unauthorized manner. Examples of loss of confidentiality are an unauthorized individual intercepting and reading a message sent across a network or the revelation of a company's trade secret. Confidentiality is related to privacy of information in that it is concerned with the unauthorized disclosure of sensitive information.

## Integrity

Integrity (as applied to data) is concerned with the unauthorized modification of data, whether by an authorized or an unauthorized individual or process. If a message is intercepted, its contents are changed, and it is then forwarded to a receiver, its integrity has been violated.

## Availability

Availability is a measure of the ability of an authorized person or process to have prompt and reliable access to computational resources when needed. For example, availability would be compromised if an individual were prevented from conducting an online financial transaction because the financial organization's website is down.

## Identification

Identification is the act of professing an identity to a system or process. A typical example of identification is a computer user typing in a user ID when logging onto a computer system.

## Authentication

Authentication is the act of verifying an identity and confirming that an entity is what it professes to be. An example of authentication is verifying that a logon user ID actually belongs to the claimed user.

## Authorization

When identification and authentication are completed, a user or process can be granted access privileges to specific computer information or resources. Individuals and processes can be granted different levels of authorization, depending on the sensitivity of the information being accessed and the level of the rights granted to those individuals or processes to access the information.

## Accountability

Accountability is the ability to identify an individual and to hold that individual responsible for his or her actions. Tools, such as audit logs, can help in determining accountability and uncovering activities that were conducted by individuals or processes.

## Auditing

Auditing is the act of reviewing logs or records at specified intervals to determine if violations of system security have occurred. Typical audit logs include information such as where and when a transaction occurred and who conducted the transaction. Information technology auditors can either be internal auditors, working for the organization under audit, or external auditors, belonging to an outside organization. A companion endeavor to auditing is monitoring, which refers to a continuing activity, such as intrusion detection on a network or host computer.

## Nonrepudiation

Nonrepudiation in digital systems refers to ensuring that the sender of a message or contract cannot later deny sending the message or contract and that, conversely, a receiver to whom the message was sent cannot deny receiving the message. Digital signatures are used to implement nonrepudiation in electronic transactions; they will be discussed later in this chapter.

## Related Terminology

In addition to the above principles, four related terms that appear throughout the information security landscape and that should be understood by all involved with protecting computer systems and networks are:

- **Asset** – An organization's important resource, whose loss or compromise could negatively impact confidentiality, integrity, or availability, and could also result in financial loss. Therefore, an asset should be protected from attack or compromise.
- **Threat** – As defined by NIST SP 800-53,[1] a threat is "any circumstance or event with the potential to adversely impact agency operations (including mission, functions, image, or reputation), agency assets, or individuals through an information system via unauthorized access, destruction, disclosure, modification of information, and/or denial of service."
- **Safeguard** – A security control or countermeasure designed to mitigate the risk associated with a specific threat or group of threats.
- **Vulnerability** – The absence or weakness of a safeguard. According to NIST SP 800-53, vulnerability is "weakness in an information system, system security procedures, internal controls, or implementation that could be exploited or triggered by a threat source."

Vulnerability can be decreased by applying controls that will mitigate the effect of an attack. Controls can be partitioned into the following categories:

- Corrective controls minimize the effect of an attack and the degree of resulting damage.
- Detective controls discern if attacks have occurred, or are occurring, and initiate control measures.
- Deterrent controls reduce the potential for an attack to occur.
- Preventive controls prevent a threat from exploiting a vulnerability.

One means that is used to prioritize which controls to apply and where to apply them is a threat matrix. The threat matrix plots the impact an attack would have if it were successful versus the likelihood that the attack will occur. In this way, resources can be applied to mitigate the threats with the highest impact and the highest likelihood of occurrence. The objective of this activity is to move all threats into the low impact and low likelihood of occurrence area, as shown in Figure 2-2.

NIST SP 800-53 provides a comprehensive list of recommended security controls.

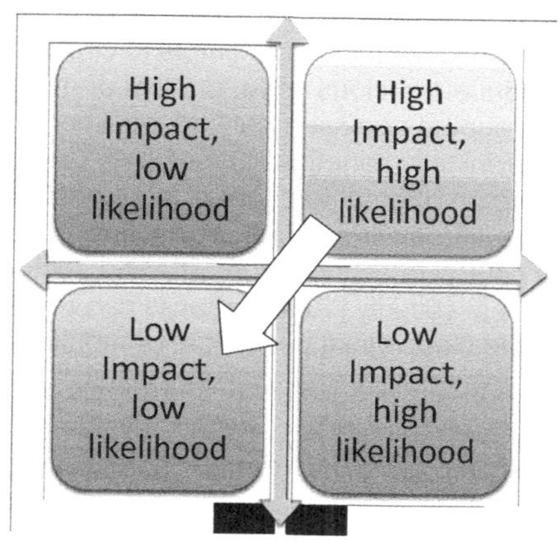


**Figure 2-2. A Threat Matrix**

## Types and Classes of Attack

The Information Assurance Technical Framework (IATF) document 3.1,[2] sponsored by the National Security Agency (NSA), identifies the following types of attacks:

- Passive
- Active
- Close-in
- Insider
- Distribution

These attacks and their descriptions are summarized in Table 2-1.

**Table 2-1. Types of Attacks**

*Source: NSA (2002)*

| Attack | Description |
|---|---|
| Passive | Passive attacks include traffic analysis, monitoring of unprotected communications, decrypting weakly encrypted traffic, and capture of authentication information (such as passwords). Passive attacks can result in disclosure of information or data files to an attacker without the consent or knowledge of the user. Examples include the disclosure of personal information such as credit card numbers and medical files. |
| Active | Active attacks include attempts to circumvent or break protection features, introduce malicious code, or steal or modify information. These attacks may be mounted against a network backbone, exploit information in transit, electronically penetrate an enclave, or attack an authorized remote user during an attempt to connect to an enclave. (An enclave is collection of computing environments connected by one or more internal networks under the control of a single authority and security policy, including personnel and physical security.) Active attacks can result in the disclosure or dissemination of data files, denial of service, or modification of data. |
| Close-In | Close-in attack consists of individuals, normally not insiders, attaining close physical proximity to networks, systems, or facilities for the purpose of modifying, gathering, or denying access to information. Close physical proximity is achieved through surreptitious entry, open access, or both. |
| Insider | Insider attacks can be malicious or nonmalicious. Malicious insiders intentionally eavesdrop, steal, or damage information; use information in a fraudulent manner; or deny access to other authorized users. Nonmalicious attacks typically result from carelessness, lack of knowledge, or intentional circumvention of security for such reasons as "getting the job done." |
| Distribution | Distribution attacks focus on the malicious modification of hardware or software at the factory or during distribution. These attacks can introduce malicious code into a product, such as a back door to gain unauthorized access to information or a system function at a later date. |

## Additional System Security Concepts

In industrial automation and control systems, as well as in information technology systems, it is important that the computer resources, networks, and associated equipment operate reliably, efficiently, and securely in performing the actions for which they were designed. A number of useful principles for designing and implementing secure IT systems have evolved and are important tools that can be applied to industrial automation and control systems. The following principles were delineated in a 1974 paper by Saltzer and Schroeder:[3]

- Complete mediation
- Defense in depth

- Economy of mechanism
- Fail-safe
- Least common mechanism
- Least privilege
- Leveraging existing components
- Open design
- Psychological acceptability
- Separation of duties
- Weakest link

The fundamental characteristics of these principles are summarized below.

## Complete Mediation

Complete mediation requires that when an entity (a user or process) requests access to an object, such as a file or document in a computer system, the entity must go through a valid authorization process that cannot be circumvented.

## Defense in Depth

Defense in depth is the act of establishing numerous layers of protection wherein a subsequent layer will provide protection if a previous layer is breached. NIST SP 800-82[4] defines a defense-in-depth architecture strategy as "the use of firewalls, the creation of demilitarized zones and intrusion detection capabilities, along with effective security policies, training programs, and incident response mechanisms." ANSI/ISA-62443-1-1 (99.01.01)-2007[5] describes defense in depth as:

> The provision of multiple security protections, especially in layers, with the intent to delay, if not prevent, an attack.

**NOTE** Defense in depth implies layers of security and detection, even on single systems, and provides the following features:

- Attackers are faced with breaking through or bypassing each layer without being detected.

- A flaw in one layer can be mitigated by capabilities in other layers.
- System security becomes a set of layers within the overall network security.

## Economy of Mechanism

Economy of mechanism advances the concept that the design and implementation of defensive mechanisms should be straightforward, understandable, and not unnecessarily complex. Following these guidelines makes it easier to detect possible unauthorized access routes into the information system that could be used to acquire sensitive data or launch an attack.

## Fail-Safe

NIST SP 800-123[6] explains the term *fail-safe* as follows: "If a failure occurs, the system should fail in a secure manner, that is, security controls and settings remain in effect and are enforced. It is usually better to lose functionality rather than security." Also, a fail-safe design should ensure that when a system recovers from a failure, it should recover in a secure state, where only authorized users have access to sensitive information.

## Least Common Mechanism

Least common mechanism refers to having the least possible sharing of common security mechanisms among users or processes. Having many users sharing common security mechanisms can result in unknown or unauthorized access paths to sensitive data.

## Least Privilege

In the principle of least privilege, a user or process is given the minimum amount of privileges, authorization, and so on, for the smallest amount of time that will permit the user or process to accomplish assigned tasks. Least privilege reduces the prospect of an individual or process having unauthorized access to sensitive information.

## Leveraging Existing Components

Leveraging existing components refers to using the security mechanisms that are already in place in the most efficient and effective manner and to their maximum capabilities. This process can be accomplished by periodically reviewing the configuration of the security devices and by optimizing their operational performance.

### Open Design

Open design proposes that making designs and security approaches available to examination and scrutiny by a variety of parties will ensure that they are robust and are performing as required. The concept is that by making the designs available for review, any flaws or weaknesses can be exposed and corrected. The alternative to open design is to keep designs proprietary and confidential in the hope that by doing so they will be more secure from compromise. In almost all cases, open design results in more effective and reliable mechanisms.

### Psychological Acceptability

Psychological acceptability is concerned with making the interface and interaction with access control mechanisms intuitive and easy for the user to understand and operate. In this way, the security controls will not be seen as interfering with the tasks at hand and as an obstacle to getting things done.

### Separation of Duties

Separation of duties requires that functions, roles, or responsibilities should be distinct and independent from each other so that no entity can solely control sensitive operations. For example, a person requesting or ordering an expensive item should not be the same person who signs the check for its purchase. Separation of duties forces multiple individuals to collude if they are going to commit fraud or obtain unauthorized access to data.

### Weakest Link

As with any group of protection mechanisms, it is important to identify the weakest element in the group. Once the weakest link is identified, actions can be taken to bolster that element and mitigate any corresponding risk.

## Policies, Standards, Guidelines, and Procedures

In addition to secure design principles, documents professing management's view of security and describing techniques of implementation are valuable tools in establishing the security of an organization. Four important categories of such tools are policies, standards, guidelines, and procedures.

## Policies

In general terms, a policy is a statement of the intent of management for the organization and compliance is mandatory. A policy is an important document because it provides top-down requirements for the organization and is necessary to protect sensitive information, prevent liability issues, and meet regulatory and legal requirements.

NIST SP 800-12[7] defines the following three types of computer system security policies:

- **Program Policy** – Creates an organization's computer security program
- **Issue-Specific Policies** – Addresses specific issues of concern to the organization
- **System-Specific Policies** – Protects a particular system through technical directives set forth by management

Table 2-2 summarizes the characteristics of the NIST policy types.

**Table 2-2. NIST Policy Types**

*Source: NIST SP 800-12 (1995)*

| Policy Type | Description | Example |
|---|---|---|
| Program policy | High-level program policy | Senior-level management statement |
| Issue-specific policy | Addresses single issue | Email privacy policy |
| System-specific policy | Single-system directives | Router access control lists |

## Standards

A standard details how specific methods must be applied in a consistent manner. For example, a standard for the types of information security controls that must be applied would ensure that all elements of the organization are incorporating technologies in a uniform manner. Conformance to standards is normally compulsory.

## Guidelines

Guidelines also detail methods to use to create secure information systems, but are considered to be recommendations and conformance is not compulsory.

### Procedures

Procedures are step-by-step actions that must be taken to implement policies and standards. Procedures describe compulsory activities.

## Malicious Code and Attacks

NIST SP 800-53[8] defines malicious code as:

> Software or firmware intended to perform an unauthorized process that will have an adverse impact on the confidentiality, integrity, or availability of an information system. A virus, worm, Trojan horse, or other code-based entity that infects a host. Spyware and some forms of adware are also examples ofmalicious code.

Malicious code (malware) is used in attacks against information systems and industrial automation and control systems. Some of the main types of attacks are reviewed in the following sections.

### Viruses and Worms

ANSI/ISA-62443-1-1 (99.01.01)-2007[9] defines a virus as a "self-replicating or self-reproducing program that spreads by inserting copies of itself into other executable code or documents," and a worm as a "computer program that can run independently, can propagate a complete working version of itself onto otherhosts on a network, and may consume computer resources destructively." Note that a worm does not require other executable software to propagate but can move across networks to infect other hosts on its own.

### Trojan Horse

A Trojan horse conceals additional code in a program that is used for a valid purpose. Then, the hidden code, which could be a virus, can perform malicious acts. Trojan horses can be transmitted through emails or downloads from websites.

### Logic Bomb

A logic bomb is a type of Trojan horse that does not execute until a preset condition is met; for example, at a specific time and date or when some activity is performed on the host computer.

## Mobile Code

Software that is obtained or downloaded over a network from a remote source onto a local computer is known as *mobile code*. This mobile code can be used for valid applications or can contain a virus that could do harm to a computer system.

## Back Door

In a back-door attack, an individual attempts to gain access to a computer system by circumventing its protection mechanisms. An example is using a dial-up modem to call into a person's desktop computer and then accessing the organization's network through that computer.

## Scanning

In scanning, an individual can locally or remotely connect to a computing system and obtain information about that system that can be used later in launching an attack. A scan can yield the following information:

- Open ports
- Services that are running
- Types of system software
- Domain names
- Existence of intrusion detection systems
- Protocols being used
- Existence of firewalls and perimeter devices

If a system is being scanned, this activity should be viewed as a prelude to a possible attack.

## Man-in-the-Middle

A man-in-the-middle (MitM) attack focuses on the authentication protocol between a claiming party and a verifying party communicating with each other. For example, a claiming party can be asked to present and transmit particular credentials while the verifying party receives those credentials. In a successful attack, the attacker can intercept data between the two parties, modify it, and then pass it on without the knowledge of the sender or receiver.

### Social Engineering

Social engineering uses "soft" methods to acquire sensitive information. Social engineering methods may be used to deceive individuals into revealing sensitive information, such as passwords or PIN numbers. Persuasive phone calls to an individual from an attacker impersonating someone in a position of authority, a phony email, or a fake call where an individual pretends to be from a help desk can elicit private information that can be used later against the organization. Social engineering threats can be minimized through effective awareness training and security policies.

### Guessing Passwords

An attacker can use a variety of means to determine a user's password to gain entry to a computer system. Guessing, searching an area for passwords written on post-it notes, and looking over someone's shoulder are all methods that can be used to determine passwords. One approach, known as a *dictionary attack*, uses a dictionary of commonly used passwords to try to gain entry to a system.

### Denial of Service/Distributed Denial of Service

A denial-of-service (DoS) attack is one that is designed to overwhelm a computer system's resources to the point where it can no longer perform its required functions. Thus, a DoS incident affects the availability of the computer system. In distributed denial of service (DDoS), the attacks on availability originate from multiple sources that work in concert to saturate a computer system's capabilities. In many instances, these sources are host computers on which the malicious DDoS software has been installed without the owner's knowledge.

### Replay

In a replay attack, the perpetrator intercepts a message and saves it to be sent at a later time, when it seems to come from the original sender. A countermeasure to this attack is to attach a timestamp to the message at the time it was sent. Then, if the message is intercepted and sent at a later time, the timestamp will show it is an old message.

### Dumpster Diving

In dumpster diving, an individual sorts through discarded material in a dumpster in the hope of finding sensitive information that can be used later to attack an organization's computer system. In many cases, user's manuals, technical manuals, correspondence, organization charts, and other related material could provide valuable information to an attacker.

## Firewalls

One of the important defenses against malware is a properly configured and implemented firewall. Firewalls implemented in industrial automation and control systems have to be "aware" of the special requirements and characteristics of such systems. The major types of firewalls are discussed in this section.

NIST SP 800-41[10] defines firewalls as "devices or programs that control the flow of network traffic between networks or hosts that employ differing security postures." ANSI/ISA-62443-1-1 (99.01.01)-2007[11] describes a firewall as:

> An internetwork connection device that restricts data communication traffic between two connected networks.

> **NOTE** A firewall may be either an application installed on a general-purpose computer or a dedicated platform (appliance) that forwards or rejects/drops packets on a network. Typically firewalls are used to define zone borders. Firewalls generally have rules restricting which ports are open.

Firewalls are categorized as to their technical implementation and their architecture. The technical categories are packet filtering, stateful inspection, application, and application proxy. The architectural categories are screened-host, dual-homed host, and screened subnet firewall. These categories are reviewed in the following sections.

### Packet-Filtering Firewall

A packet-filtering firewall resides at the network layer of the OSI model and determines which packets can be allowed to continue to their destination addresses by inspecting the packet source and destination addresses. It filters the packets based on firewall access control lists (ACLs), which specify the packets that can be sent to particular destination addresses or to specific application port destinations. Figure 2-3 illustrates a packet-filtering firewall.

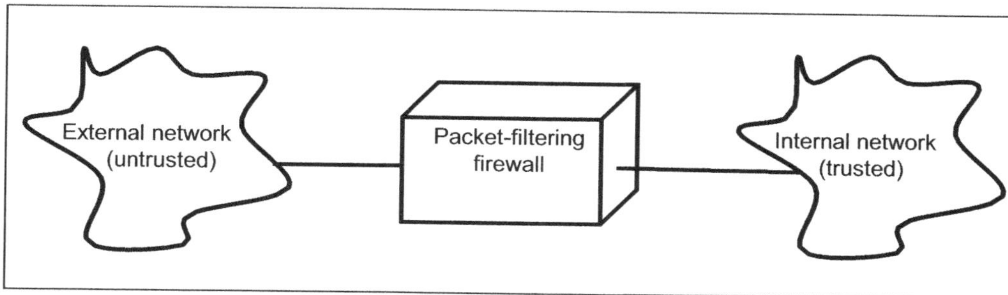


**Figure 2-3. A Packet-Filtering Firewall**

## Stateful Inspection

A stateful inspection firewall improves on a packet-filtering firewall by determining the state of incoming data connections at the network layer and blocking data packets that do not conform to the anticipated state. A state refers to one of the three primary states associated with TCP communications, namely, initiating the connection, establishing and using the connection, and terminating the connection. This information is contained in TCP headers and can indicate if a packet is not in its proper state. An attacker might place information in the header, indicating that the packet is in one state when it is actually in another. In this instance, the packet will not be passed through the firewall.

In addition to state information, the packet header also includes the source IP address, destination IP address, and port numbers. This information is stored in a dynamic state table for processing. An example of a state table is shown in Table 2-3.

**Table 2-3. Example State Table**

*Source: NIST SP 800-41 (2009)*

| Source Address | Source Port | Destination Address | Destination Port | Connection State |
|---|---|---|---|---|
| 192.168.1.100 | 1030 | 192.0.2.71 | 80 | Initiated |
| 192.168.1.102 | 1031 | 10.12.18.74 | 80 | Established |
| 192.168.1.101 | 1033 | 10.66.32.122 | 25 | Established |
| 192.168.1.106 | 1035 | 10.231.32.12 | 79 | Established |

As an example of the firewall use of the state table, the source address (192.168.1.100) in the first row of the state table is an internal address behind the firewall and is initiating a connection to an IP address (192.0.2.71) outside of the firewall. This connection would have to be allowed under the firewall security policy.

## Application Firewall

An application firewall performs what is sometimes referred to as deep packet inspection by incorporating stateful protocol analysis. This analysis is performed at the application layer of the OSI model and compares stored profiles of normal protocol usage against actual usage that might indicate malicious activity in an application. If malicious activity is detected, the firewall could inhibit specific connections to the information system. Potential malicious activity could include email with harmful attachments, unexpected command sequences, and the downloading of active mobile code.

## Application-Proxy Gateway

An application-proxy gateway firewall serves as an intermediary between networks and masks the source of a message. The application-proxy gateway operates at the application layer of the OSI model and provides a higher level of security at the expense of additional time being required for processing the data.

This type of firewall can be implemented by executing proxy software on a workstation that masks an internal network from external entities attempting to gain information about that network. Thus, the connection consists of communication between an internal computer and the proxy server and a connection between the proxy server and an external network. Most application proxy firewalls operate with HTTP. Figure 2-4 is an example of an application-proxy gateway.

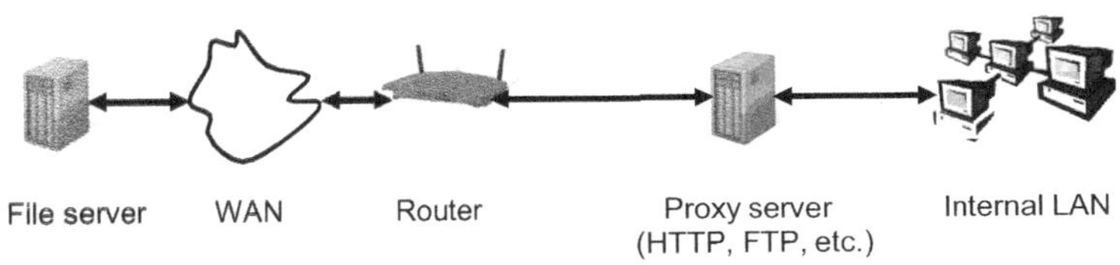


**Figure 2-4. Application-Proxy Gateway**

## Screened-Host Firewall

A screened-host firewall uses a host to connect two networks together. This connection is accomplished through two network interface cards (NICs) that respectively connect to each one of the networks, with a screening router between the host and the untrusted network. This arrangement offers both proxy and routing capabilities. In this architecture, the firewall is facing outward toward the untrusted network and serves to protect the host and the trusted network. Figure 2-5 illustrates a screened-host firewall.

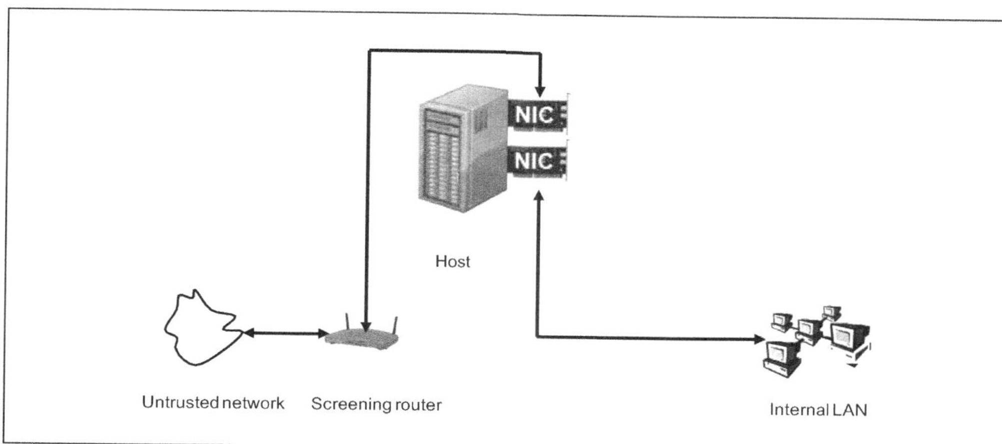


**Figure 2-5. Screened-Host Firewall**

## Dual-Homed Host Firewall

An architecture similar to the screened-host firewall is the dual-homed host firewall. In this configuration, a host with two NICs connects a trusted network to an untrusted network, but does not employ a screening router. The host acts as a filter to packets passing between the networks and is configured as a strong defense against attacks. This type of host is known as a *bastion host*. Figure 2-6 shows a dual-homed host firewall configuration with two NICs.

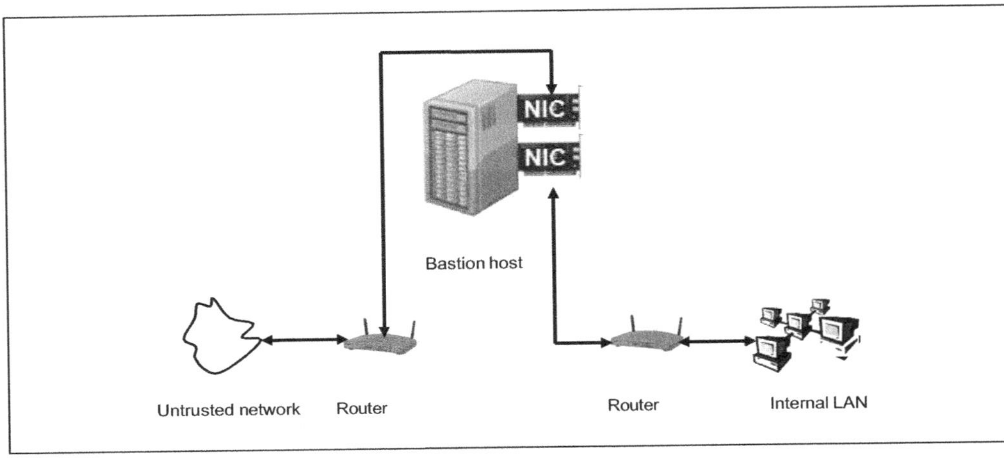


**Figure 2-6. Dual-Homed Host Firewall**

## Screened-Subnet Firewalls

A screened-subnet firewall employs another variation of two NICs with a bastion host. It uses packet-filtering routers on the trusted and untrusted network sides, and the host provides proxy services. The screened-subnet architecture also defines a demilitarized zone (DMZ) or perimeter network, which is a third network that lies between the internal trusted network and the external untrusted network as an added layer of protection. Figure 2-7 is an example of a screened-subnet firewall.

Additional means to protect information systems include scrambling the data so that it cannot be read or compromised by an attacker. This approach employs cryptography, which will be discussed in the next section.

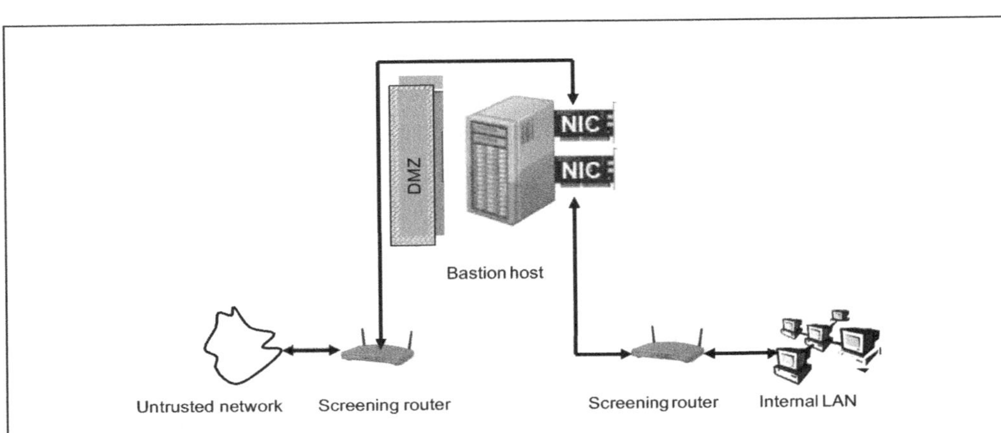


**Figure 2-7. Screened-Subnet Firewall**

## Cryptography

The purpose of cryptography is to protect transmitted information from being read and understood by anyone other than the intended recipient. Cryptography can be used to implement confidentiality, integrity, authentication, and nonrepudiation.

The most familiar form of cryptography is secret key or symmetric key cryptography, where the sender and receiver both know a secret key. The sender encrypts a plaintext message with the secret key, and the receiver decrypts the message with the secret key. Thus, the sender and receiver have a shared secret. One of the problems with secret key cryptography is how to securely distribute the secret key used by the sender to the receiver. This distribution might be accomplished with a messenger or by another out-of-band communication means, such as by fax or mail. The term *out-of-band communication* refers to using another means of communication that is different from the means used to send the main message.

A solution to the key distribution problem of secret key encryption came in the development of asymmetric or public key encryption. In asymmetric key cryptography, the receiver holds a private key and the sender uses a public key to encrypt the message. The public and private keys are related mathematically such that a message encrypted with the public key can be decrypted only with the private key held by the receiver. The public key is made available to anyone who wants to use it, and the private key of the pair is held secretly by the receiver. The private and public keys are related, but in the ideal case, they share the characteristic that the private key cannot be derived using knowledge about the public key.

Public key cryptography requires more computations than secret key cryptography and is therefore much slower in data transmission than secret key cryptography. Symmetric and asymmetric key cryptography are discussed in more detail in the following sections.

### Symmetric Key Cryptography

In symmetric key cryptography, it is critical that the secret key be kept secure and uncompromised. In an ideal situation, a secret key would be used only once and would not be used in subsequent transmissions. In general, the symmetric key encryption algorithm is known and is available to the public, but the cryptographic system is secure if the secret key (or *cryptovariable*, as it is sometimes called) is kept private and secure. A symmetric key cryptographic system (*cryptosystem*) block diagram is shown in Figure 2-8.

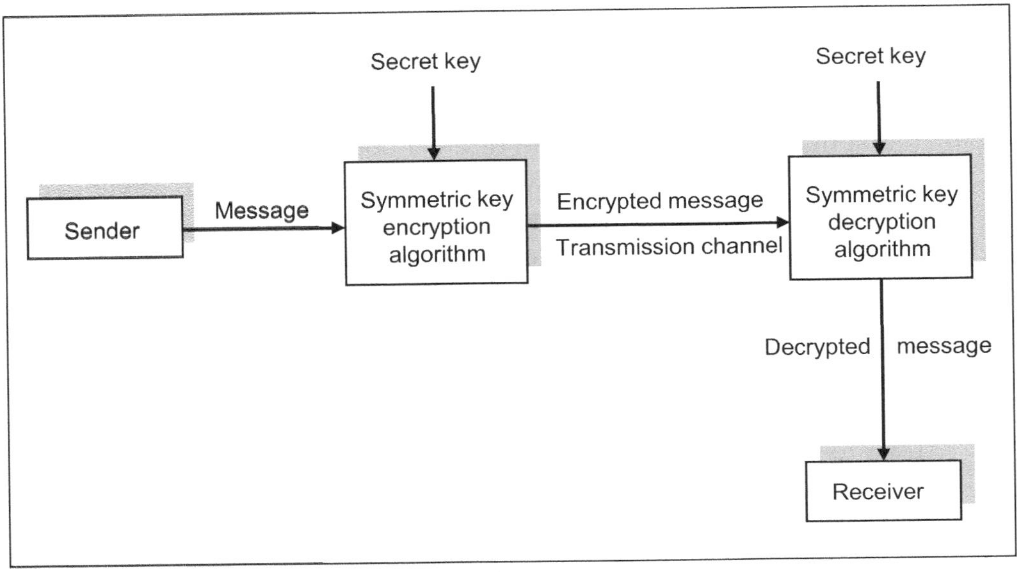


**Figure 2-8. A Symmetric Key Cryptosystem**

Symmetric key encryption algorithms are fast compared to asymmetric key algorithms and are widely used to securely transmit large volumes of data. One popular symmetric key algorithm is the Advanced Encryption Standard (AES), which is specified in NIST Federal Information Processing Standards (FIPS) Publication 197.[12] The standard specifies three symmetric key strengths of 128, 192, or 256 bits. AES has been adopted by private and public organizations inside and outside the United States and is widely used in smart cards, routers, and a variety of other applications.

## Asymmetric Key Cryptography

As mentioned, asymmetric or public key cryptography uses a public and a private key. The public key can be accessed by anyone who desires to send an encrypted message to the holder of the corresponding private key. The important properties of a properly generated public-private key pair are summarized as follows:

- A message encrypted by one of the keys can be decrypted by the other key in the pair.

- The private key is kept confidential.
- The public key cannot be used to determine the private key.
- The public key cannot decrypt the message that it encrypted.

A typical asymmetric key cryptosystem is illustrated in Figure 2-9.

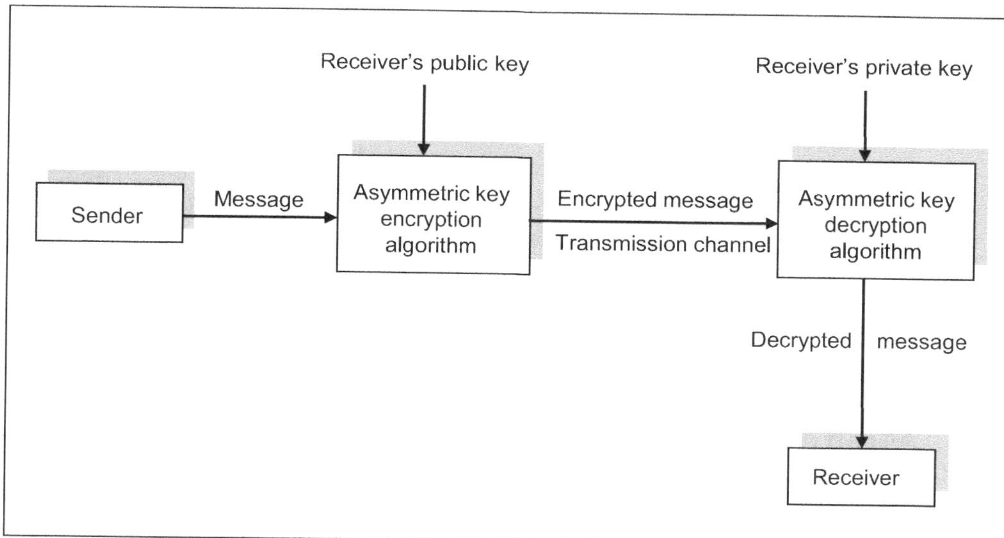


**Figure 2-9. An Asymmetric Key Cryptosystem**

One of the most popular asymmetric key algorithms is the RSA algorithm, named after its developers, Rivest, Shamir, and Addleman.[13] The strength of the RSA algorithm is based on the difficulty of finding the prime number factors of a large number, which would be required to generate the private key from the corresponding public key.

Asymmetric cryptographic key sizes have to be much larger than symmetric key sizes to obtain the same level of security. Table 2-4 lists the equivalent key strengths in bits of symmetric and asymmetric cryptographic systems to obtain the same level of protection through encryption.

An important application of asymmetric cryptography is in generating digital signatures.

**Table 2-4. Symmetric vs. Asymmetric Key Sizes**

*Source: NIST SP 800-57 (2011)*

| Symmetric Key Size | Asymmetric Key Size |
|---|---|
| 80 | 1024 |
| 112 | 2048 |
| 128 | 3072 |
| 192 | 7680 |
| 256 | 15360 |

## Digital Signatures

The NIST Digital Signature Standard (DSS)[14] defines a digital signature as "the result of a cryptographic transformation of data that, when properly implemented, provides a mechanism for verifying origin authentication, data integrity, and signatory nonrepudiation." A digital signature uses asymmetric key encryption to achieve nonrepudiation of a transmitted message and to authenticate the sender and receiver.

As part of the implementation of a digital signature, a variable-length message is fed into a type of encryption system, known as a *hash function*, which produces a shorter, fixed-length output digital stream—a message digest. The message digest is uniquely bound to the input message and any change in the input message will result in a change in the hash function output. This characteristic provides integrity checking and is also used to implement nonrepudiation. The ideal characteristics of a hash function are:

- It is a one-way function because the original input file cannot be created from the message digest.
- Two files should not have the same message digest.
- The message digest should be calculated by using all the original input file's data.
- It should not be feasible to find another file with the same message digest, given a file and its corresponding message digest.

The digital signature of a message, M, is accomplished through the following steps:

1. Generate the message digest of M using a valid hash function.

2. Encrypt the message digest with the private key of the sender, and attach it to M.
3. Transmit M with the attached message digest.
4. Decrypt the message digest at the receiver's end by using the sender's public key. Because the public key of the sender is the only key that can open a file encrypted with the sender's private key, the sender is authenticated.
5. Compute the message digest of M at the receiving end, using the same hash function as the sender.
6. Compare the message digest computed at the receiving end with the message digest transmitted by the sender together with message M. If the two message digests are identical, this verifies that the message was not modified after it left the sender.

Note that in this process, the message itself is not encrypted. Figure 2-10 summarizes the digital signature process.

An important issue that has to be addressed in using asymmetric cryptographic systems is the possibility that an individual, A, would falsely have the name of another person, B, attached to A's public key. Then, any messages sent to B could only be opened and read by A using A's private key. Thus, A would be reading all of B's messages. In order to thwart this kind of attack, an independent third party could be used to certify that a person's public key was actually the key associated with that person. This independent third party is known as a *certificate authority* and, if presented with the proper credentials, can warrant that an individual's public key is his or her valid public key and can generate a digital certificate to that effect. The certificate contains the subject's name, the subject's public key, the name of the certificate authority, and the period in which the certificate is valid. This certificate is then sent to a repository, which holds the certificates and certificate revocation lists (CRLs) that denote any revoked certificates.

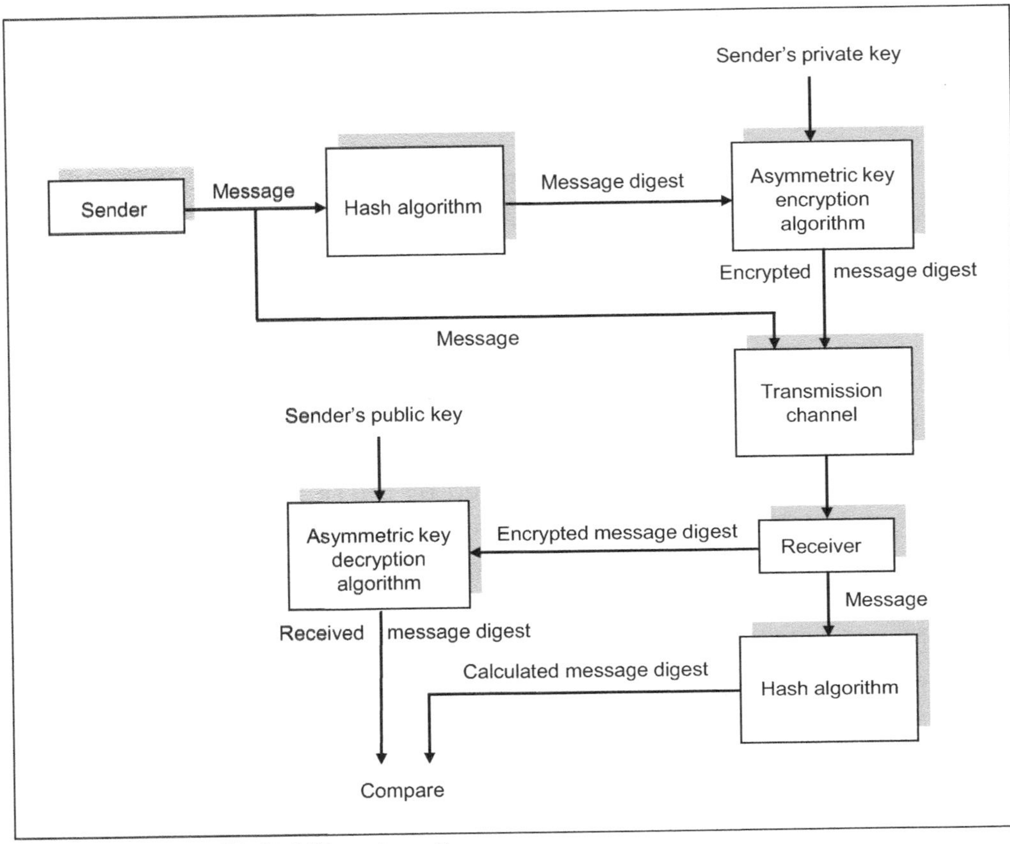


**Figure 2-10. The Digital Signature Process**

## Attacks Against Cryptosystems

The formal term for attacking cryptographic systems is cryptanalysis. In cryptanalysis, an attacker attempts to obtain the original, unencrypted message (plaintext) from the encrypted message (ciphertext). A number of approaches have been developed to perform cryptanalysis, from using mathematic formulations to measuring electronic emissions from chips. The most common attacks are summarized asfollows:

- **Known Plaintext Attack** – The attacker has access to samples of the unencrypted and encrypted versions of messages and uses this information to attempt to determine the key.
- **Ciphertext-Only Attack** – The attacker has access to multiple samples of encrypted messages that have been encrypted withthe same algorithm and attempts to find the key.

- **Man-in-the-Middle Attack** – The attacker intercepts messages being sent between two parties and attempts to decipher them and, possibly, to forward false messages to the parties.
- **Replay Attack** – The attacker intercepts a message and sends it at a later time.
- **Chosen Plaintext Attack** – The attacker has the ability to select plaintext messages, have them encrypted, and then analyze the results to determine the key.
- **Chosen Ciphertext Attack** – The attacker has the ability to select ciphertext messages for trial decryption while having access to the corresponding decrypted plaintext.
- **Adaptive Chosen Ciphertext** – The attacker conducts a chosen ciphertext attack but uses the results of one iteration to select the ciphertext to be used in the next iteration.
- **Adaptive Chosen Plaintext** – The attacker conducts a chosen plaintext attack but uses the results of one iteration to select the plaintext to be used in the next iteration.
- **Differential Cryptanalysis** – The attacker analyzes ciphertext pairs generated by the encryption of plaintext pairs and attempts to use any differences in the ciphertext pairs to determine the key.
- **Linear Cryptanalysis** – The attacker analyzes ciphertext pairs and their corresponding plaintext pairs to develop a linear approximation to portions of the key.
- **Brute Force** – The attacker tries all possible combination of keys in the hope of finding the correct key.

Encryption and digital signatures can be used effectively to transmit sensitive information securely over the Internet. One of the major means of accomplishing this type of transmission is the use of a virtual private network.

## Virtual Private Network

NIST SP 800-82[15] defines a virtual private network (VPN) as:

> A restricted-use, logical (i.e., artificial or simulated) computer network that is constructed from the system resources of a relatively public,

> physical (i.e., real) network (such as the Internet), often by using encryption (located at hosts or gateways), and by tunneling links of the virtual network across the real network.

ANSI/ISA-62443-1-1 (99.01.01)-2007[16] gives the following definition of a VPN:

> A VPN is a private network that operates as an overlay on a public infrastructure. It contains three components, which are handled at the recipient end of the VPN:
>
> – **Authenticity and Authentication** – This component consists of security measures designed to establish the validity of a transmission, message, originator, or a means of verifying an individual's authorization to receive specific categories of information.
>
> – **Integrity** – In a formal security mode, integrity is interpreted more narrowly to mean protection against unauthorized modification or destruction of information.
>
> – **Confidentiality** – This component provides assurance that information is not disclosed to unauthorized persons, processes, or devices.

A secondary component of VPNs is authorization, which encompasses:

- The rights granted to a user to access, read, modify, insert, or delete certain data, or to execute certain programs
- Access privileges granted to a user, program, or process

In other words, the purpose of a VPN is to provide a user with the ability to use a public network, such as the Internet, to transmit and receive information that must be protected from compromise. A typical VPN is illustrated in Figure 2-11.

There are three general types of VPNs:

- **Gateway-to-Gateway** – This VPM provides secure communications over a public network between two trusted networks, such as between two organizations.

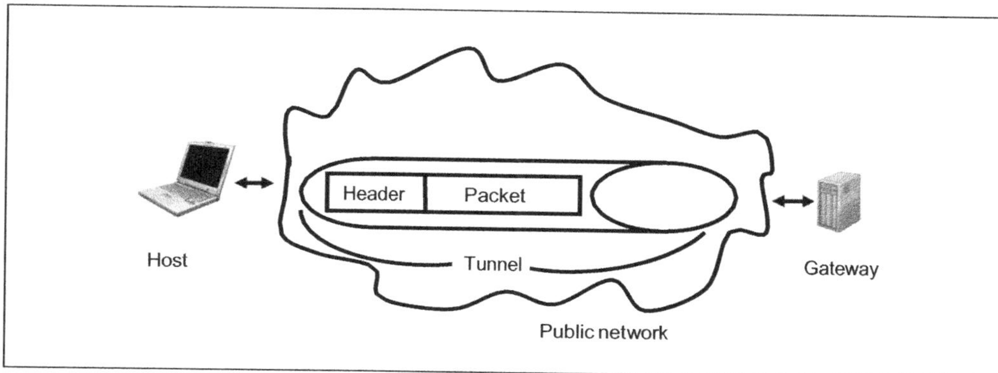


**Figure 2-11. A Typical Virtual Private Network**

- **Host-to-Gateway** – Also known as a *remote access VPN*, it provides secure communication between a host computer and an organization's network. An example of this type of VPN is a traveling salesman connecting his laptop to the organization's network from a remote location.
- **Host-to-Host** – This VPM enables secure communication between two computers over an untrusted network.

The characteristics of VPNs lend themselves to their use in industrial automation and control systems through their ability to establish secure communications between host systems and controllers over the Internet.

A VPN uses a tunnel through the public network to transmit the data to be protected. Establishing a tunnel, or tunneling, consists of encapsulating packets of a particular VPN protocol inside headers and transmitting the packets over another carrier protocol. The transmission takes place over a logical network connection established between the VPN client and server. The information being sent may be encrypted or not encrypted. For maximum security, the packets should be encrypted using an approved algorithm and a strong key. At the destination host, the reverse procedure takes place and the packets are unencapsulated and decrypted, if necessary. Two of the most popular VPN technologies are the Internet Security Protocol (IPsec) and Secure Sockets Layer (SSL).

### IPsec

IPsec operates at the OSI network layer and provides the following two modes of encryption:

- **Transport Mode** – Encrypts the data segment of the packet to be transmitted and does not encrypt the header
- **Tunnel Mode** – Adds an additional header to the packet and encrypts the entire packet, including the original header

IPsec operates with two security protocols, Authentication Header (AH) and Encapsulating Security Payload (ESP). AH can provide authentication of the source and integrity of packets and data; however, it does not use encryption. ESP supports encryption and also provides authentication and integrity. Of the two protocols, ESP is more commonly used.

### Secure Sockets Layer

The Secure Sockets Layer (SSL) protocol was developed in 1994 by Netscape to provide for secure Internet transactions. SSL encrypts the packet contents and authenticates the server to the client using digital signatures and digital certificates. The Internet Engineering Task Force (IETF)[17] modified SSL 3.0 to produce the Transaction Layer Security (TLS) protocol, which has evolved to version 1.2.[18] SSL and TLS are popularly used to secure Internet HTTP communications and are known as HTTP secure (HTTPs). Information on applying TLS is given in NIST SP 800-52.[19]

## Summary

Chapter 2 has reviewed some of the basic tenets of information system security with the objective of developing a foundation for applying these methods, with appropriate adjustments, to harden and protect industrial automation and control systems. Many of the attack types reviewed in this chapter can be used against industrial automation and control systems in order to cause malfunctions, cause physical and economic damage, or even threaten lives. The countermeasures and techniques presented in Chapter 2, when properly applied, have the ability to improve the security posture of the systems controlling the nation's infrastructure and manufacturing facilities, thereby preventing or mitigating disruptive and damaging occurrences.

With the security concepts of this chapter in mind, Chapter 3 will explore the similarities and differences between industrial automation and control system characteristics and requirements, as well as those of IT systems.

## Review Questions for Chapter 2

2.1 Information system security is defined as comprising which of the following three basic elements?

A. Confidentiality, integrity, and authorization

B. Confidentiality, integrity, and availability

C. Audit, integrity, and availability

D. Security, integrity, and availability

2.2 Protecting documents and messages from unauthorized disclosure refers to which of the following?

A. Confidentiality

B. Availability

C. Authorization

D. Integrity

2.3 An attack that overloads the resources of a computing system is an attack against which of the following?

A. Integrity

B. Availability

C. Confidentiality

D. Authentication

2.4 Which of the following items refers to the act of verifying a user's identity and confirming that the user is who he or she professes to be?

A. Authentication

B. Authorization

C. Registration

D. Accountability

2.5 Which of the following is a detective activity used to determine if violations of information system security have occurred?

A. Confirming

B. Authenticating

C. Auditing

D. Authorizing

2.6 Ensuring that the sender of a message or contract cannot later deny sending the message or contract is known as which of the following?

A. Nonrepudiation

B. Denial

C. Validation

D. Confirmation

2.7 A security control that minimizes the effect of an attack and the degree of resulting damage is known as which type of control?

A. Corrective

B. Preventive

C. Deterrent

D. Detective

2.8 The Information Assurance Technical Framework Forum (IATF) Document (Table 2-1) defines which of the following attacks as an "attempt to circumvent or break protection features, introduce malicious code, or steal or modify information"?

A. Distribution

B. Close-in

C. Active

D. Passive

2.9 The act of establishing numerous layers of protection wherein a subsequent layer will provide protection if a previous layer is breached is known as which of the following?

A. Defense in depth

B. Complete mediation

C. Least privilege

D. Open design

2.10 Which of the following is noncompulsory?

A. Standards

B. Guidelines

C. Policies

D. Procedures

2.11 A self-replicating or self-reproducing program that spreads by inserting copies of itself into other executable code or documents is known as which of the following?

A. Worm

B. Mobile code

C. Back door

D. Virus

2.12 An attack that focuses on the authentication protocol between a claiming party and a verifying party communicating with each other is known as which of the following?

A. Meet-in-the-middle

B. Man-in-the-middle

C. Social engineering

D. Brute force

2.13 NIST SP 800-41 defines which of the following as "devices or programs that control the flow of network traffic between networks or hosts that employ differing security postures"?

A. Firewalls

B. Demilitarized zones

C. Trap doors

D. Browsers

2.14 What type of firewall filters packets based on firewall access control lists (ACLs), which specify the packets that can be sent to particular destination addresses or on specific application port destinations?

A. Stateful inspection

B. Packet filtering

C. Application proxy

D. Screened-host

2.15 What type of firewall serves as an intermediary between networks and masks the source of a message?

A. Application

B. Packet filtering

C. Stateful inspection

D. Application proxy

2.16 What type of firewall defines a demilitarized zone (DMZ) or perimeter network, which is a third network that lies between the internal trusted network and the external untrusted network as an added layer of protection?

A. Dual-homed host

B. Screened-subnet

C. Screened-host

D. Application

2.17 What type of cryptography uses public and private key pairs?

A. Symmetric

B. Asymmetric

C. Secret key

D. Transposition

2.18 The Advanced Encryption Standard (AES) is an example of which of the following types of cryptography?

A. Symmetric key

B. Asymmetric key

C. Public key

D. Dual key

2.19 The result of a variable length message being fed into a hash function is a shorter, fixed-length output digital stream, known as which of the following?

A. Ciphertext

B. Plaintext

C. Message digest

D. Cryptovariable

2.20 A digital signature uses asymmetric key encryption to achieve which of the following?

A. Nonrepudiation

B. Nonrepudiation and authentication

C. Authorization

D. Confidentiality

2.21 A cryptographic attack in which the attacker uses multiple samples of encrypted messages that have been encrypted with the same algorithm to attempt to find the key is known as which of the following?

A. Chosen ciphertext

B. Ciphertext only

C. Adaptive chosen ciphertext

D. Known plaintext

2.22 A private network that operates as an overlay on a public infrastructure is known as which of the following?

A. Virtual private network

B. Dual-band network

C. 4G network

D. Demilitarized zone

2.23 IPsec provides which of the following two modes of operation?

A. Symmetric key mode and transport mode

B. Authentication mode and tunnel mode

C. Transport mode and tunnel mode

D. Transport mode and transparent mode

2.24 A VPN that provides secure communications over a public network between two trusted networks is known as which of the following?

A. Host-to-host

B. Gateway-to-gateway

C. Host-to-gateway

D. Host-to-demilitarized zone

## References

1. NIST SP 800-53. *Recommended Security Controls for Federal Information Systems*. Revision 4. Washington, DC: NIST (National Institute of Standards and Technology), 2013.

2. NSA (National Security Agency). *Information Assurance Technical Framework (IATF)*. Release 3.1. Fort Meade, MD: NSA, 2002.

3. Saltzer, J. H., and M. D. Schroeder. "The Protection of Information in Computer Systems." Paper presented at the 4th ACM Symposium on Operating Systems Principles. October 1974.

4. NIST SP 800-82. *Guide to Industrial Control Systems (ICS) Security*. Revision 2. Washington, DC: NIST (National Institute of Standards and Technology), 2015.

5. ANSI/ISA-62443-1-1 (99.01.01)-2007. *Security for Industrial Automation and Control Systems – Part 1: Terminology, Concepts, and Models*. Research Triangle Park, NC: ISA (International Society of Automation).

6. NIST SP 800-123. *Guide to General Server Security*. Washington, DC: NIST (National Institute of Standards and Technology), 2008.

7. NIST SP 800-12. *An Introduction to Computer Security: The NIST Handbook*. Washington, DC: NIST (National Institute of Standards and Technology), 1995.

8. NIST SP 800-53. *Recommended Security Controls for Federal Information Systems*. Revision 4. Washington, DC: NIST (National Institute of Standards and Technology), 2013.

9. ANSI/ISA-62443-1-1 (99.01.01)-2007. *Security for Industrial Automation and Control Systems – Part 1: Terminology, Concepts, and Models*. Research Triangle Park, NC: ISA (International Society of Automation).

10. NIST SP 800-41. *Guidelines on Firewalls and Firewall Policy*. Revision 1. Washington, DC: NIST (National Institute of Standards and Technology), 2009.

11. ANSI/ISA-62443-1-1 (99.01.01)-2007. *Security for Industrial Automation and Control Systems – Part 1: Terminology, Concepts, and Models*.

Research Triangle Park, NC: ISA (International Society of Automation).

12. FIPS 197 (Federal Information Processing Standard Publication 197). *Advanced Encryption Standard (AES)*. Washington, DC: NIST (National Institute of Standards and Technology), 2001.

13. Rivest, R. L., A. Shamir, and L. M. Addleman. "A Method for Obtaining Digital Signatures and Public-Key Cryptosystems." *Communications of the ACM 21* no. 2 (1978): 120–126.

14. FIPS 186-3 (Federal Information Processing Standard Publication 186-3). *Digital Signature Standard (DSS)*. Washington, DC: NIST (National Institute of Standards and Technology), 2009.

15. NIST SP 800-82. *Guide to Industrial Control Systems (ICS) Security*. Revision 2. Washington, DC: NIST (National Institute of Standards and Technology), 2015.

16. ANSI/ISA-62443-1-1 (99.01.01)-2007. *Security for Industrial Automation and Control Systems – Part 1: Terminology, Concepts, and Models*. Research Triangle Park, NC: ISA (International Society of Automation).

17. The Internet Engineering Task Force. www.ietf.org.

18. The Internet Engineering Task Force, Network Working Group. *RFC 5246. The Transport Layer Security (TLS) Protocol*. Version 1.2. August 2008.

19. NIST SP 800-52. *Guidelines for the Selection and Use of Transport Layer Security (TLS) Implementations*. Washington, DC: NIST (National Institute of Standards and Technology), 2005.

# 3

# Industrial Automation and Control System Culture versus IT Paradigms

*Some of the basic principles of information system security were presented in Chapter 2 as a prelude to selectively and properly applying them to securing industrial automation and control systems. As a prerequisite to this adaptation, it is important to examine the differences in culture, requirements, and operational issues between automation and control systems and IT systems. Critical areas that have to be addressed include safety, real-time demands, maintenance, productivity, training, and personnel mindsets. These topics and related subject areas are discussed in this chapter to help the reader better understand how to apply security principles to automation and control systems without negatively impacting their primary mission and in full acknowledgement of their special requirements.*

## Differences in Culture, Philosophy, and Requirements

The major advances in securing computer systems and networks have come through the information system technology route, with origins in computer science and software engineering. The principal players are IT system administrators, systems analysts, database administrators, software engineers, network administrators, chief information officers (CIOs), and so on. On the other hand, a large number of the personnel populating the industrial automation and control system field come from engineering backgrounds, with training in such areas as electrical engineering, chemical engineering, mechanical engineering, systems engineering, and control engineering.

The motivation, requirements, and focus of each of the groups are, in many instances, largely divergent, with some overlapping common areas. For example, software quality and process improvement methods widely used in the IT environment are often foreign to control engineers and in fact may be viewed as cumbersome in implementing SCADA and process control algorithms. In addition, the performance of a process in a plant is critical, and inadequate performance in production areas can result in huge financial losses, equipment damage, and personnel injuries. These severe consequences of operational errors are not usually a common occurrence in IT facilities. Similarly, safety is a critical concern in a production environment, and control system malfunctions can result in fires or explosions in some instances. Thus, in a production environment, safety and performance usually take precedence over information security, which is not the case in an IT system.

Some of the major differences between IT and industrial automation and control system requirements are listed in Table 3-1.

**Table 3-1. Comparison between IT and Industrial Control and Automation Systems Issues**

| Issue | IT Systems | Industrial Automation and Control Systems |
|---|---|---|
| Application of encryption | Advanced encryption systems used. | Encryption used sometimes or not at all. Delays caused by encryption software are a consideration. |
| Degree of employment of patch management | Implementation of software patches performed routinely under formal procedures. | Installation of software patches usually done infrequently, with deliberation, and with involvement of vendors. The impact of software patches should be tested off-line to ensure no harmful effects are introduced into the process or plant. In some instances, adding patches can produce some dangerous consequences. |
| Degree of vulnerability testing | Penetration testing by ethical hackers is widely used to expose vulnerabilities. | Penetration testing must be conducted sparingly and at appropriate times in order not to inadvertently cause malfunctions in monitoring and control systems. |

## Table 3-1. Comparison between IT and Industrial Control and Automation Systems Issues

| | | |
|---|---|---|
| Employment and impact of security software | Antivirus software is effective and commonly used to scan for malware. | Antivirus software can be used but issues, such as delays due to scanning and processing and lack of spare computing cycles, might cause problems in plant systems. |
| Employment of change management | Change management is used extensively to document and confirm system changes. | Change management is employed inconsistently and sporadically in many organizations. Changes might also result in hazardous situations. |
| Impact of equipment upgrades | Equipment modifications, upgrades, and replacements occur almost continuously in various parts of organizations. | Many legacy systems in use with few changes over periods of many years. |
| Level of security awareness training | Awareness training is actively conducted in accordance with security policy. | Security awareness training is conducted sporadically and inconsistently in many cases. |
| Tolerance to data loss | Recovery from data loss can usually be accomplished from backups without severe consequences in most situations. | Loss of data can result in loss of product, unsafe conditions, and disruption of plant operations. |
| Tolerance to loss of availability | Recovery follows standard procedures and unless availability is lost for extended periods of time, effects are not disastrous. However, availability is critical and can affect mission objectives. | Systems should be designed with resiliency; however many are not, particularly legacy systems. Loss of availability can have serious consequences, including safety of personnel and equipment. |
| Tolerance to loss of confidentiality | Depending on the data involved, the consequences might be minimal or extremely harmful, such as the loss of trade secrets, personnel data, credit card numbers, and health information. | Consequences not usually severe. |
| Tolerance to loss of integrity | Consequences can range from minimal to extremely harmful if critical data has been altered and is acted on as being valid. | Consequences can be very severe because, in many instances, critical decisions are made based on data assumed to be correct. |

**Table 3-1. Comparison between IT and Industrial Control and Automation Systems Issues**

| | | |
|---|---|---|
| Tolerance to system delays | In most cases, delays might be considered a nuisance, but can be tolerated. | In situations where critical control decisions are based on deterministic response times, delays might result in serious or dangerous consequences. |
| Use of auditing | Based on organizational polices, internal and external audits are normally conducted and provide detective controls for breaches of information system security. | Audits of a systems' security posture are conducted irregularly and are a function of the particular organization and regulatory requirements. |
| System performance | Performance can be suboptimal at times. | Proper performance is critical. |
| Safety | Safety is critical, particularly that of personnel. | Safety is critical, particularly that of personnel. |

Figure 3-1 summarizes the important issues listed in Table 3-1 and emphasizes some of the common areas between IT and automation and control systems.

The lesson to be learned from these comparisons is that traditional information system security knowledge and methods provide a solid basis for addressing industrial automation and control system security, albeit with deliberate, appropriate, and intelligent modifications required to address the unique characteristics of automation and control systems.

One important starting point in incorporating these modifications is education. In general, most universities and certification programs addressing computer and network security have been heavily focused on IT security. Automation and control systems, which are typically sitting on isolated networks and are relatively few in number compared to IT systems, have not been considered to be interesting targets. With the advent of the terrorism threat, this situation is no longer the case. In addition, SCADA and plant process control systems are now being connected to large networks and the Internet.

The Certified Information System Security Professional (CISSP) and related certifications do not address the security of industrial automation and control

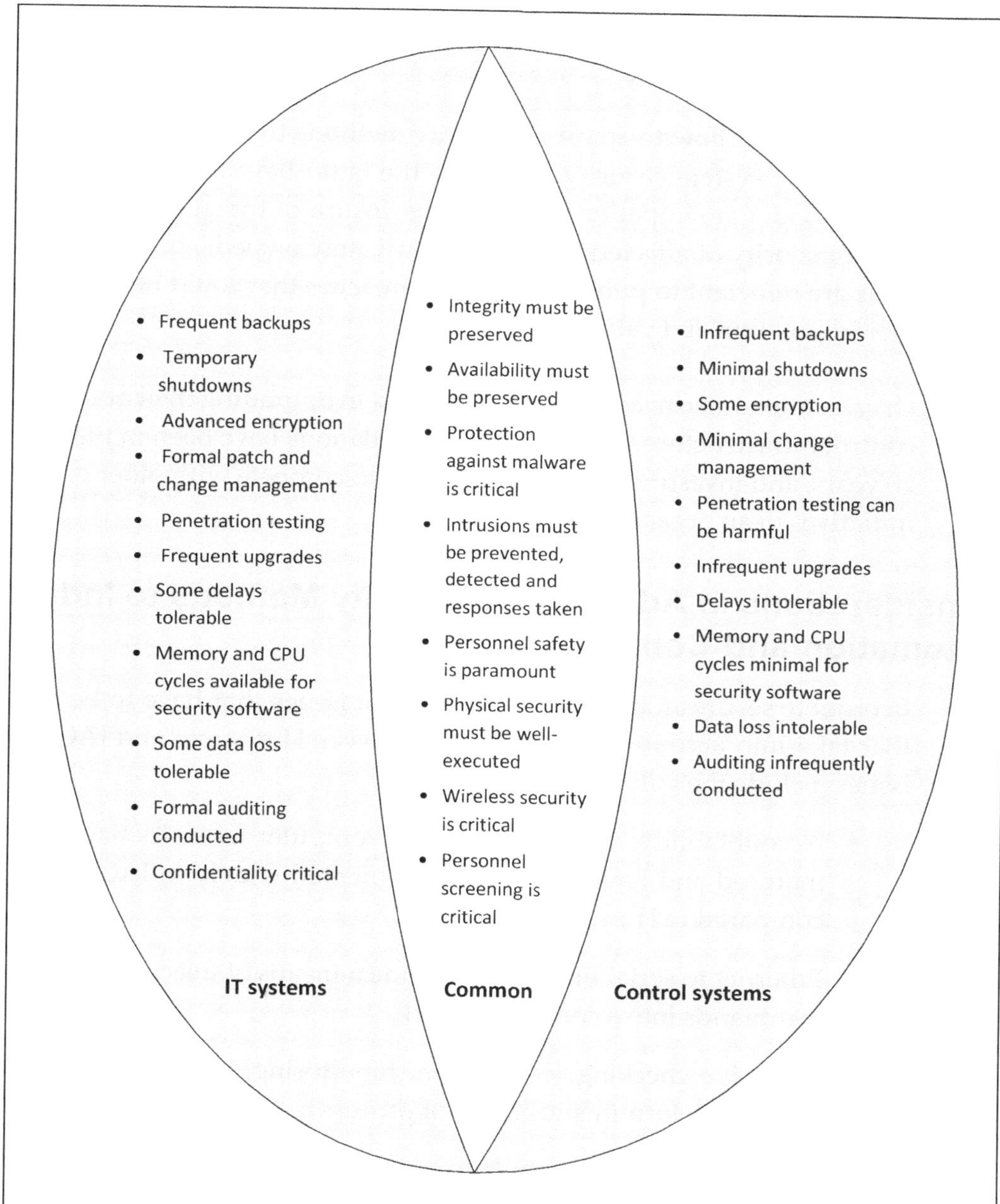


**Figure 3-1. IT and Automation and Control System Issue Comparisons**

systems. Organizations, such as ISA, have addressed this problem and are filling a critical need. NIST has generated special publications that directly address industrial automation and control systems. However, it is important that security training related to the control of production lines, industrial processes, electrical transmission and distribution, pipelines, chemical plants, and so on moves to the fore in universities, technical institutes, and certification organizations.

To understand how to adapt IT security methods to industrial automation and control system security, threats to the latter have to be identified and understood. One impediment to full disclosure of threats realized is the fact that a majority of affected facilities are privately owned, and these organizations are reluctant to publicize security breaches that could negatively affect their reputation and value.

Organizations also need incentives to invest in upgrading their automation and control infrastructure. Many existing installations have been in place for 10 or 20 years, and investments in security have to compete with other compelling initiatives in an organization.

## Considerations in Adapting IT Security Methods to Industrial Automation and Control Systems

In order to secure an IACS, there are specific issues that have to be addressed that take into account the differences between IT systems and IACSs. These issues include the following:

- Accountability, authorization, and computer forensics have not matured and have not been implemented widely in IACSs as compared to IT systems.
- Ethernet to serial line paths provide a means of injecting malicious commands into a control network.
- Excessive checking, encryption, monitoring, and so on can interfere with the deterministic nature of process control systems.
- In many IACS environments, control engineers have multiple responsibilities that, in many instances, violate the security principle of separation of duties.
- Installing patches and upgrades in process control systems can lead to serious and sometimes dangerous situations in production facilities.
- Life-cycle design disciplines common in the IT field are not widely used in industrial automation and control systems.

- Maintenance hooks and trap doors installed in automation and control systems for remote maintenance can be easy entry points to modify critical software and firmware with negative consequences.
- Many IACS vendors combine safety mechanisms with security mechanisms, leading to single points of failure and less resiliency than separating these two functions logically and physically.
- Many manufacturing facilities and SCADA installations house legacy systems with outdated technology, minimal memory and computing power, and little thought to security.
- Port scanning of automation and control systems can result in blockages and lack of system availability.
- Remote access into automation and control systems via older modems or newer wireless devices poses a serious threat to security.
- There is a trend to apply protocols used for IT systems to industrial control and automation systems because of their wide availability, their lower cost, and the existence of trained personnel. However, in most instances, these protocols were not designed for deterministic process control systems, and they are vulnerable to many existing attacks.
- There is heavy reliance on suppliers who provide modified software and hardware for IACSs, resulting in nonstandard implementations that are difficult to maintain without support from these suppliers.
- Weak authentication mechanisms in many SCADA systems and networked plant control systems leave them vulnerable to attack.

A variety of additional items must be considered when discussing comparisons between IT and industrial automation and control systems. The concepts related to risk management and the means to protect industrial automation and control systems will be discussed in detail in Chapters 5 and 6, respectively. However, it is important to now examine some related critical subject areas to provide a basis for developing more specific security solutions.

## Threats

Threats to IT and industrial automation and control systems come from different sources, with different motivations. It is important to understand these threat sources and their characteristics in order to counter any malicious activities on their part. NIST SP 800-30[1] summarizes the various types of threat sources and some of their driving factors, as shown in Table 3-2. Table 3-3, also from NIST SP 800-30, provides a listing of some general threat sources, including environmental ones, which can also cause disruptions to industrial automation and control systems.

The categories of terrorists, industrial espionage, and insiders are of particular interest in connection with industrial automation and control systems. Traditionally, insider threats have been considered one of the most dangerous because they give insiders the ability to bypass protective measures. However, external threats are increasing and are also of grave concern, particularly relating to our nation's critical infrastructure and resource processing plants. In addition, threats to automation systems can materialize from environmental and structural sources, as illustrated in the next section.

## Sensitivity of Industrial Automation and Control Systems to Upgrades and Modifications

One area that is not usually considered when discussing the relative sensitivities of IT systems and industrial automation and control systems is the effects of equipment upgrades and modifications. A particularly relevant example concerns the consequences of converting analog controls to digital controls. Digital systems transfer information via pulses, which inherently generate high frequency electromagnetic radiation that can interfere with control system operations. An article in the journal *Interference Technology*[2] describes the electromagnetic radiation emission environment in a nuclear plant that was being changed from analog to digital controls. The authors obtained measurement data in the range of 100 Hz to 6 GHz in instances before and after the conversion.

**Table 3-2. Threats and Motivations for Attackers**

*Source: NIST SP 800-30 (2012)*

| Threat-Source | Motivation | Threat Actions |
|---|---|---|
| Hacker, cracker | Challenge Ego Rebellion | • Hacking<br>• Social engineering<br>• System intrusion, break-ins<br>• Unauthorized system access |
| Computer criminal | Destruction of information Illegal information disclosure<br>Monetary gain<br>Unauthorized data alteration | • Computer crime (e.g., cyber stalking)<br>• Fraudulent act (e.g., replay, impersonation, interception)<br>• Information bribery<br>• Spoofing<br>• System intrusion |
| Terrorist | Blackmail Destruction<br>Exploitation Revenge | • Bomb/terrorism<br>• Information warfare<br>• System attack (e.g., distributed denial of service)<br>• System penetration<br>• System tampering |
| Industrial espionage (companies, foreign governments, other government interests) | Competitive advantage<br>Economic espionage | • Economic exploitation<br>• Information theft<br>• Intrusion on personal privacy<br>• Social engineering<br>• System penetration<br>• Unauthorized system access (access to classified, proprietary, and/or technology-related information) |
| Insiders (poorly trained, disgruntled, malicious, negligent, dishonest, or terminated employees) | Curiosity Ego Intelligence<br>Monetary gain<br>Revenge<br>Unintentional errors and omissions (e.g., data entry error, programming error) | • Assault on an employee<br>• Blackmail<br>• Browsing of proprietary information<br>• Computer abuse<br>• Fraud and theft<br>• Information bribery<br>• Input of falsified, computed data<br>• Interception<br>• Malicious code (e.g., virus, logic bomb, Trojan horse)<br>• Sale of personal information<br>• System bugs<br>• System intrusion<br>• System sabotage<br>• Unauthorized system access |

The testing followed guidelines in U.S. Nuclear Regulatory Commission Regulatory Guide, NUREG 1.180,[3] and Electric Power Research Institute (EPRI) document TR-102023-2004.[4] In the tests, antennas were installed next to three cabinets housing control electronics, and radiation emission measurements were taken from the analog and digital control installations. Some of the results obtained are summarized in Table 3-4, showing frequencies at which peak amplitudes occur at antennas 1 and 2.

**Table 3-3. Listing of General Threat Sources**

*Source: NIST SP 800-30 (2012)*

| Type of Threat Source | Description | Characteristics |
|---|---|---|
| **Adversarial**<br>• Individual<br>• Outsider<br>• Insider<br>• Trusted insider<br>• Privileged insider<br>• Group<br>• Ad hoc<br>• Established<br>• Organization<br>• Nation-state | Individuals, groups, organizations, or states that seek to exploit the organization's dependence on cyber resources (i.e., information in electronic form, information and communications technologies, and the communications and information-handling capabilities provided by those technologies). | Capability, intent, targeting |
| **Accidental**<br>• Ordinary user<br>• Privileged user/ administrator | Erroneous actions taken by individuals in the course of executing their everyday responsibilities. | Range of effects |
| **Structural**<br>• IT Equipment<br>• Storage<br>• Processing<br>• Communications<br>• Display<br>• Sensor<br>• Controller<br>• Environmental controls<br>• Temperature/humidity controls<br>• Power supply<br>• Software<br>• Operating system<br>• Networking<br>• General-purpose application<br>• Mission-specific application | Failures of equipment, environmental controls, or software due to aging, resource depletion, or other circumstances that exceed expected operating parameters. | Range of effects |

### Table 3-3. Listing of General Threat Sources

*Source: NIST SP 800-30 (2012)*

| | | |
|---|---|---|
| **Environmental**<br>• Natural or man-made disaster<br>• Fire<br>• Flood/tsunami<br>• Windstorm/tornado<br>• Hurricane<br>• Earthquake<br>• Bombing<br>• Overrun<br>• Unusual natural event (e.g., sunspots)<br>• Infrastructure failure/outage<br>• Telecommunications<br>• Electrical power | Natural disasters and failures of critical infrastructures on which the organization depends, but which are outside the control of the organization.<br><br>NOTE: Natural and man-made disasters can also be characterized in terms of their severity and/or duration. However, because the threat source and the threat event are strongly identified, severity and duration can be included in the description of the threat event (e.g., Category 5 hurricane causes extensive damage to the facilities housing mission-critical systems, making those systems unavailable for three weeks). | Range of effects |

### Table 3-4. Radiation Emission Measurements

*Source: Keebler and Berger (2011)*

| Antenna | Analog Measurement | | Digital Measurement | |
|---|---|---|---|---|
| | Frequency | Significant Peak Amplitudes (dBμV/m) | Frequency | Significant Peak Amplitudes (dBμV/m) |
| 1 | 1.34 MHz | 99.2 | 468 MHz | 71.6 |
| 1 | 928 MHz | 76.6 | 826 MHz | 94.5 |
| 1 | | | 928 MHz | 113.5 |
| 1 | | | 1.35 GHz | 49.9 |
| 1 | | | 1.88 GHz | 50.5 |
| 1 | | | 1.92 GHz | 53.4 |
| 1 | | | 2.41 GHz | 76.3 |
| 1 | | | 2.46 GHz | 54.4 |
| 1 | | | 5.82 GHz | 60.6 |
| 2 | 1.04 MHz | 99.3 | 2 MHz | 84 |
| 2 | 4.55 MHz | 88.5 | 10 MHz | 80 |
| 2 | | | 1 GHz | 109 |
| 2 | | | 1.17 GHz | 48.8 |
| 2 | | | 1.92 GHz | 48.9 |
| 2 | | | 2.42 GHz | 57.7 |
| 2 | | | 5.82 GHz | 50.9 |

This data is plotted in Figures 3-2 and 3-3 for antennas 1 and 2, respectively. Note that the digital electronics generate more peak radiation generally and more at high frequencies compared to the analog equipment. These peak emissions have the potential to interfere with control system signals and cause malfunctions if proper shielding and isolation are not applied.

The sample electromagnetic emanations collected illustrate the necessity to ensure electromagnetic compliance (EMC) when equipment upgrades are made to plant control systems. These actions will serve to protect against interruptions of control systems' operation due to electromagnetic emissions from digital systems.

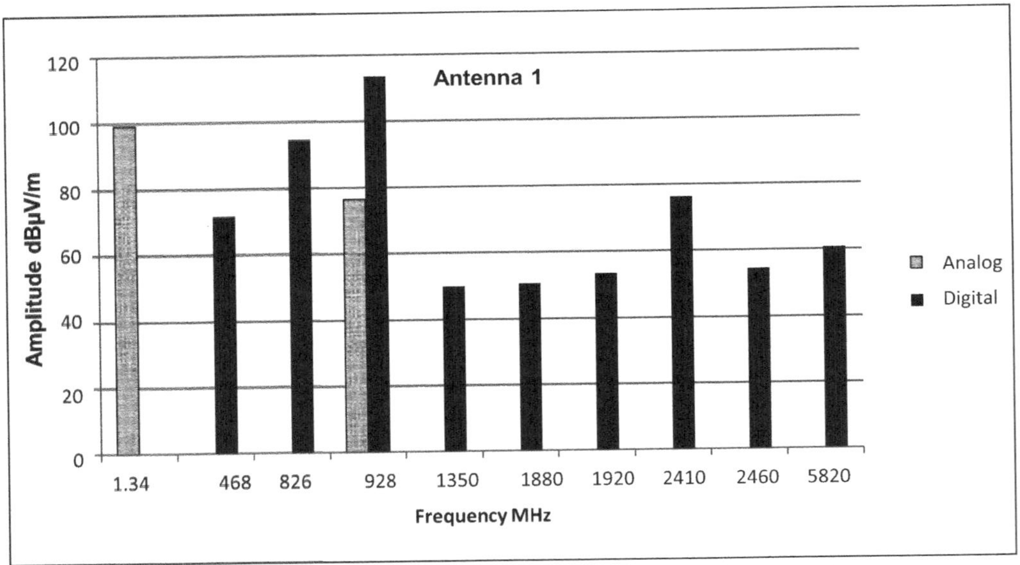


**Figure 3-2. Analog and Digital Radiation Emissions Received at Antenna 1**

## IT and Industrial Automation and Control Systems Comparisons from a Standards Perspective

Valuable insight into the contrasts and similarities between IT systems security focus areas and those of industrial automation and control systems can be obtained from an example using standards that represent each of the areas. In this example, ISO/IEC 27002, *Code of Practice for Information Security Management*,[5] will be used to represent IT systems security areas of emphasis while ANSI/ISA-62443-2-1 (99.02.01)-2009, *Security for Industrial Automation and Control Systems Part 2-1: Establishing an Industrial Automation and Control Systems*

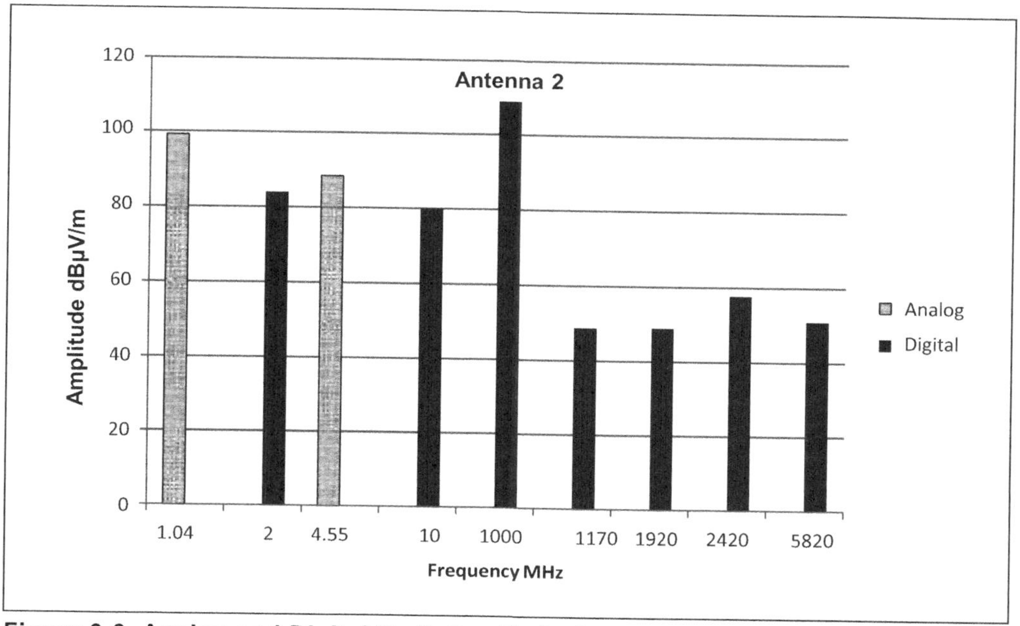


**Figure 3-3. Analog and Digital Radiation Emissions Received at Antenna 2**

*Security Program*,[6] will be used to illustrate the major concerns of automation and control system security.

In each document, there are common areas addressed by both standards and other areas addressed by one standard and not the other. Figure 3-4 summarizes the main characteristics of each standard and identifies common areas addressed by both, as well as topics that are addressed mainly by one document and not the other.

Figure 3-4 shows that topics, such as change management, email security, access control policies, digital signatures, compliance, and business continuity planning are among the areas considered critical for IT systems that are not emphasized in automation and control system standards. Conversely, for automation and control systems, the significant domains not covered include security architecture analysis, quantitative and qualitative analysis, information security management, and information security testing. Areas of common emphasis include information security policy, risk assessment, training, media physical security, remote access, event logging, and protection against malware.

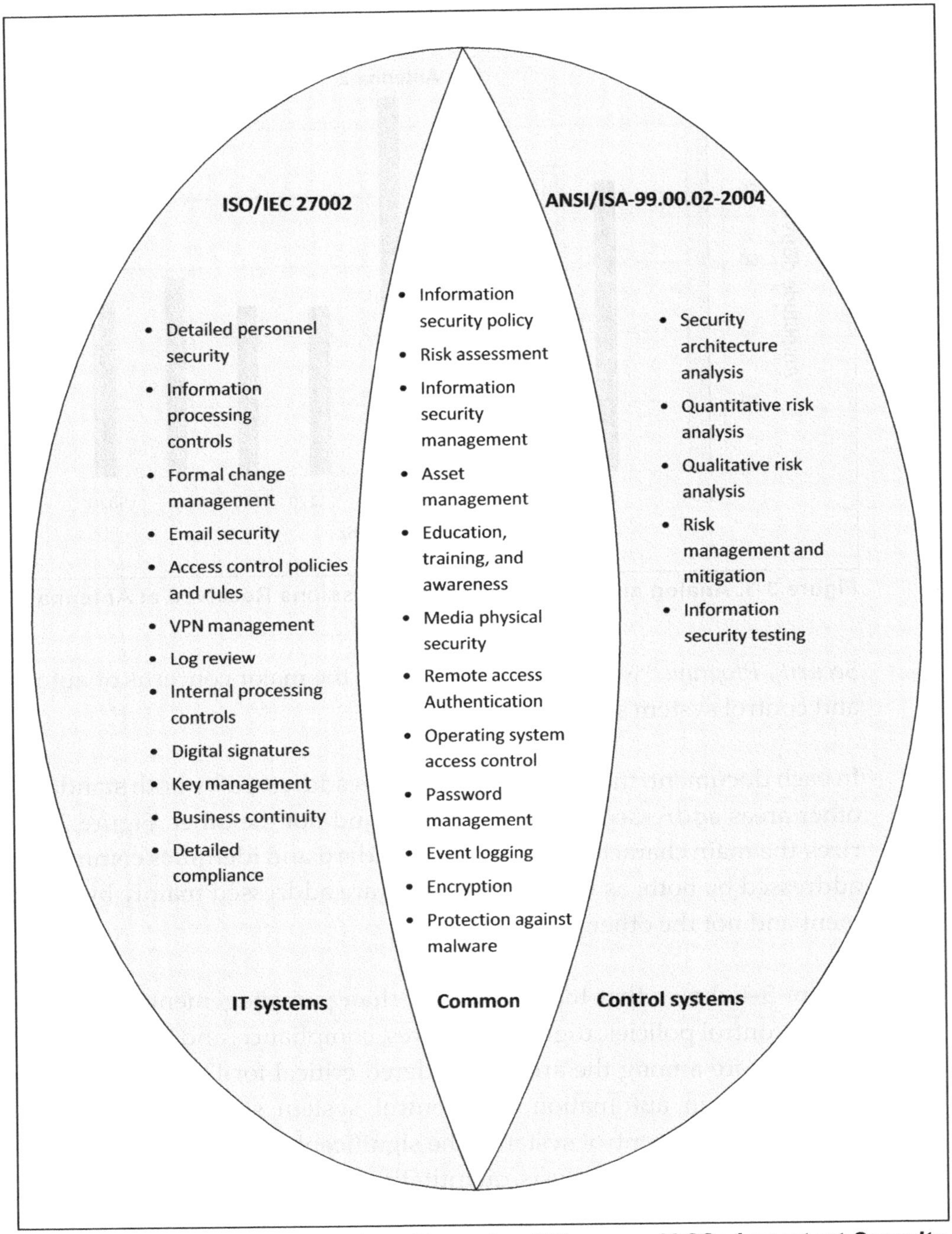


**Figure 3-4. Standards Comparison Example of IT versus IACSs Important Security**

## Summary

Understanding the requirements of industrial automation and control systems security and how they relate to IT systems requires a mapping of these requirements onto the emerging technologies being employed in the control of production processes, as well as the critical infrastructure (represented primarily by the electrical generation and distribution grid). The advances in capability and sophistication of industrial automation and control systems require a tailored approach to security. Some of the factors pushing the industrial automation and control systems security envelope include:

- The Smart Grid
- Advanced cryptography and key management applications
- Advanced PLCs and PACs
- Advanced protective relaying
- Advanced wireless networks
- Alarm processing
- Availability of real-time energy information
- Multisphere security among IT, transportation, and power systems
- Redundancy in networks, equipment, and sensors
- Fiber communication
- Use of GPS tracking

Chapter 4 will investigate the technical evolution taking place in elements of the critical infrastructure and in key production facilities to identify the important risk factors and areas of maximum potential impact in the event of an attack.

## Review Questions for Chapter 3

3.1 Which of the following statements is generally TRUE regarding an industrial automation and control system?

A. Installation of software patches can be performed routinely and frequently.

B. Encryption of data can sometimes lead to problematic delays.

C. Penetration testing can be conducted routinely and frequently.

D. Confidentiality is a key concern in automation systems as opposed to integrity and availability.

3.2 In both IT and automation and control systems, which of the following is the primary concern in the event of an emergency or malicious event?

A. Equipment safety

B. Preservation of documentation

C. Personnel safety

D. Facility protection

3.3 Which of the following statements is FALSE?

A. Flash drives and other portable memory devices can be sources of malware injections into control systems.

B. Maintenance hooks and trap doors installed in automation and control systems for remote maintenance can be easy entry points to modify critical software and firmware with negative consequences.

C. In many control system environments, control engineers, in general, do not have multiple responsibilities, such that the security principle of separation of duties is not normally violated.

D. Many facilities house legacy systems with outdated technology, minimal memory and computing power, and little thought to security.

3.4 Which of the following actions is the most likely to result in blockages and lack of system availability in automation and control systems?

A. Remote access

B. Life-cycle design

C. Accountability

D. Port scanning

3.5 Which threat source is motivated by revenge, ego, and dissatisfaction?

A. Insider

B. Espionage

C. Criminal

D. Hacker

3.6 What is a source of a possible disruption of control system functions that is not normally considered?

A. Changes from digital to analog systems

B. Upgrades from analog to digital systems

C. Malware

D. Attacks

3.7 In general, what distinguishes analog control equipment from digital control equipment?

A. Analog controls generate more high-frequency peak voltages than digital controls.

B. Digital controls generate more high-frequency peak voltages than analog controls.

C. Analog controls generate essentially the same number of high-frequency peak voltages as digital controls.

D. Digital controls generate essentially the same number of high-frequency peak voltages as analog controls.

3.8 Which of the following is more likely to be performed in an IT environment than in an automation and control system environment?

A. Security architecture analysis

B. Information security testing

C. Quantitative risk analysis

D. Change management

3.9 Which of the follow statements is FALSE?

A. In many IACS environments, control engineers have multiple responsibilities that, in many instances, violate the security principle of separation of duties.

B. Installing patches and upgrades in process control systems can lead to serious and sometimes dangerous situations in production facilities.

C. Life-cycle design disciplines common in the IT field are not widely used in industrial automation and control systems.

D. Many IACS vendors combine safety mechanisms with security mechanisms, which results in increased system reliability.

3.10 Which of the following threat sources is motivated by economic exploitation and competitive advantage, and uses social engineering?

A. Insider

B. Terrorist

C. Industrial espionage

D. Computer criminal

3.11 What is a detective control that is more frequently applied in IT systems than in control and automation systems?

A. Firewall

B. Separation of duties

C. Biometrics

D. Auditing

3.12 Which of the following is NOT a usual reason for an organization's reluctance to disclose successful attacks against it?

A. Hope the attack will harm competitors

B. Embarrassment

C. Effect on reputation

D. Possible loss of customers

3.13 Which of the following can lead to a single point of failure in an industrial automation and control system?

A. Separation of duties

B. Disk redundancy

C. Combination of safety and security mechanisms

D. Use of authentication with identification

3.14 What is a typical characteristic of industrial automation and control systems?

A. Have excess computing cycles

B. Have limited extra computing cycles

C. Have excess memory

D. Computational speed is not an issue

3.15 What is a typical characteristic of an automation and control system supplier?

A. Usually ensures maintenance hooks are never left enabled without the customer's approval

B. Usually ensures default passwords are never duplicated from one customer to another

C. Usually provides unmodified off-the-shelf hardware and software

D. Usually provides modified hardware and software

## References


1. NIST SP 800-30. *Guide for Conducting Risk Assessments*. Revision 1. Washington, DC: NIST (National Institute of Standards and Technology), 2012.
2. Keebler, P., and S. Berger. "Going from Analog to Digital." *Interference Technology, 2011 EMC Directory and Design Guide*, 2011.
3. U.S. Nuclear Regulatory Commission (NRC) Regulatory Guide NUREG 1.180. *Guidelines for Evaluating Electronic and Radio Frequency Interference in Safety-Related Instrumentation and Control Systems*. Revision 1. Washington, DC: U.S. Nuclear Regulatory Commission, 2003.
4. EPRI TR-102323-2004. *Guidelines for Electromagnetic Interference Testing in Power Plants*. Revision 3. Palo Alto, CA: EPRI (Electric Power Research Institute), 2004.
5. ISO/IEC Standard 27002-2005. *Information Technology – Security Techniques – Code of Practices for Information Security Management*. Geneva 20 – Switzerland: IEC (International Electrotechnical Commission) and ISO (International Organization for Standardization).
6. ANSI/ISA-62443-2-1 (99.02.01)-2009. *Security for Industrial Automation and Control Systems – Part 2-1: Establishing an Industrial Automation and Control Systems Security Program*. Research Triangle Park, NC: ISA (International Society of Automation).

# 4

# The Continuing Technological Evolution Affecting IAC Systems

*Chapter 3 presented the important differences between IT systems and industrial automation and control systems, along with some areas that are common to both of them. However, it is important to recognize that these differences are based on elements in both technologies that are constantly evolving. Thus, in order to truly understand and safeguard automation and control systems, these technology drivers and their potential effects on the components of the critical infrastructure have to be identified and examined. Chapter 4 investigates these areas and focuses on technological issues that relate directly to the security posture of the critical infrastructure, using the Smart Grid as an example.*

## Important Technological Trends

At any point in time, as technological advances occur and the technologies are adopted, related issues often appear. In this chapter, some of the major technological trends and associated concerns that directly and indirectly affect (or will affect) the industrial automation and control system landscape are examined. Then, the Smart Grid initiative will be explored relative to these technological developments as an example in the automation and control systems field.

Some of the emerging technological trends and associated concerns that will have a significant impact on both business and industry are home area networks, energy storage, analytics, cloud computing, privacy, social networks, mobile technology, and interoperability.

## Home Area Networks

An increasing number of homes have a home area network (HAN) to some extent. The most common implementation is connecting devices, such as PCs, printers, laptops, tablet devices, Internet telephones, smart phones, and game computers. Additional items that are being connected and that will be more commonly connected in the future are digital video recorders, TV sets, and network-attached storage (NAS) devices.

Relative to the Smart Grid, home area networks are being viewed as important network elements to connect to smart meters. It is anticipated that the home area network will host smart thermostat controls, user information displays, and interfaces to remotely control such items as appliances, lighting, and security systems. With this capability, a smart meter could communicate with appliances over the home area network to reduce energy consumption during periods of peak demand while also supplying information to customers, service providers, suppliers, and utilities. Customers could use this information to alter usage patterns and reduce energy costs.

## Energy Storage

Energy storage device technology is advancing because of the needs of hybrid and electric vehicles. In addition, finding a means to capture and store intermittent, renewable energy is driving such technologies as "super" batteries and ultracapacitors as well as other devices. The advent of the Smart Grid has put a new emphasis on efficient energy storage methods that can be applied at various points in the Grid to supplement conventional electricity generation methods.

## Analytics

Analytics is the application of data mining and analysis to large amounts of data for the purposes of discovering knowledge and making intelligent predictions and decisions. Because large volumes of data are being generated and captured in very short amounts of time in automation and control systems, being able to analyze this data and make sound decisions based on this analysis is becoming increasingly important and necessary. A typical example is using analytic tools to visualize data within databases, networks, and applications, and to make critical decisions in real time. A related issue is the ability to meaningfully display acquired information and share it among colleagues to support collaboration and innovative decision-making.

Traditionally, analytics has been applied in electrical power systems for condition monitoring of high voltage electrical equipment. However, analytics hold promise for more sophisticated support of industrial automation and control systems to support rapid decision-making and to provide an instinctively understandable visualization of critical information. Specifically, data collected on the Smart Grid through wireless, optical fiber, and other communication means can provide data on the health of Grid components, such as transformers, as well as voltage monitoring and energy consumption information. This data can be used to identify areas of potential failure and to initiate corrective action before actual outages occur.

Analytics can also be applied to secure industrial automation and control systems in the form of cyber analytics. In cyber analytics, data from a variety of sources can be analyzed in real time or near real time to perform predictive modeling and thereby support physical, logical, and administrative security.

## Cloud Computing

The cloud computing paradigm is being embraced by many organizations, and this will affect how data is collected, stored, and processed in these organizations, including in automation and control systems. NIST[1] defines cloud computing as:

> A model for enabling ubiquitous, convenient, on-demand network access to a shared pool of configurable computing resources (e.g., networks, servers, storage, applications, and services) that can be rapidly provisioned and released with minimal management effort or service provider interaction. This cloud model promotes availability and is composed of five essential characteristics, three service models, and four deployment models.

The NIST document defines the essential characteristics of cloud computing as:

- **On-Demand Self-Service** – A consumer can unilaterally provision computing capabilities, such as server time and network storage, as needed automatically without human interaction with each service's provider.
- **Broad Network Access** – Capabilities are available over the network and accessed through standard mechanisms that promote use by

heterogeneous thin or thick client platforms (e.g., mobile phones, laptops, and PDAs).

- **Resource Pooling** – The provider's computing resources are pooled to serve multiple consumers, using a multitenant model, with different physical and virtual resources dynamically assigned and reassigned according to consumer demand. There is a sense of location independence in that the consumer generally has no control over or knowledge about the exact location of the provided resources but may be able to specify a location at a higher level of abstraction (e.g., country, state, or data center). Examples of resources include storage, processing, memory, network bandwidth, and virtual machines.
- **Rapid Elasticity**– Capabilities can be rapidly and elastically provisioned—in some cases automatically—to quickly scale out and rapidly released to quickly scale in. To the consumer, the capabilities available for provisioning often appear to be unlimited and can be purchased in any quantity at any time.
- **Measured Service** – Cloud systems automatically control and optimize resource use by leveraging a metering capability at some level of abstraction appropriate to the type of service (e.g., storage, processing, bandwidth, and active user accounts). Resource usage can be monitored, controlled, and reported, providing transparency for both the provider and consumer of the utilized service.

The three NIST cloud service models are:

- **Cloud Software as a Service (SaaS)** –This service provides the customer with the capability to use the provider's applications running on a cloud infrastructure.
- **Cloud Platform as a Service (PaaS)** – This service provides the customer with the capability to deploy onto the cloud infrastructure consumer-created or acquired applications created using programming languages and tools supported by the provider.
- **Cloud Infrastructure as a Service (IaaS)** – This service provides the customer with the capability to provision processing, storage, networks, and other fundamental computing resources where the consumer is able to deploy and run arbitrary software, which can include operating systems and applications.

NIST defines the cloud deployment models as:

- **Private Cloud** –The cloud infrastructure is operated solely for an organization. It may be managed by the organization or a third party and may exist on premises or off premises.
- **Community Cloud** –The cloud infrastructure is shared by several organizations and supports a specific community that has shared concerns (e.g., mission, security requirements, policy, and compliance considerations). It may be managed by the organizations or a third party and may exist on premises or off premises.
- **Public Cloud** – The cloud infrastructure is made available to the general public or a large industry group and is owned by an organization selling cloud services.
- **Hybrid Cloud** – The cloud infrastructure is a composition of two or more clouds (private, community, or public) that remain unique entities but are bound together by standardized or proprietary technology that enables data and application portability.

Cloud computing emphasizes service and can result in lower costs, agility, and the ease of integration of business and industrial processes. It can also increase availability through the distribution of capabilities to areas both inside and outside of the organization. Availability can also be enhanced by the cloud's ability to automatically manage and transport virtual applications dynamically. This approach supports automatic failover, provisioning to adapt to load demands, enforcement of security policies, and monitoring of security violations.

The cloud can also provide exceptional resources in the implementation of analytics solutions.

## Privacy

Privacy is becoming more and more important in the conduct of business and computer-related activities. Special precautions have to be taken to protect personally identifiable information (PII) in all types of environments to avoid violation of regulatory and compliance requirements, such as the Health Insurance Portability and Accountability Act (HIPAA). PII is information that can be used to identify an individual and discover sensitive information about that individual. It is imperative that the management of an organization ensures that proper and adequate controls are in place to protect this sensitive information. These controls must include robust identification and authentication mechanisms to protect both individuals and physical equipment. Disclosure of PII (such as a user's password, social security number, address, and health situation) can be used to compromise the person and eventually gain access to critical organizational information and equipment. Threats to computer systems continually evolve and are becoming more sophisticated. Intentional or unintentional employee actions can bypass protection mechanisms and cause serious physical and financial damage and loss or compromise of sensitive information.

An organization should have privacy policies in place that are a statement of management intent and commitment to protect PII. A privacy policy usually addresses what information is collected, email rules, what information is shared with partners, how information is protected, disclosure rules, and authentication mechanisms.

Most privacy policies embody the following five basic principles:

1. **Access** – Individuals should be able to review and correct their PII.
2. **Choice** – Individuals should have the choice to opt out or opt in on permitting the disclosure of their PII to other entities.
3. **Enforcement** – Privacy rules should be enforced.
4. **Notice** – Individuals should be kept informed of the disclosure and use of PII.
5. **Security** – Unauthorized disclosure of PII should not occur.

The European Union (EU) has codified privacy principles that apply to their citizens. These principles are:

- Data should be collected in accordance with the law.
- Data should be used only for the purposes for which it was collected, and it should be used only for a reasonable period of time.
- Individuals are entitled to receive a report on the information that is held about them.
- Individuals have the right to correct errors contained in their personal data.

- Information collected about an individual cannot be disclosed to other organizations or individuals unless authorized by law or by consent of the individual.
- Records kept on an individual should be accurate and up to date.
- Transmission of personal information to locations where equivalent personal data protection cannot be assured is prohibited.

Personal data will be collected, transmitted, and stored in the Smart Grid paradigm; as such, privacy protection should be inherent in the Grid. The principles expounded in the previous paragraphs can serve as guidance in privacy implementations for the Smart Grid.

As discussed earlier in this chapter, cyber analytics can be useful in determining patterns and threat vectors from large and diverse sources of data to predict possible future attacks or compromises of PII.

## Social Networks

Social networks are becoming widely used by individuals in all walks of life. According to Facebook statistics,[2] there are more than 500 million active users who spend over 700 billion minutes per month on Facebook. Social sites, such as Facebook, which many consider hobbies, can be sources of vital information and serve as platforms for collaboration, innovation, and rapid transactions. Incorporating such platforms and media into a corporate environment can provide benefits, such as providing business intelligence, conducting training, and enhancing communications among colleagues, customers, and partners. On the other hand, social networks can be points of vulnerability for attackers and serve as avenues for the compromise of sensitive information. Participation in social networks at all levels of an organization can impact the security and privacy of the organization by providing information that can lead to the disclosure of proprietary data.

## Mobile Technology

Interactivity is becoming the norm in dealing with digital devices and systems. Users can communicate with systems through social media, touchscreens, and mobile devices. Accessing data and controlling automation and control systems easily through iPad-like interfaces from a variety of physical locations offer both opportunities and concerns. The proliferation of 4G WiMAX will make remote communications from mobile devices more common and allow messaging and controls to be available to a variety of personnel and applications. Transactions on these networks will have to be managed, and policies redesigned, to accommodate such communications.

## Interoperability

Interoperability is the ability of an entity to exchange data with another entity. In the computer realm, this data exchange can consist of software applications communicating with each other through common protocols and file formats. With the advent of technologies such as cloud computing and mobile platforms, it is important not to tie applications to specific types of platforms and not to discourage moving applications among different computers, storage, and networks.

Historically, industrial automation and control systems were developed on hardware and software modified by vendors for targeted applications and could not be ported to other platforms.

In implementing advanced industrial automation and control systems for programs, such as the Smart Grid, it is critical that hardware and software are interoperable and are verified for interoperability through some of the following means:

- Use independent testing groups to evaluate interoperability of software and hardware products from different sources
- Statistically test samplings of combinations of applications for interoperability and project the probability of other similar combinations being interoperable
- Test all software items against a standard reference to determine interoperability
- Test hardware items against applicable standards to ensure interoperability

The security of interoperable software components poses a different set of issues. The question to be answered is how to determine the security posture of application software comprised of software components whose security has been evaluated. It is not always the case that the features and characteristics of the resultant composite model reflect the security characteristics of the component software elements. These issues are explored in detail in a paper by Oladimeji and Chung from the University of Texas.[3] In the paper, the authors demonstrate that using secure software components to build a large system does not always guarantee the security of the large system. They also present an approach that can be used to ensure that the security requirements of the composite system are fulfilled when using secure components or to show that some components might not satisfy the security specifications of the composite system.

## The Smart Grid and Technological Trends

It is important to identify the effects of technological developments on industrial automation and control systems. Technological advances have been incorporated into the elements of the critical infrastructure and in doing so, have increased performance and efficiency. Networks that were once isolated are now interconnected and in many instances, manage data from a variety of organizations. A natural progression of this technological evolution is to use advanced automation and networking methods to optimize system operations. However, the improvements in performance obtained by these enhancements have to be considered in light of the impact of the new technologies on safety, security, cost, and long-term consequences. To provide an illustration of these issues, the Smart Grid technologies will be overlaid with the technological trends described in the first part of this chapter. The following discussion is not meant to be an exhaustive exploration of the Smart Grid, but it is presented to provide enough detailed information to serve as a basis for exploring the consequences of technological evolution and applications on an advanced automation and control system.

In the NIST document *Framework and Roadmap for Smart Grid Interoperability Standards,*[4] the Smart Grid is characterized as "adding and integrating many varieties of digital computing and communication technologies and services with the power-delivery infrastructure. Bidirectional flows of energy and two-way communication and control capabilities will enable an array of new functionalities and applications that go well beyond 'smart' meters for homes and businesses." The NIST document also cites the following distinguishing characteristics of the Smart Grid as:

- Increased use of digital information and controls technology to improve the reliability, security, and efficiency of the electric grid
- Dynamic optimization of grid operations and resources with full cybersecurity

- Deployment and integration of distributed resources and generation, including renewable resources
- Development and incorporation of demand response, demand-side resources, and energy efficiency resources
- Deployment of "smart" technologies for metering, communications concerning grid operations and status, and distribution automation
- Integration of "smart" appliances and consumer devices
- Deployment and integration of advanced electricity storage and peak-shaving technologies, including plug-in electric and hybrid electric vehicles and thermal-storage air conditioning (e.g., peak shaving can refer to a hybrid electric vehicle sending power back to the grid when demand is high)
- Provision to consumers of timely information and control options
- Development of standards for communication and interoperability of appliances and equipment connected to the electric grid, including the infrastructure serving the grid
- Identification and lowering of unreasonable or unnecessary barriers to the adoption of Smart Grid technologies, practices, and services

In summary, the idea of the Smart Grid is to automate the electric power system by providing the ability to monitor and optimize operations across the grid from bulk generation to the consumer through two-way transmission of electricity and information.

Some of the benefits of the Smart Grid are put forth as:

- Built-in cybersecurity
- Higher grid resiliency
- Higher power quality
- Improved energy delivery efficiency
- Improved generation efficiency
- Reduced emissions and pollutants
- Reduced operating costs

With the increased interconnectivity and linking of Smart Grid networks, there are increased risks. These risks are summarized as follows from the *NIST Guidelines for Smart Grid Cybersecurity*:[5]

- Increasing the complexity of the grid could introduce vulnerabilities and increase exposure to potential attackers and unintentional errors.
- Interconnected networks can introduce common vulnerabilities.
- Increasing vulnerabilities to communication disruptions and introducing malicious software/firmware or compromised hardware could result in denial of service (DoS) or other malicious attacks.
- An increased number of entry points and paths are available for potential adversaries to exploit.
- Interconnected systems can increase the amount of private information exposed and increase the risk when data is aggregated.
- Expanding the amount of data that will be collected can lead to the potential for compromise of data confidentiality, including the breach of customer privacy.

Relative to the C-I-A triad of confidentiality, integrity, and availability, availability will be the most important of the three principles in the Smart Grid environment. In general, data integrity is also important while confidentiality ranks third in the list. For confidentiality, encryption keys will be used in the Smart Grid and management of the large numbers of keys involved will be an important task.

NIST has developed a framework for the Smart Grid as shown in Figure 4-1.

In Figure 4-1, customers have the historical role of consuming electricity. However, in the Smart Grid architecture of Figure 4-1, the customer can also be a source of electricity that can be generated and stored at the customer's site. Operations are responsible for managing the movement of electricity in the Smart Grid, and the term *markets* refers to the markets for selling and buying electricity. Bulk generation produces and stores electricity and uses transmission systems to move the electricity over large distances. Then the electricity is taken from the transmission system and provided to the customer through the distribution system. A summary of the roles of these domains is given in the following sections.

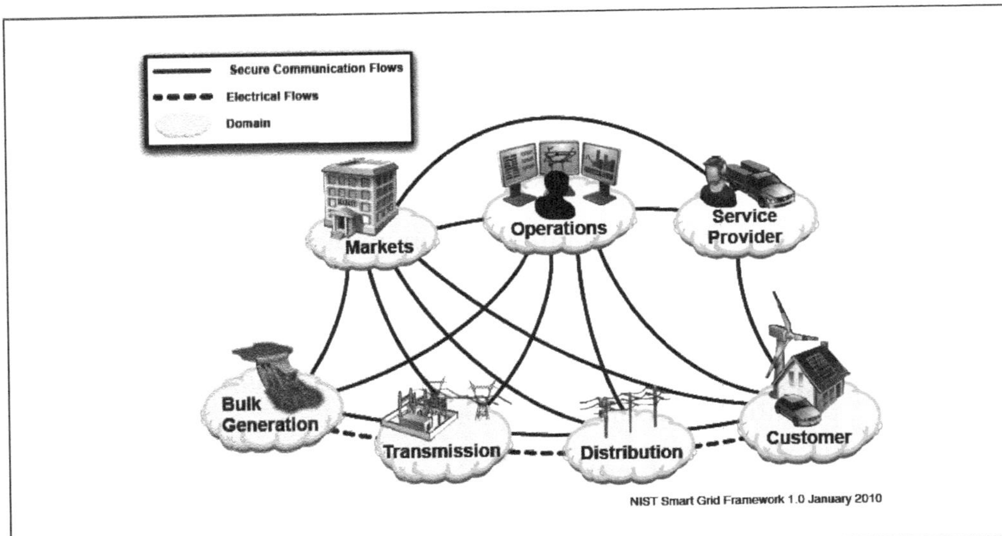


**Figure 4-1. Smart Grid Interactions and Electrical Flow**

## The Bulk Generation Domain

The bulk generation domain principally connects to and communicates with the transmission and operations domains, but it also communicates with some of the other domains, as shown in Figure 4-1. Bulk generation provides energy generated from sources such as coal, gas, water from dams, nuclear reactions, geothermal, the sun, and wind. In the Smart Grid environment, bulk generation must also be able to store energy from sources that are not continuously available.

SCADA systems are employed in the bulk generation domain to control and measure its subsystems and exchange data with the operations domain.

## The Transmission Domain

The transmission domain delivers electrical power from the bulk generation domain to the distribution domain and balances the required electrical load between these two entities. It also provides energy storage capabilities. There is close coordination between the transmission domain and the operations domain. The transmission domain incorporates transformer substations that step up the voltage for transmission and then step down the voltage for the distribution domain.

Additional energy supplies required by the transmission domain are obtained through the marketing domain. Transmission networks are usually run by a regional transmission operator (RTO) or an independent system operator (ISO).

### The Distribution Domain

The distribution domain serves as the intermediary between the transmission and customer domains and is normally connected in a radial configuration with customers and their metering systems. However, with the capabilities of the Smart Grid, the interaction of the distribution domain with the customer can also be a mesh or loop through wired or wireless networks. This domain also will connect to energy storage facilities.

### The Operations Domain

The operations domain provides reliable and high-quality power delivery on the Grid and interacts with all other domains of the Grid. Historically, operations have been the responsibility of a regulated utility, but they can also be the function of a service provider. This domain provides for locating faults, monitoring equipment (such as breakers and switches), evaluating security, keeping inventory, scheduling maintenance, monitoring intelligent field devices, gathering business intelligence, and optimizing Grid performance.

### The Service Provider Domain

The service provider domain communicates with the operations domain, if they are separate entities, to provide the basis for controlling critical elements of the Grid. It also interacts with the markets and customer domains to provide account management and billing. Additional functions performed by the service provider domain include:

- Handling customer problems
- Maintenance and installation of customer premise devices
- Managing business accounts
- Providing building and home energy management

## The Markets Domain

The markets domain buys and sells energy services and communicates with the other domains to balance supply and demand. Additional functions provided by this domain include:

- Establishing pricing mechanisms
- Managing regulation issues
- Providing roaming billing information
- Providing secure communication with other domains
- Supporting interoperability across suppliers of market data

## The Customer Domain

The customer domain is where the electricity is used and where customers can manage their electricity usage and electricity generation, if applicable. A customer might be a home user, an industrial plant, or a building. In addition to the meter, the customer domain interacts with the utility through the Energy Services Interface (ESI), which might communicate with additional domains through the Advanced Metering Infrastructure (AMI). A customer might also generate and store energy for transmission back into the Grid.

## Advanced Metering Infrastructure

An important subsystem of the Smart Grid that is part of the customer domain in Figure 4-1 is the AMI. The AMI comprises hardware and meter data management (MDM) software that provide the normal meter reading functions and support two-way communications that can exchange energy information and commands with customers' devices through a home area network. The AMI can be used to implement dynamic electricity pricing and load management, including voluntary programs where a customer can schedule load reductions with prior notice in return for pricing reductions. There are analogous networks that apply to the industrial customer.

In addition to the smart meter, another critical component of the AMI is the ESI, which acts as a gateway to the customer's network. While the smart meter provides functions, such as determining and communicating energy usage, connecting or disconnecting service, measuring the amount of power fed into the grid from distributed power sources, and communicating with the home area network or equivalent, the ESI serves as the connection to the energy service provider. The service provider is usually a utility, but it can also be a third party that provides energy management services to the customer.

Some of the functions that can be supported by the AMI include:

- Providing the ability for external, third-party clients to communicate with and control customer devices to allow customers to take part in load reduction options
- Supporting voluntary load reduction at the customer's location by providing the customer with instantaneous kWhr electricity pricing
- Measuring the effectiveness of the voluntary load reduction incentives to the customer
- Optimizing building loads through building automation systems for large customers
- Providing the ability to detect, map, and restore outages to the customer

NIST[6] identifies the following Smart Grid elements as participating in the AMI:

- **Regional Transmission Organization (RTO)** – An independent organization that coordinates, controls, and monitors the operation of the electrical power system and supply in a particular geographic area; it is similar to an independent system operator.
- **Independent System Operator (ISO)** – An independent entity that controls a power grid to coordinate the generation and transmission of electricity and ensure a reliable power supply.
- **Energy Market Clearinghouse** – Wide-area energy market operation system providing high-level market signals for operations.
- **Service Provider** – A person or company combining two or more customers into a single purchasing unit to negotiate the purchase of electricity from retail electric providers or the sale of electricity to these entities; provides consolidated bills to customers.
- **Utility Energy Management S ystems (EMS)** – The entities that continue to provide regulated services for the distribution of

electricity (and gas and water) to customers and to serve customers that do not choose direct access. Regardless of where a customer chooses to purchase power, the customer's current utility will deliver the power to the customer's home or business.

- **Operations** – The calculation and analysis of power flow/state estimation results in the inclusion of distribution automation capabilities, demand response signaling, distributed energy resources, electric storage, plug-in electric vehicles (PEVs), and load management. Providing a website enables customer access to usage and billing records.
- **Meter Data Management System (MDMS)** – A system that provides meter data validation and verification so that reliable information can be used at bill settlement.
- **Metering System** – The system used for collecting metering information.
- **Load Management System (LMS)** – This includes controlling demand responses (DR), distributed energy resources (DER), PEVs, energy storage (ES) charging/discharging; processing and storing data on load management programs; storing information on contracts and relevant historic information; creating behavioral models; collecting, processing, and storing customer-specific power quality and reliability characteristics; and so on.
- **Customer Information System (CIS)** – Repository of customer information related to distribution company services. CIS contains load data for customers that are estimated for each nodal location on a feeder, based on billing data and time-of-day and day-of-week load shapes for different load categories.
- **SCADA** – The SCADA database is updated via remote monitoring and operator inputs. Required scope, speed, and accuracy of real-time measurements are provided; supervisory and closed-loop control is supported.
- **Field Devices** – Field equipment with local intelligence for monitoring and controlling automated devices in the distribution domain, which communicates with SCADA systems. This equipment also includes distributed intelligence capabilities for automatic

operations in a localized distribution area based on local information and on data exchange between members of the group.

- **Meter** – Unless otherwise qualified, a device of the utility used to measure watts, vars, var-hours, volt-amperes, or volt-ampere-hours. (The term *var* stands for volt-ampere reactive power, which is the power in an AC circuit where there is a capacitive or inductive load.)
- **Energy Services Interface (ESI)** – Provides cybersecurity and (often) coordination functions that enable secure interactions between relevant home area network devices and the utility. Permits applications such as remote load control, monitoring and control of distributed generation, in-home display of customer usage, reading of nonenergy meters, and integration with building management systems. Provides auditing/logging functions that record transactions to and from home area networking devices. Can also act as a gateway.
- **Customer EMS** – A customer energy management system that can receive pricing and other signals for managing customer devices, including appliances, DER, electric storage, and PEVs.
- **Customer Appliances, DER, PEV, and Electric Storage** – Equipment and systems at the customer site that could participate in demand response and other programs.
- **Submeter** – A meter that measures a subload, such as a plug-in electric vehicle.

## Energy Storage and Management of Stored Energy

Energy storage (ES) is an important element of the Smart Grid concept. It can support the use of intermittent renewable energy devices and energy storage devices at the generating site, as well as allow for the use of electricity generated at a customer site. The connection of ES devices to the Grid must be done according to specified standards and guidelines to ensure the safe and efficient connection of external generation and storage devices to the Grid. These standards and guidelines have to address the following issues:

- Accommodation of intermittent renewable energy sources by storage of energy that is generated
- Effective deployment of ES-distributed energy resource (DER) systems
- ES-DER transmission implementations

- Interfaces to rotating machines
- Interoperability among ES-DER systems
- Modification of existing standards such as IEEE 1547[7] to address various types of ES-DER systems and the DER connections with the Grid
- New Smart Grid storage and power electronics technologies

There are quite a few energy storage technologies being developed and/or improved that can be used in the Smart Grid architecture. Some of the more promising approaches are:

- **Compressed Air** – Energy is stored by pumping compressed air into a storage container or underground cavity and then releasing it when needed to power a turbine that generates electricity. An advantage is that it is a well-known technology, but a disadvantage is that it does impact the environment and requires a lengthy approval process. Compressed-air projects are used in large utility applications, but can be used in smaller applications.
- **Flywheels** – Energy is stored in large, high-mass, rapidly rotating discs, which can impart energy to generators to provide back-up or storage power and also support frequency regulation.
- **Fuel Cells** – Fuel cells produce electricity through conversion of chemical energy. A fuel cell requires a fuel supply and an oxidizing agent. A hydrogen fuel cell uses hydrogen as the fuel and oxygen as an oxidizer. Fuel cells can rapidly recharge.
- **Lithium-Ion Batteries** – Modern technology battery with higher charging rates and higher energy densities than lead acid batteries. Development is focused on the automotive and electrical grid markets. These batteries have fast responses and are best used at the generation project level.
- **Pumped Hydro** – Water is pumped from a lower-level storage basin to one at a higher level. Electricity can then be generated at a later time by having the water flow downhill through a turbine to generate electricity. This technology is used for centralized, utility- scale projects and is able to handle load with quick response.

- **Sodium Sulfur Batteries** – Batteries using sodium and liquid sulfur. These batteries are more efficient in larger sizes but have to be properly sealed and protected from water as sodium spontaneously explodes when it comes in contact with water.
- **Ultracapacitors** – Large amounts of electrical charge can be stored on new technology capacitors and then tapped when energy is needed. Capacitors typically have short charging times and provide high quantities of energy when discharged.

## Smart Grid Protocols

The Smart Grid is a system of systems with two types of flow: power and communications. For data to be exchanged among all the domains of the Smart Grid, standard protocols must be used. There are viable candidate protocols, each with advantages and disadvantages. One major candidate protocol is the Internet Protocol (IP), which was discussed in Chapter 1.

IP was originally designed to allow dissimilar networks to communicate with each other and to provide flexibility in the types of physical connections used. It is widely deployed and has a base of support tools and security mechanisms. IP and the Internet protocols provide the advantages of interoperability with other applications that have been developed and deployed over the years relating to the Internet.

To be effective in the Smart Grid, a protocol must meet the requirements of different domains of the Grid. For example, in the AMI, data collection does not have to be instantaneous and data can be collected every 15 or 20 minutes. Conversely, in bulk generation facilities, operations have to be performed on the order of milliseconds. Thus, IP would be a good fit for the Smart Grid in general, but where deterministic elements are involved, other protocols more appropriate to control systems should be employed.

An additional standard that has been developed for interfacing to data communication networks, like the one used in the Smart Grid, is defined in ANSI C12.22-2008.[7] ANSI C12.22 specifies a protocol for communicating with devices, such as data modules and meters. It provides security through the use of AES encryption methods. For Smart Grid purposes, ANSI C12.22 is an improvement over previous standards C12.18 and C12.21, which addressed meter reading through optical ports and modems, respectively, and are more suited to high-volume commercial and industrial customers.

ANSI C12.22 is more appropriate for reading residential meters than C12.18 and C12.21. The main features of C12.22 include a protocol for the data link, physical layers for meter communications, and an application-level protocol that is independent of the transport method used.

Another important protocol suite that is widely used in Europe in conjunction with smart meters is the IEC 62056-XX series. It is based on the Device Language Message Specification (DLMS)/Companion Specification for Energy Metering (COSEM) suite. DLMS uses the abstract modeling of communication elements, and COSEM establishes rules for smart meter data exchanges. DLMS/COSEM provides interoperability among a variety of meters and other modules. The components of the IEC 62056 series are summarized as follows:

- **IEC 62056-21** – Defines data communications through electrical and optical ports
- **IEC 62056-31** – Provides guidance on the use of local area networks on twisted pair
- **IEC/TS 62056-41** – Addresses data communications over wide area networks
- **IEC 62056-42** – Addresses physical layer services for asynchronous data interchanges
- **IEC 62056-46** – Covers the High-Level Data Link Control (HDLC) protocol for use in the data link layer
- **IEC 62056-47** – Focuses on the COSEM IPv4 transport layer and communications from the application layer
- **IEC/TS 62056-51** – Addresses protocols in the application layer
- **IEC/TS 62056-52** – Discusses protocols for the management distribution line message specification (DLMS) server
- **IEC 62056-53** – Provides guidance on establishing application associations among peer applications and messaging services
- **IEC 62056-61** – Supports data identification for various types of data, including abstract data
- **IEC 62056-62** – Defines the COSEM interface classes for modeling the range of meter functions

## Mapping of Emerging Technology Issues onto an Example Automation System – The Smart Grid

The technologies associated with home area networks, energy storage, analytics, cloud computing, privacy, social networks, mobile technology, and interoperability will have a significant impact on industrial automation and control systems. Figure 4-2 summarizes the domains of the Smart Grid that will be most affected by these technologies.

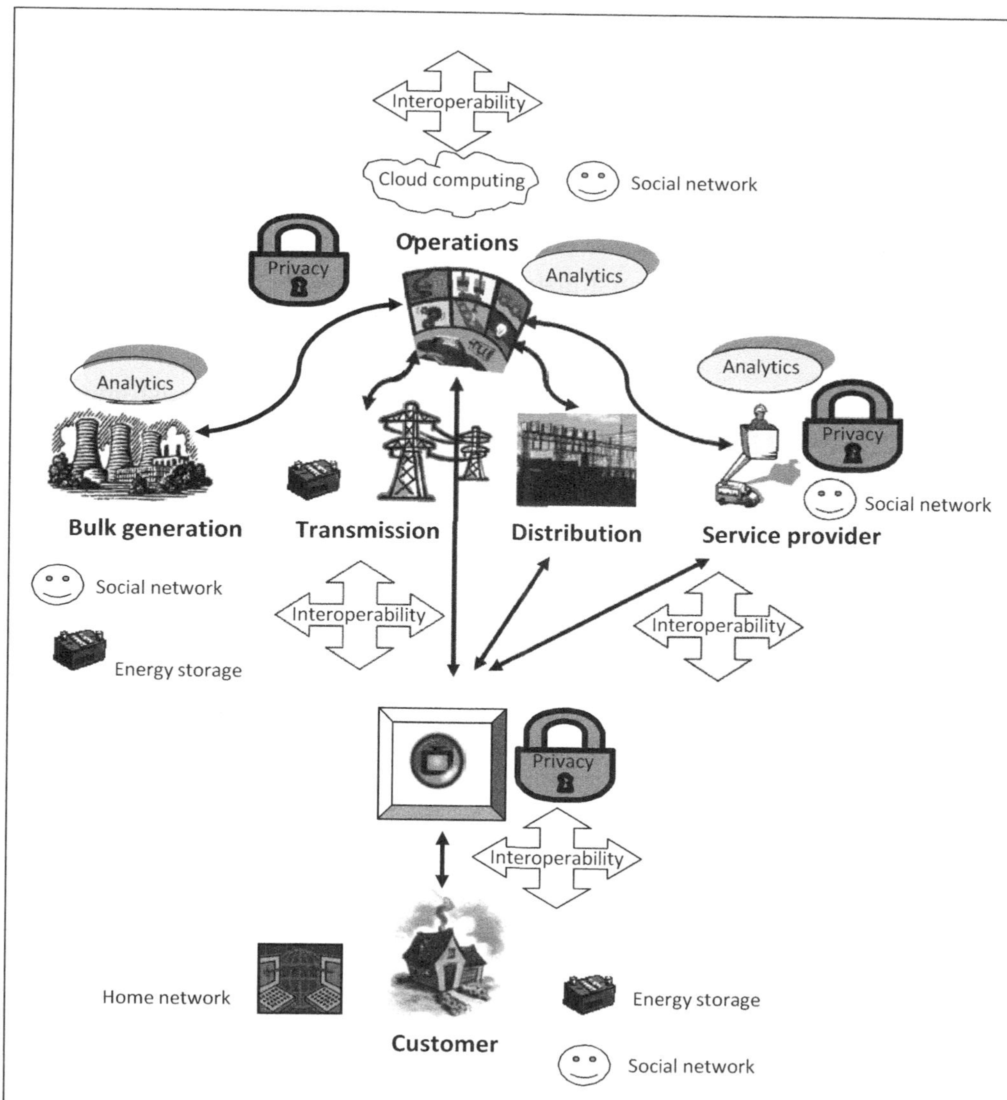


**Figure 4-2. Smart Grid Domains**

To recap the earlier discussion:

Applying analytics in bulk generation can lead to improved and more efficient maintenance strategies by alerting operators before outages occur. In addition, the information obtained from applying analytics can lead to significant reductions in fuel costs, thus increasing profitability. Bulk generation facilities also require efficient and cost-effective energy storage, particularly for intermittent renewable energy sources, such as wind farms and solar collectors.

Social networking by operators with other domains can improve cooperation and understanding; however, it also holds the potential for compromising sensitive information.

The transmission domain must maintain close communications with the operations domain and provide energy management and storage as needed. The distribution domain must also communicate with the operations domain and effectively provide electricity to the customer.

The service provider domain handles large amounts of customer PII and must ensure that adequate privacy protections are in place. Social networking used in this environment offers benefits but might also be a source of security and privacy breaches.

The operations domain must communicate effectively with the other domains and provide efficient and reliable power management. In performing its functions, the operations domain can apply analytics to provide insight into potential problem areas before they become critical, this insight can then be used to initiate proactive solutions instead of reacting to system failures. With the large amount of data being gathered by the operations domain, cloud computing can offer significant value and savings by making on-demand applications and computing capacity available as needed. However, because social networking might also be involved in this domain, privacy protections have to be adequately implemented.

The AMI is a critical link to the customer domain and must ensure efficient transfer of information and offer privacy protections for PII.

The customer domain is the focus of all the other domains and can be a source or sink of electricity. It also provides communications with home area networks and smart meters to implement dynamic pricing and customer opt-in for energy savings. The customer domain, as a source of electricity to the Grid, can have energy storage capacity for intermittent sources of electricity. The customer domain can also have the highest volume of social networking and can thus be a source of compromise and loss of sensitive information.

All the domains must have high degrees of interoperability or the benefits of the Smart Grid will be significantly reduced.

## Summary

Technological advancements pose a Cornelian dilemma in that choosing to use them or not to use them can both yield negative consequences. However, in most instances, the benefits of using them outweigh the risks. The key is to apply the new technologies in an intelligent and effective manner while implementing protections against the vulnerabilities introduced by these advancements.

Home area networks, energy storage, analytics, cloud computing, privacy, social networks, mobile technology, and interoperability can provide significant improvements in the performance, safety, security, and cost-effectiveness of industrial automation and control systems, if appropriate safeguards are employed.

Chapter 5 will explore the threats and risks to industrial automation and control systems and provide methods to quantify, evaluate, and manage the risks.

## Review Questions for Chapter 4

4.1 Which of the following is an important element in connecting to smart meters?

   A. Home area network

   B. Energy storage devices

   C. Cloud computing

   D. Printers

4.2 Energy storage devices are especially useful in which of the following cases?

A. Fossil fuel energy sources

B. Nonrenewable energy sources

C. Intermittent renewable energy sources

D. Constant energy sources

4.3 The application of data mining and analysis to large amounts of data for the purposes of discovering knowledge and making intelligent predictions and decisions is known as which of the following?

A. Artificial intelligence

B. Analytics

C. Data aggregation

D. Predictive calculus

4.4 Which of the following techniques is used for condition monitoring of high-voltage electrical equipment and identifying areas of potential failure?

A. Analytics

B. Artificial intelligence

C. Robotics

D. Scanning

4.5 Data from a variety of sources that is analyzed in real time or near real time to perform predictive modeling and support physical, logical, and administrative security is known as which of the following?

A. Data aggregation

B. Cyber analytics

C. Cybersecurity

D. Sensitivity analysis

4.6 What of the following is NOT a characteristic of cloud computing?

A. Measured service

B. Resource pooling

C. On-demand self-service

D. Minimal elasticity

4.7 Which of the following is NOT one of the three cloud service models?

A. Cloud Application as a Service

B. Cloud Software as a Service

C. Cloud Platform as a Service

D. Cloud Infrastructure as a Service

4.8 Which of the following is NOT one of the five basic privacy principles?

A. Access

B. Choice

C. Preemption

D. Notice

4.9 Which of the following items is NOT one of the European Union (EU) privacy principles?

A. Data should be collected in accordance with the law.

B. Data should be used only for the purposes for which it was collected, and it should be used only for a reasonable period of time.

C. Transmission of personal information to locations where equivalent personal data protection cannot be assured is permitted.

D. Individuals have the right to correct errors contained in their personal data.

4.10 The ability of an entity to exchange information with another entity is known as which of the following?

A. Compatibility

B. Commonality

C. Interchangeability

D. Interoperability

4.11 Which of the following is NOT a distinguishing characteristic of the Smart Grid?

A. Development and incorporation of demand response, demand-side resources, and energy efficiency resources

B. Deployment and integration of distributed resources and generation, excluding renewable resources

C. Provision to consumers of timely information and control options

D. Increased use of digital information and controls technology to improve reliability, security, and efficiency of the electric grid

4.12 Which of the following is NOT an expected result of Smart Grid implementation?

A. Higher grid resiliency

B. Improved energy delivery efficiency

C. Reduced operating costs

D. Possible loss of customers

4.13 Which of the following is NOT a domain in the NIST Smart Grid framework?

A. Control

B. Customer

C. Operations

D. Bulk generation

4.14 Which domain of the Smart Grid framework provides energy generated from sources such as coal, gas, water from dams, nuclear reactions, geothermal, the sun, and wind?

A. Distribution

B. Transmission

C. Bulk generation

D. Service provider

4.15 Which domain of the Smart Grid delivers electrical power to the distribution domain and balances the required electrical load?

A. Transmission

B. Service provider

C. Operations

D. Bulk generation

4.16 Which domain of the Smart Grid handles customer problems and manages business accounts?

A. Markets

B. Service provider

C. Customer

D. Operations

4.17 What subsystem of the Smart Grid comprises hardware and meter data management (MDM) software applications that provide the normal meter reading functions as well as supporting two-way communications that can exchange energy information and commands with customer's devices through a home area network?

A. Advanced Metering Infrastructure (AMI)

B. Energy Management System (EMS)

C. Energy Services Interface (ESI)

D. Smart Grid Interface (SGI)

4.18 In Smart Grid terminology, DER stands for which of the following terms?

A. Delayed energy reduction

B. Determined energy requirements

C. Distributed energy requirements

D. Distributed energy resources

4.19 Which domain of the Smart Grid framework has a home area network and also might have a requirement for energy storage for intermittent sources of electricity?

A. Customer

B. Distribution

C. Operations

D. Markets

4.20 Which domain of the Smart Grid framework handles large amounts of customer personally identifiable information (PII) and must ensure that adequate privacy protections are in place?

A. Customer

B. Markets

C. Service provider

D. Operations

4.21 Which of the following evolving technologies can increase communications, collaboration, and innovation, but also be a source of compromise of PII and other sensitive information?

A. Social networks

B. Interoperability

C. Cloud computing

D. Analytics

4.22 Which domain of the Smart Grid framework does not have an energy storage requirement for intermittent sources of electricity?

A. Customer

B. Transmission

C. Bulk generation

D. Markets

4.23 Which domain of the Smart Grid framework has the most potential for benefiting from the use of cloud computing?

A. Customer

B. Transmission

C. Distribution

D. Operations

## References

1. NIST SP 800-145. *The NIST Definition of Cloud Computing*. Washington, DC: NIST (National Institute of Standards and Technology), 2011.

2. Facebook Newsroom. http://newsroom.fb.com/Key-Facts/Statistics-8b.aspx.

3. Oladimeji, E. A., and L. Chung. "Analyzing Security Interoperability during Component Integration." In the proceedings of the 5th IEEE/ACIS International Conference on Computer and Information Science and 1st IEEE/ACIS International Workshop on Component-Based Software Engineering, Software Architecture and Reuse (July 10–12, 2006). Washington, DC: ICIS-COMSAR, IEEE Computer Society: 121–129.

4. NIST SP 1108. *NIST Framework and Roadmap for Smart Grid Interoperability Standards*. Release 1.0. Washington, DC: Office of the National Coordinator for Smart Grid Interoperability. NIST (National Institute of Standards and Technology), 2010.

5. Smart Grid Interoperability Panel – Cybersecurity Working Group. NISTIR 7628. *Guidelines for Smart Grid Cybersecurity: Volume 1, Smart Grid Cybersecurity Strategy, Architecture, and High-Level Requirements*.

Washington, DC: NIST (National Institute of Standards and Technology), 2010.

6. NIST SP 1108. *NIST Framework and Roadmap for Smart Grid Interoperability Standards*. Release 1.0. Washington, DC: Office of the National Coordinator for Smart Grid Interoperability and NIST (National Institute of Standards and Technology), 2010.

7. IEEE 1547. Standard for Interconnecting Distributed Resources with Electric Power Systems. 2003. http://ieeexplore.ieee.org/xpl/articleDetails.jsp?arnumber=1225051&filter=AND(p_Publication_Number:8676.

# 5 Risk Management for Industrial Automation and Control Systems

*Chapter 4 explored the technological changes affecting industrial automation and control systems and their impact on securing such systems. Chapter 5 will develop the methodologies for identifying and mitigating risks in automation and control systems and provide the foundation for implementing the security controls discussed in Chapter 6.*

## Risk Management

Identifying and managing risks to industrial automation and control systems is a critical undertaking and a fundamental activity that is necessary to ensure the safety and security of these systems. ANSI/ISA-62443-1-1 (99.01.01)-2007[1] defines risk as the "expectation of loss expressed as the probability that a particular threat will exploit a particular vulnerability with a particular consequence." Important definitions from the standard are:

- **Threat** – Potential for violation of security, which exists when there is a circumstance, capability, action, or event that could breach security and cause harm
- **Vulnerability** – Flaw or weakness in a system's design, implementation, or operation and management that could be exploited to violate the system's integrity or security policy
- **Residual Risk** – Remaining risk after the security controls or countermeasures have been applied

- **Risk Assessment** – Process that systematically identifies potential vulnerabilities to valuable system resources and threats to those resources, quantifies loss exposures and consequences based on probability of occurrence, and (optionally) recommends how to allocate resources to countermeasures to minimize total exposure

- **Risk Management** – Process of identifying and applying countermeasures commensurate with the value of the assets protected based on a risk assessment

To obtain another perspective, NIST SP 800-39[2] defines risk as "a measure of the extent to which an entity is threatened by a potential circumstance or event and typically a function of: (1) the adverse impacts that would arise if the circumstance or event occurs; and (2) the likelihood of occurrence." Below are NIST SP 800-39 definitions for the terms listed above:

- **Threat** – Any circumstance or event with the potential to adversely impact organizational operations (including mission, functions, image, or reputation), organizational assets, individuals, other organizations, or the nation through an information system via unauthorized access, destruction, disclosure, modification of information, and/or denial of service.

- **Vulnerability** – Weakness in an information system, system security procedures, internal controls, or implementation that could be exploited by a threat source.

- **Risk Assessment** – The process of identifying risks to organizational operations (including mission, functions, image, and reputation), organizational assets, individuals, other organizations, and the nation, resulting from the operation of an information system.

- **Risk Management** – The program and supporting processes to manage information security risk to organizational operations (including mission, functions, image, and reputation), organizational assets, individuals, other organizations, and the nation, and includes: (1) establishing the context for risk-related activities; (2) assessing risk; (3) responding to risk once determined; and (4) monitoring risk over time. Part of risk management incorporates threat and vulnerability analyses, and considers mitigations provided by security controls planned or in place.

The *Common Criteria*[3] document describes the relationships among threats, risk, countermeasures, and assets to be protected, as shown in Figure 5-1.

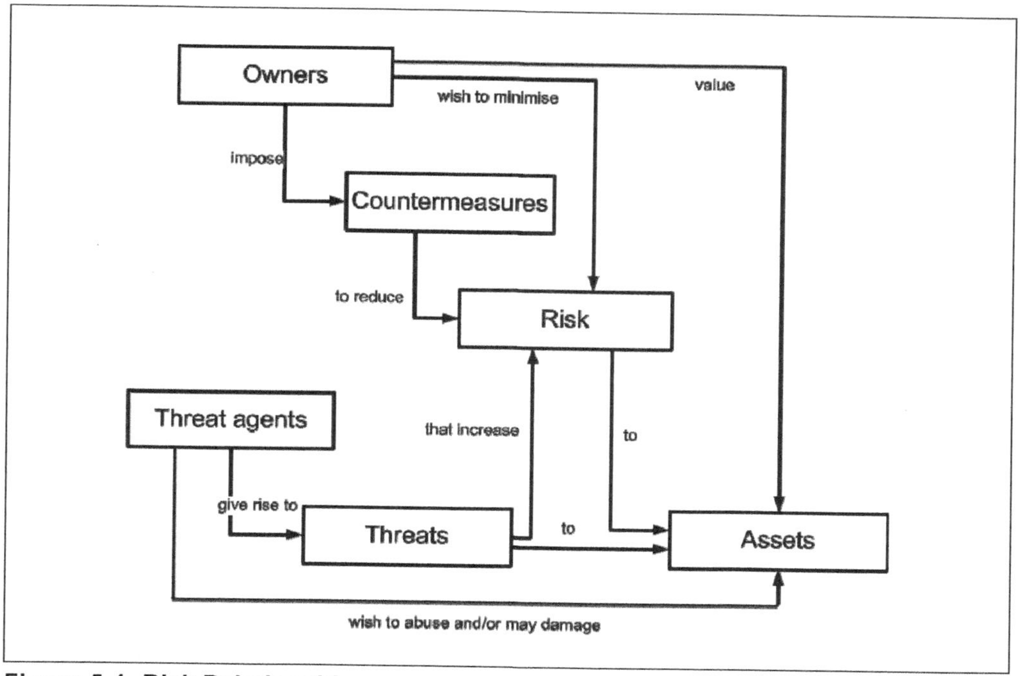


**Figure 5-1. Risk Relationships**
*Source: ISO/IEC 15408(2009)*

Assets can include physical elements, such as computer hardware, control systems, or specialized equipment, in addition to intangible assets, such as intellectual property, compliance information, or process expertise. Another important category of asset is the individual or professional who possesses unique knowledge and skills that, for example, are critical in the operation and maintenance of industrial automation and control systems.

Three valuable approaches to risk management are presented in ANSI/ISA-62443-2-1 (99.02.01)-2009, NIST SP 800-39, and NIST SP 800-37, which will be reviewed in the following sections.

## ANSI/ISA-62443-2-1 (99.02.01)-2009 Cyber Security Management System

To determine and mitigate risk, ANSI/ISA-62443-2-1 (99.02.01)-2009[4] defines a Cyber Security Management System (CSMS) as shown in Figure 5-2.

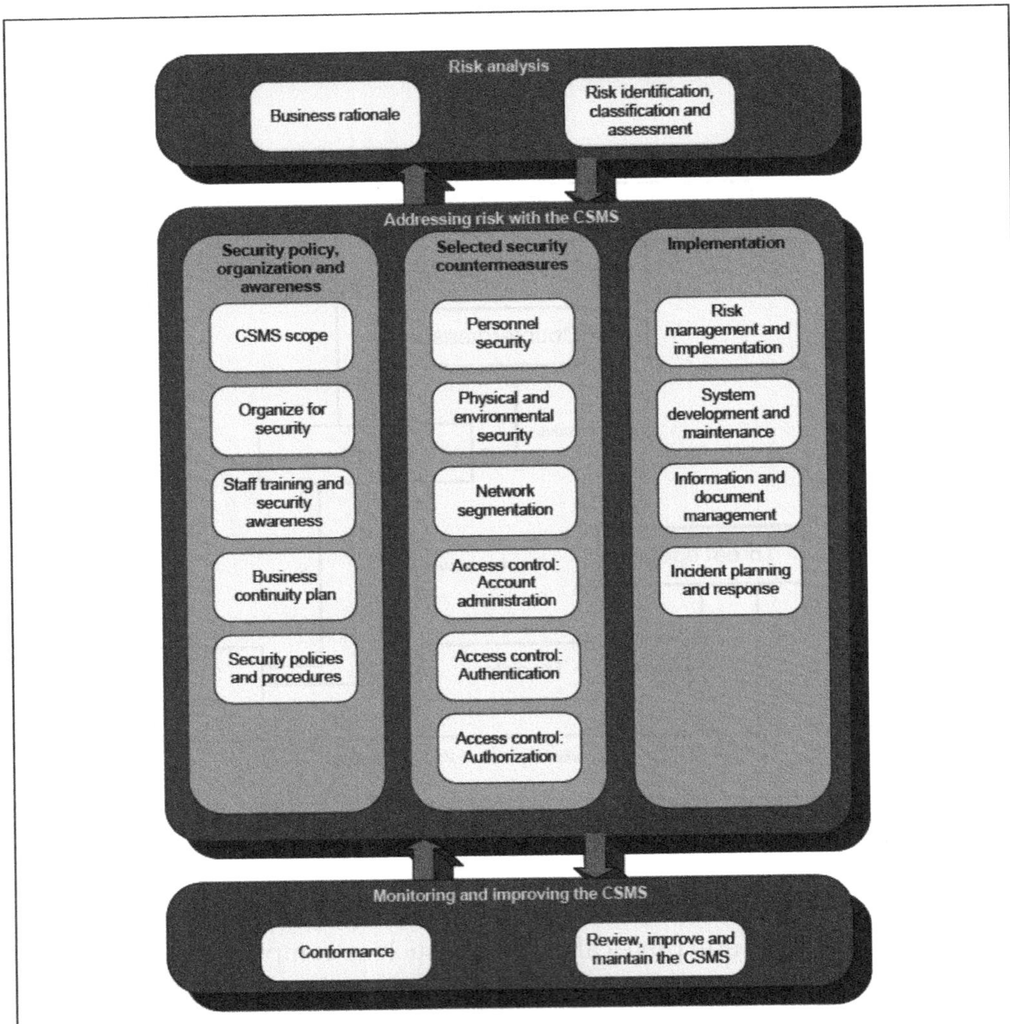


**Figure 5-2. Cyber Security Management System (CSMS)**
*Source: Data from ANSI/ISA-62443-2-1 (99.02.01)-2009*

The CSMS comprises three main categories: risk analysis, addressing the risk, and monitoring and improving the CSMS.

## Risk Analysis

In the risk analysis category, a business rationale is established, and risk identification, classification, and assessment are conducted. The business rationale includes the following:

- Establishing a basis for securing the industrial automation and control system

- Obtaining management support for securing the automation and control system
- Identifying and understanding the consequences of a successful attack on the automation and control system

Risk identification, classification, and assessment include ascertaining and prioritizing the risks to industrial automation and control systems and estimating the likelihood and impact of those risks. Specifically, this area of the CSMS includes the following activities:

- Select a risk assessment methodology
- Conduct a high-level risk assessment
- Identify the industrial automation and control systems to be addressed
- Develop simple network diagrams
- Prioritize the systems for mitigation
- Perform a detailed vulnerability assessment
- Conduct a detailed risk assessment
- Maintain vulnerability assessment records

## Addressing Risk

The next major CSMS category, addressing risk, comprises the following areas:

- Security policy, organization, and awareness
- Selected security countermeasures
- Implementation

As shown in Figure 5-2, security policy, organization, and awareness include documenting the elements that fall within the scope of the CSMS and assigning responsibilities for overall cybersecurity. In addition, this category incorporates providing staff and contractor personnel with adequate training and information to be aware of threats and weaknesses, and to have the ability to take remedial action if necessary. Further activities include developing a business continuity plan for reestablishing essential business operations when an event that causes a business disruption occurs and developing security policies and procedures applicable to automation and control systems that are compatible with general organizational policies.

The second group under "Addressing risk with the CSMS" is selected security countermeasures. The activities in this group include personnel security, physical and environmental security, network segmentation, and access control of account administration, authentication, and authorization.

- **Personnel security** is concerned with evaluating personnel to ensure they exhibit appropriate behaviors and will be able to preserve the security of the automation and control systems.
- **Physical and environmental security** are focused on protecting industrial automation and control systems from physical threats and malfunctions in environmental systems, such as power supplies and temperature controls.
- In **network segmentation**, networks are subdivided into zones that have similar security requirements and characteristics, and they are separated by protective devices to screen traffic from one network segment to another.
- **Access control** determines which entities can have access to resources; it is broken down into account administration, authentication, and authorization. In account administration, access to accounts is controlled and limited to entities that have privileges associated with those accounts. Authentication requires strong verification of a claimed user's identity prior to granting access to specific resources, and authorization provides access privileges to specific accounts upon successful authentication.

The third CSMS group under addressing risk is implementation (comprised of risk management and implementation), system development and maintenance, information and document management, and incident planning and response.

- **Risk management and implementation** consists of identifying and putting in place countermeasures that are technically and financially warranted to address determined risks.
- **System development and maintenance** emphasizes the inclusion of cybersecurity considerations throughout the system development life

cycle, including in the maintenance phase when system changes might frequently occur.

- **Information and document management** is concerned with applying information and document management policies in order to protect sensitive information from unauthorized disclosure.
- The last item in this group, **incident planning and response**, focuses on having plans and procedures in place to effectively respond to incidents, including reporting and investigative and follow-up activities.

## Monitoring and Improving the CSMS

The last category of the CSMS, monitoring and improving the CSMS, consists of the elements of conformance and reviewing, improving, and maintaining the CSMS.

- **Conformance** verifies that an automation and control system is compliant with appropriate policies, procedures, and regulations.
- The **review, improve, and maintain the CSMS** element requires continuous supervision of the CSMS to make sure responses to changes or threats are incorporated into the CSMS.

An additional benefit of implementing a CSMS, as cited in ANSI/ISA-62443-1-1 (99.01.01)-2007,[5] is maintaining the security level of industrial automation and control systems over time. Many security efforts are undertaken on a project basis, and system security deteriorates over time with the evolution of new threats and changes in the system architecture. Figure 5-3 illustrates the relative security levels of project-oriented industrial automation and control systems (IACS) and CSMS-based IACS.

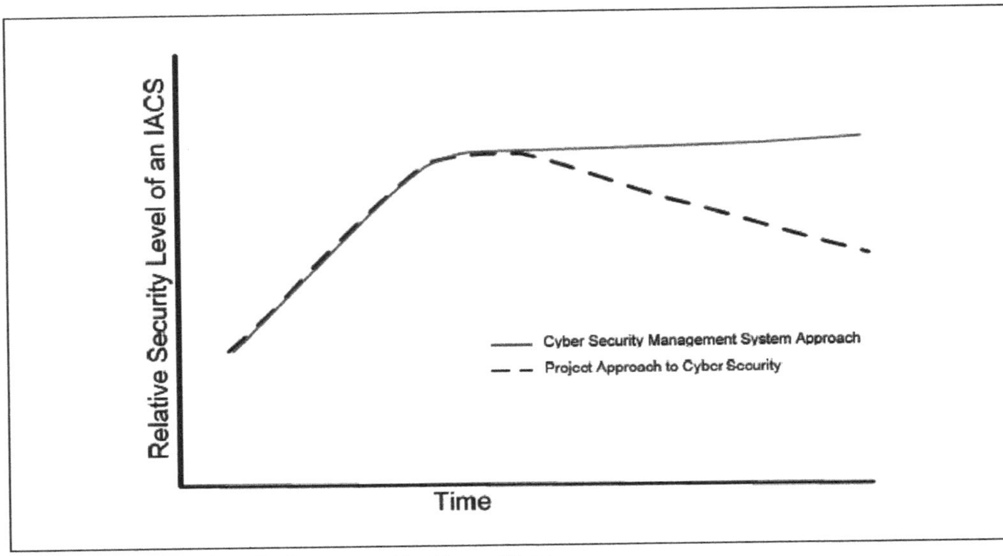


**Figure 5-3. Relative Cybersecurity Over Time**
*Source: ANSI/ISA-62443-1-1 (99.01.01)-2007*

## NIST SP 800-39 Integrated Enterprise Risk Management

NIST SP 800-39,[6] *Managing Information Security Risk: Organization, Mission, and Information System View*, replaces NIST SP 800-30,[7] *Risk Management Guide for Information Technology Systems*, which has been revised to provide guidance on risk assessment as a supporting document to SP 800-39. NIST SP 800-39 views risk management as:

> A comprehensive process that requires organizations to: (1) frame risk (i.e., establish the context for risk-based decisions); (2) assess risk; (3) respond to risk once determined; and (4) monitor risk on an ongoing basis, using effective organizational communications and a feedback loop for continuous improvement in the risk-related activities of organizations.

Toward these ends, it takes a global view of risk management and does not limit the risk management process to elements of information systems or automation and control systems alone. The approach views risk management as a hierarchical structure that has to involve the organization's mission and business goals, as well as business processes and the system development life cycle of information and automation system components. The document formalizes these subject areas into the following three tiers:

- **Tier 1 – Organization**: This tier establishes and implements governance structures consistent with the organization's mission and goals.
- **Tier 2 – Mission/Business Processes:** This tier designs, develops, and implements mission/business processes to support the mission defined in Tier 1.
- **Tier 3 – Information Systems:** The final tier integrates risk management activities into the system development life cycle (SDLC) of organizational information systems and addresses the resilience of organizational information systems. Tier 3 risk is addressed through NIST SP 800-53, which incorporates security controls for automation systems. These controls will be discussed in detail in Chapter 6.

This multi-tiered approach is summarized in Figure 5-4.

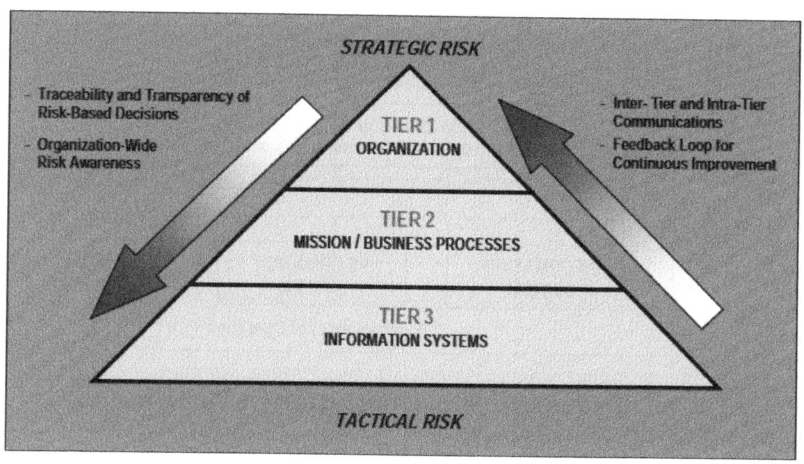


**Figure 5-4. Multi-Tiered Risk Management**
*Source: NIST SP 800-39 (2011)*

In Figure 5-4, Tier 1 risk management prioritizes mission/business functions, evaluates strategic and funding alternatives, provides risk management guidance to authorizing officials, and establishes the order of recovery of critical information systems.

Tier 2 execution follows from the output of Tier 1 activities. Tier 2 functions include defining the necessary and critical mission/business processes of the organization, prioritizing these functions based on organizational objectives, and determining the specific types of information necessary to carry out the organization's mission/business processes. An important concept in Tier 2 is the information security architecture, which is that part of the enterprise architecture that is concerned with the robustness of the information or automation and control system and its resistance to attacks. The information security architecture should incorporate the best elements of defense in depth and defense in breadth (illustrated in Figure 5-5). Defense in breadth is a planned, systematic set of multidisciplinary activities that seek to mitigate risk at every stage of the system development life cycle.

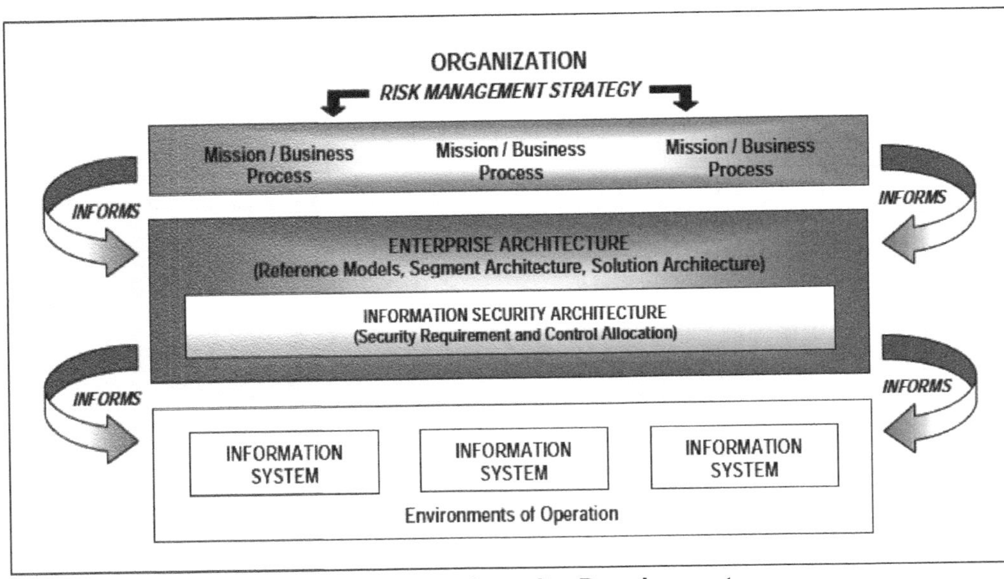


**Figure 5-5. Integration of Information Security Requirements**
*Source: NIST SP 800-39 (2011)*

Figure 5-5 shows that risk management can be addressed as an element of the enterprise architecture. The process is summarized in NIST SP 800-39 as:

- Developing a segment architecture linked to the strategic goals and objectives of organizations, defined missions/business functions, and associated mission/business processes
- Identifying where effective risk response is a critical element in the success of organizational missions and business functions

- Defining the appropriate, architectural-level information security requirements within organization-defined segments based on the organization's risk management strategy
- Incorporating an information security architecture that implements architectural-level information security requirements
- Translating the information security requirements from the segment architecture into specific security controls for information systems/ environments of operation as part of the solution architecture
- Allocating management, operational, and technical security controls to information systems and environments of operation as defined by the information security architecture
- Documenting risk management decisions at all levels of the enterprise architecture

Tier 3 focuses on information systems or automation and control systems and follows from the information passed down from Tiers 1 and 2. Some Tier 3 functions include categorizing organizational information systems, properly allocating security controls to those systems to support the organization's mission/business processes, and managing and monitoring security controls on an ongoing basis. Tier 3 addresses risk management in the following system development life-cycle phases, which includes that of industrial automation and control systems:

- Initiation
- Development/acquisition
- Implementation
- Operations/maintenance
- Disposal

As defined above, NIST SP 800-39 defines the components of risk management as risk framing, risk assessment, risk response, and risk monitoring. The relationship of these components to the three tiers is given in Figure 5-6.

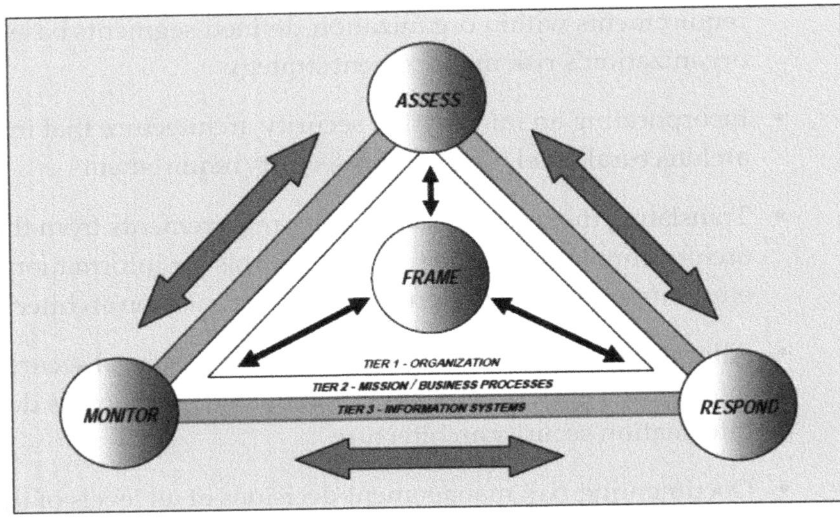


**Figure 5-6. Risk Management and the Tiers**
*Source: NIST SP 800-39 (2011)*

Risk framing (Figure 5-6) includes the following tasks:

- Making assumptions about threats, vulnerabilities, consequences/ impact, and likelihood of occurrence
- Determining restraints on risk assessment, response, and monitoring
- Determining risk tolerance levels
- Identifying priorities and trade-offs

Risk assessment includes the following activities:

- Identifying threats to the organization
- Identifying vulnerabilities
- Determining the impact of threats realized
- Determining the likelihood of harm occurring

Risk response includes:

- Developing alternative courses of action
- Evaluating these alternative courses of action

- Selecting the best courses of action
- Implementing selected responses

Risk monitoring is concerned with:

- Confirming that risk response measures have been implemented
- Evaluating the effectiveness of implemented risk response measures
- Determining changes to organizational information systems that would affect risk

## NIST SP 800-37 Risk Management Framework

NIST SP 800-37,[8] *Guide for Applying the Risk Management Framework to Federal Information Systems*, provides additional risk management guidance through the Risk Management Framework (RMF), which "provides a disciplined and structured process that integrates information security and risk management activities into the system development life cycle. The RMF operates primarily at Tier 3 in the risk management hierarchy, but can also have interactions at Tiers 1 and 2." The RMF comprises the following elements:

- **Categorize** the information system and the information processed, stored, and transmitted by that system based on an impact analysis.
- **Select** an initial set of baseline security controls for the information system based on the security categorization, tailoring and supplementing the security control baseline as needed based on an organizational assessment of risk, and on local conditions.
- **Implement** the security controls and describe how the controls are employed within the information system and its environment of operation.
- **Assess** the security controls using appropriate assessment procedures to determine the extent to which the controls are implemented correctly, operating as intended, and producing the desired outcome with respect to meeting the security requirements for the system.
- **Authorize** information system operation based on a determination of the risk to organizational operations and assets, individuals, other organizations, and the nation, resulting from the operation of the information system and the decision that this risk is acceptable.

- **Monitor** the security controls in the information system on an ongoing basis, including assessing control effectiveness, documenting changes to the system or its environment of operation, conducting security impact analyses of the associated changes, and reporting the security state of the system to designated organization officials.

## Threats

Threats fall into the two major categories of insider and external threats. Insider threats are insidious in that they originate from sources in the organization and can intentionally or unintentionally cause harm to a control system. There are a variety of external threats but, in this section, we will focus on those threats that are particularly relevant to IACS.

### The Insider Threat

As discussed in the previous ANSI and NIST documents in this chapter, one of the major efforts in risk management is to determine the threats to an industrial automation and control system. Most conventional threats have been reviewed extensively in many books and papers. One important conventional threat that deserves attention is the insider attack. This type of attack, from individuals within an organization, is of particular concern because it can bypass many institutional safeguards. Some of the common characteristics of insider attacks are:

- Many attacks are conducted by a disgruntled insider.
- In common with other types of attacks, insider attacks make systems unavailable and can result in serious losses and harm to reputation.
- Insiders who initiate attacks against computers usually give some indication of dissatisfaction beforehand.
- Many insider attacks are conducted remotely.
- Most inside attackers have privileged access to computer systems.
- Many insider attacks are planned in advance.

### Relevant IACS External Threats

A number of threats are so potentially harmful that it is important to look at them as learning examples and as a means to understand how to counteract threats in general.

There are a number of threats to industrial automation and control systems that are out of the ordinary and worthy of review. These threats include malware and electromagnetic pulse (EMP) emanations.

*Malware*

The following list, compiled in the *Oil and Gas IQ publication*[9] summarizes the prominent attacks and threats against control systems:

- **2000, Gazprom Network** – OAO Gazprom, the world's largest extractor of natural gas, announced that hackers within the Russian state had managed to seize control of the company's entire natural gas pipeline network.
- **2009, Operation Aurora** – A sophisticated cyber espionage operation based out of the People's Republic of China attempted to siphon intellectual property from major multinationals including Exxon Mobil, ConocoPhillips, and Marathon Oil.
- **2009, Night Dragon** – Beginning in November 2009, a series of ongoing cyberattacks based out of China were targeted at the global oil and gas sector. Involving a host of tools and techniques from spear-phishing attacks to the exploitation of Microsoft Windows operating system vulnerabilities and the use of remote access tools (RATs), these attacks were calculated not to maim but to purloin. Night Dragon sought to harvest sensitive information about everything, from specific field operations to upcoming licensing bids and project tenders.
- **2010, Stuxnet** – The Stuxnet worm attacked automation systems at Iran's Natanz nuclear facility, halting progress on the country's uranium enrichment program. The computer worm is considered the world's first cyber weapon, infecting up to 12 million computers in China. Stuxnet could attack SCADA and industrial control systems (ICS) and reprogram them to gradually self-destruct, which is exactly what it did to centrifuges at the heart of the Iranian nuclear program.
- **2012, Shamoon** – A Trojan horse stole data and wiped files in an attack that disabled 30,000 terminals at Saudi Aramco, the national oil company of Saudi Arabia and the world's largest oil producing corporation. Shamoon took RasGas, the second largest producer of liquefied natural gas (LNG) in Qatar completely offline.

- **2012, Flame** – Modular computer malware that burrowed into Iran's oil ministry, severing Internet links to rigs and the hub for nearly all the country's crude exports, caused widespread data loss. Flame was described as one of the most advanced pieces of malware found up to that time.
- **2014, Norway** – The Norwegian government announced that more than 50 Norwegian oil and energy companies were hacked by unknown assailants. The authorities advised another 250 firms that their networks and systems might have been compromised in the largest cyberattack to take place in the country's history.
- **2014, Havex** – A hacking group based out of Russia and calling itself the Dragonfly group (a.k.a. Energetic Bear), which appears to be a "state-sponsored" undertaking, focused on espionage with sabotage as a "definite secondary capability." It launched a campaign to infect energy and industrial firms around the world with the malicious RAT codenamed Havex. This malware had the ability to shut down major power grids, and oil and gas pipelines.

Because Stuxnet was such a significant and revolutionary worm aimed at control systems, it is useful to examine it, and countermeasures to it, in more detail.

***Stuxnet and Defensive Approaches***

Stuxnet is a worm that was designed to change control outputs on specific PLCs and conceal its existence from control room observers. It is of such complexity that it is probably the result of a team of programmers who worked many months to develop, debug, and test it. The sophistication of Stuxnet leads many to believe it is the product of one or more nations working together. Some of the salient characteristics of Stuxnet are:

- Discovered in June 2010, but suspected to have been spread many months before
- Infects Windows PCs
- Modifies WinCC databases, which provide process visualization HMI functions
- Attacks both networked and non-networked PCs
- Accomplishes infection through USB flash drives

- Uses zero-day exploits, which take advantage of previously unidentified vulnerabilities
- Installs device drivers using valid, digital certificates "appropriated" from trusted sources
- Specifically targets Siemens SIMATIC STEP 7 PLCs through a Windows PC
- Manipulates output bits on PLCs that control processes
- Focuses, it is speculated, on disrupting the operation of centrifuges at the Iranian Natanz uranium enrichment facility by changing values on turbine controls and frequency converters
- Receives updates through communications with servers in Denmark and Malaysia and through peer-to-peer communications
- Conceals its actions from the network control center so that operators do not have any indication of improper operations or malfunctions

From the information known about Stuxnet, it is not a great leap to imagine it being targeted at elements of a nation's critical infrastructure, particularly the Smart Grid, in addition to refineries, chemical plants, pipelines, and so on.

Preventive measures against Stuxnet include scanning and using firewalls on both incoming and outgoing connections to control the spread of the worm to other systems. However, because Stuxnet launched a zero-day attack, which exploited heretofore-unknown Microsoft vulnerabilities, it will be difficult to detect other possible intrusions taking advantage of future, unknown vulnerabilities.

A promising countermeasure against worms, such as Stuxnet, is the use of whitelisting of valid, executable code. In whitelisting, software that is "clean" and not compromised is placed on a "whitelist" and only that code can be run on the control system. The code is validated by generating a cryptographic hash or message digest from the code and the value is stored for future reference. Then, whenever a program is scheduled for execution, its hash is generated and compared against the hash of the original "clean" program. If the hashes are identical, the code has not been modified and is permitted to run. If the hashes are different or there is no such hash on the valid hash list, then the code is marked as malware and is not executed. Thus, whitelisting is effective against "zero-day" attacks in that this countermeasure is not dependent on knowing the signature of malicious code in advance.

One problem with whitelisting is that, in an environment where there are frequent changes and patches to software, the whitelist has to be constantly updated to stay current. Fortunately, most automation and control systems do not require as many updates as IT systems.

Yet another, albeit more restrictive, approach to protecting industrial automation and control systems is to limit the transmission of data to only one direction, from the automation and control system to the higher corporate IT system. In this manner, no malicious code can be sent from the IT-connected Internet to control systems. This method is known as a *data diode* in that it allows data to be transmitted in one direction only. The unilateral transmission of messages has obvious advantages, but its use is limited to architectures that do not rely on two-way transmission of data and commands. A number of data diode products are available that support a variety of automation and control system protocols.

Worms, such as Stuxnet, raise the specter of cyber warfare among nations. This situation raises a number of difficult questions for dealing with such attacks. For example:

- Does the military have a role in defending against cyberattacks?
- Should the military role be offensive, defensive, or both?
- How does one determine if the attack was initiated by a state or by rogue individuals operating in a state?
- How may the origin of a cyberattack be determined and with what accuracy?
- Is international law up-to-date with the technical progress of digital systems?
- Should preemptive attacks be used?
- What part does international law play in cyber warfare?

### *Electromagnetic Pulse*

A threat not commonly planned for in industrial automation and control systems is electromagnetic pulse (EMP) interference. William Radasky, in a paper titled "High Power Electromagnetic (HPEM) Threats to the Smart Grid,"[10] analyzes this threat and its potential effects on power distribution systems. Because the Smart Grid depends more on electronic communication and processing resources, it can be more vulnerable to EMP than older power systems.

EMP can originate from the following three primary sources:

1. A high-altitude nuclear detonation in space at heights above 30 km – high-altitude EMP (HEMP)
2. Solar activity – Geomagnetic storms
3. Electromagnetic emanations created by terrorists or criminal elements – Intentional electromagnetic interference (IEMI)

*HEMP*

As discussed in the "High Power Electromagnetic (HPEM) Threats to the Smart Grid" paper, HEMP has the following three time-based components:

1. **Early-time (E1)** within 10 nanoseconds of the burst – Produces an electric field pulse in the range of 2.5 to 25 nanoseconds at levels of approximately 50 kV/m that travels at the velocity of light. This field can penetrate substation walls and induce voltages of around 20 kV into substation control circuits.
2. **Intermediate-time (E2)** between 1 microsecond and 1 second after the burst – Produces field levels reaching 100 V/m.
3. **Late-time (E3)** between 1 second and several hundred seconds after the burst – Produces field levels reaching 40 V/km.

It is estimated that the environments at times E1 and E3 have the potential of causing the most damage to electric power systems because of the combination of high coupling efficiency to control lines and the peak field level values. Also, microwave towers with their extensive cabling are ideal induction targets for E1 HEMP, providing a path to sensitive communications and control equipment. E1 HEMP is also a serious threat to SCADA systems in power generation facilities, including controls and communications for wind turbines, solar cells, and other variable generation sources. E3 HEMP pulses are typically on the order of 1 second to several hundred seconds and can generate fields that are capable of coupling 4000 V between transformer grounded neutrals and inducing currents on the order of 800 A. This activity can also generate harmonics that would propagate through the power system and cause large amounts of damage, particularly to the equipment of the Smart Grid.

### *Solar*

Regarding solar activity, extreme geomagnetic storms on the sun can generate electric fields in the range of 20 V/km and cause distortions in Earth's surface geomagnetic field. Such changes cause fluctuations in Earth's electric field; these fluctuations induce currents in the neutral lines and generate damaging harmonics in the power grids, similar to E3 HEMP. On August 6, 2011, U.S. government scientists warned users of satellite, telecommunications, and electrical equipment to prepare for possible disruptions occurring during the week. "The magnetic storm that is soon to develop probably will be in the moderate to strong level," said Joseph Kunches, a space weather scientist at the Space Weather Prediction Center, a division of the U.S. National Oceanic and Atmospheric Administration (NOAA). He stated that solar storms occurring during this period could affect communications and global positioning system (GPS) satellites and might even produce an aurora visible as far south as Minnesota and Wisconsin. Tom Bogdan, director of the center, said the sun is going from a quiet period into a busier cycle for solar flares, and an increase in the number of such blasts is expected over the next 3 to 5 years. Similarly, an article titled "Effects of Solar Flares Arriving on Earth" in the Sunday, August 7, 2011 edition of *The New York Times*, stated that "operators of electrical grids are working to avoid outages, but the National Oceanic and Atmospheric Administration says some satellite communications and global positioning systems could face problems."

### *IEMI*

IEMI is categorized into wideband and narrowband fields above 100 MHz with peaks at 10 kV/m or higher. IEMI poses significant threats to substation electronics, smart meters, communications systems, and so on.

### *Protection*

Protective measures cited by Radasky[11] against EMP threats include placing control rooms and electronics in basements surrounded by soil, which offers some protection against EMP fields. Cabling should also be placed in the center of the building, if possible, to minimize EMP inductive fields. Additional procedures that are useful in defending against EMP include:

- EM shielding should be applied to protect vulnerable electronics in control centers and power-generating facilities.

- Fiber-optic cabling can minimize coupling if properly applied.
- High-frequency grounding with filters and surge arrestors can be used to protect sensors, smart meters, communications systems, and so on.

Protection against EMP should take into account the following field characteristics:

- High-frequency pulses can induce currents in floating wires.
- High-frequency pulses can penetrate through windows and gaps in metal shields.
- High-frequency fields do not have high energy content and therefore, surge protection devices can be effective.
- Low-frequency electromagnetic fields, such as those generated by E3 HEMP and extreme geomagnetic storms, can be mitigated by incorporating good grounding techniques and by using capacitors on neutral connections to provide a fast bypass mechanism for surges.

Some organizations and standards that address EMP threats include:

- IEC SC 77C, *EMC: High Power Transient Phenomena and Smart Grid Implications*
- IEEE P1642, The IEEE EMC Society/TC-5 (High-power EM), *Recommended Practice for Protecting Public Accessible Computer Systems*
- IEEE/ANSI C62.41, *Product Application Location, Test Methods, Guide to Using the Standard*
- MIL-STD 188-125-1, *HEMP Hardening (Fixed Facilities)*
- MIL-STD 188-125-2, *HEMP Hardening (Transportable Systems)*
- MIL-STD 202, *Environmental Requirements Component Level*
- MIL-STD 461, *EMI Requirements (Subsystems)*
- MIL-STD 464, *EMI Requirements (Systems)*
- MIL-STD 810, *Environmental Requirements Box Level*
- RTCA DO-160F, *Aircraft Protection*

- The International Council on Large Electric Systems, WG C4.206, *Protection of the High Voltage Power Network Control Electronics against Intentional Electromagnetic Interference (IEMI)*
- The International Telecommunications Union (ITU-T) Study Group 5
- UL1449, *Safety Standards for Low Voltage Surge Protective Devices* (Steering Committee)

## Summary

Chapter 5 discussed and summarized three important risk management methodologies and their application to industrial automation and control systems. A key component of risk management is risk mitigation, and in developing this topic, this chapter provides a basis for Chapter 6. In Chapter 6, approaches and means to secure industrial automation and control systems are developed along with guidelines for maintaining that security.

## Review Questions

5.1 In ANSI/ISA-62443-1-1 (99.01.01)–2007, a "potential for violation of security, which exists when there is a circumstance, capability, action, or event that could breach security and cause harm," is which of the following?

A. Threat

B. Vulnerability

C. Weakness

D. Risk

5.2 The "expectation of loss expressed as the probability that a particular threat will exploit a particular vulnerability with a particular consequence" is which of the following?

A. Consequence

B. Threat source

C. Weakness

D. Risk

5.3 The program and supporting processes to manage information security risk to organizational operations (i.e., mission, functions, image, reputation), organizational assets, individuals, other organizations, and the nation are defined as which of the following?

A. Risk assessment

B. Risk management

C. Risk mitigation

D. Risk association

5.4 Which of the following is considered an asset in risk management?

A. Computer hardware

B. A control system

C. A professional with unique knowledge

D. All of the above

5.5 The ANSI/ISA-62443-2-1 (99.02.01)-2009 Cyber Security Management System (CSMS) comprises which of the following three main categories?

A. Risk analysis, addressing the risk, and monitoring and improving the CSMS

B. Risk mitigation, addressing the risk, and monitoring and improving the CSMS

C. Risk analysis, addressing the risk, and monitoring and improving the automation system

D. Risk analysis, eliminating the risk, and monitoring and improving the CSMS

5.6 In the CSMS, a business rationale is established. Which of the following is NOT one of the elements of the business rationale?

A. Establishing a basis for securing the industrial automation and control system

B. Obtaining management support for securing automation systems

C. Identifying and understanding the consequences of a successful attack on automation systems

D. Confirmation of the value of the automation system

5.7 The risk identification, classification, and assessment activities of the CSMS do NOT include which of the following?

A. Conduct a high-level risk assessment

B. Develop simple network diagrams

C. Mitigate the risk

D. Perform a detailed vulnerability assessment

5.8 Which of the following is NOT one of the areas included in "Addressing risk with the CSMS?"

A. Security policy, organization, and awareness

B. Correction

C. Selected security countermeasures

D. Implementation

5.9 Subdividing a network into zones that have similar security requirements and characteristics and are separated by protective devices to screen traffic from one network segment to another is known as which of the following?

A. Network segmentation

B. Network isolation

C. Network masking

D. Network fragmentation

5.10 In the CSMS, access control is subdivided into which of the following three areas?

A. Account verification, authentication, and authorization

B. Account administration, authentication, and authorization

C. Account administration, audit, and authorization

D. Account administration, authentication, and activation

5.11 Risk management, system development and maintenance, and incident planning are part of which of the following areas of addressing risk in the CSMS?

A. Security policy, organization, and awareness

B. Selected security countermeasures

C. Implementation

D. Scope

5.12 In the monitoring and improving category of the CSMS, which element verifies that an automation system is compliant with appropriate policies, procedures, and regulations?

A. Review

B. Conformance

C. Improve

D. Maintain

5.13 NIST SP 800-39 views which of the following as a comprehensive process that requires organizations to: (1) frame risk (i.e., establish the context for risk-based decisions); (2) assess risk; (3) respond to risk once determined; and (4) monitor risk on an ongoing basis?

A. Risk management

B. Risk assessment

C. Risk monitoring

D. Risk evaluation

5.14 NIST SP 800-39 defines three tiers. Which of the following includes those three?

A. Policy, mission/business processes, information systems

B. Organization, policy, information systems

C. Organization, mission/business processes, objectives

D. Organization, mission/business processes, information systems

5.15 Categorizing organizational information systems and properly allocating security controls to those systems to support the organization's mission/business processes is performed in which tier of the multi-tiered risk management architecture?

A. Tier 1

B. Tier 2

C. Tier 3

D. Tier 4

5.16 According to NIST SP 800-39, risk framing, risk assessment, risk response, and risk monitoring are components of which of the following?

A. Risk valuation

B. Risk determination

C. Risk management

D. Risk acceptance

5.17 Making assumptions about threats, vulnerabilities, consequences/impact, and likelihood of occurrence are part of which of the following, according to NIST SP 800-39?

A. Risk assessment

B. Risk framing

C. Risk response

D. Risk monitoring

5.18 Which of the following is NOT an element of responding to risk, according to NIST SP 800-39?

A. Evaluating the effectiveness of implemented risk response measures

B. Developing alternative courses of action

C. Selecting the best courses of action

D. Implementing selected responses

5.19 According to NIST SP 8000-37, which of the following "provides a disciplined and structured process that integrates information security and risk management activities into the system development life cycle?"

A. Risk Monitoring Framework (RMF)

B. Risk Assessment Framework (RAF)

C. Risk Management Framework (RMF)

D. Risk Evaluation Framework (REF)

5.20 Which of the following groups of elements is presented in NIST SP 800-37 to manage risk?

A. Categorize, select, implement, authorize

B. Categorize, evaluate, implement, monitor

C. Determine, select, implement, authorize

D. Determine, evaluate, implement, monitor

5.21 Which of the following is NOT characteristic of an insider threat?

A. Many insider attacks are conducted by disgruntled insiders.

B. Most insider attacks do not result in serious losses or harm.

C. Many insider attacks are conducted remotely.

D. Many inside attackers have privileged access to computer systems.

5.22 Which of the following best describes Stuxnet?

A. A worm that was designed to change control outputs on specific PLCs and conceal its existence from control room observers

B. A worm that was designed to modify email messages and conceal its existence from control room observers

C. A virus that was designed to modify disk storage and send random messages to users' computers

D. A virus that was designed to crash computers when email attachments were opened

5.23 Which of the following is NOT true regarding Stuxnet?

A. Attacks both networked and non-networked PCs

B. Installs device drivers using valid, digital certificates

C. Infects Apple Macintosh computers

D. Infection is accomplished through USB flash drives

5.24 The process that only uses software that is "clean" and not compromised and is verified before running is known as which of the following?

A. Confirming

B. Whitelisting

C. Cleanlisting

D. Prechecking

5.25 Modular computer malware that burrowed into Iran's oil ministry, severing Internet links to rigs and the hub for nearly all the country's crude exports, causing widespread data loss is known as which of the following?

A. Night Dragon

B. Stuxnet

C. Flame

D. Havex

5.26 An electromagnetic pulse generated by a high-altitude nuclear detonation in space at heights above 30 km is known as which of the following?

A. Intentional electromagnetic interference (IEMI)

B. Space generated EMP (SGEMP)

C. Intensive electromagnetic interference (IEMI)

D. High-altitude EMP (HEMP)

5.27 Which one of the following is NOT a component of a high-power electromagnetic (HPEM) burst?

A. Final-time (E4) – Several hundred seconds after the burst

B. Early-time (E1) – Within 10 nanoseconds of the burst

C. Intermediate-time (E2) – Between 1 microsecond and 1 second after the burst

D. Late-time (E3) – Between 1 second and several hundred seconds after the burst

5.28 Solar-generated EMP is the result of which of the following actions?

A. Changes in the sun's orbit

B. Solar eclipses

C. Geomagnetic storms on the sun

D. Changes in Earth's orbit

5.29 Which of the following statements is NOT true regarding EMP?

A. High-frequency pulses can induce currents in floating wires.

B. High-frequency pulses can penetrate through windows and gaps in metal shields.

C. High-frequency grounding with filters and surge arrestors can be used to protect sensors, smart meters, and communications systems.

D. Fiber-optic cabling cannot minimize coupling.

## References

1. ANSI/ISA-62443-1-1 (99.01.01)-2007. *Security for Industrial Automation and Control Systems Part 1-1: Terminology, Concepts, and Models.* Research Triangle Park, NC: ISA (International Society of Automation).

2. NIST SP 800-39. *Managing Information Security Risk: Organization, Mission, and Information System View.* Washington, DC: NIST (National Institute of Standards and Technology), 2011.

3. ISO/IEC 15408. *Common Criteria for Information Technology Security Evaluation Part 1: Introduction and General Model.* Version 3.1, Revision 3. Geneva: ISO/IEC, 2009.

4. ANSI/ISA-62443-2-1 (99.02.01)-2009, *Security for Industrial Automation and Control Systems: Establishing an Industrial Automation and Control Systems Security Program.* Research Triangle Park, NC: ISA (International Society of Automation).

5. ANSI/ISA-62443-1-1 (99.01.01)-2007. *Security for Industrial Automation and Control Systems Part 1-1: Terminology, Concepts, and Models.* Research Triangle Park, NC: ISA (International Society of Automation).

6. NIST SP 800-39. *Managing Information Security Risk: Organization, Mission, and Information System View.* Washington, DC: NIST (National Institute of Standards and Technology), 2011.

7. NIST SP 800-30. *Guide for Conducting Risk Assessments.* Revision 1. Washington, DC: NIST (National Institute of Standards and Technology), 2012.

8. NIST SP 800-37. *Guide for Applying the Risk Management Framework to Federal Information Systems.* Washington, DC: NIST (National Institute of Standards and Technology, 2010.

9. Oil and Gas IQ. "Cyber 9/11: Is The Oil & Gas Industry Sleepwalking Into A Nightmare?" White paper. http://www.oilandgasiq.com/strategy-management-and-information/white-papers/cyber-9-11-is-the-oil-gas-industry-sleepwalking-in/.

10. Radasky, W. "High Power Electromagnetic (HPEM) Threats to the Smart Grid." *Interference Technology, 2011 EMC Directory and Design Guide* (2011): 49–58.

11. Ibid.

# 6

# IAC Systems Security Methodologies and Approaches

*Chapter 5 presented important risk management methodologies and their application to industrial automation and control systems. Chapter 6 will build on the results of risk management and develop in detail the foundation of security controls that are proven, effective, and applicable to industrial automation and control systems.*

## Automation and Control System Security Standards and Guidelines

The fundamental principles for protecting computer systems and networks are valid across all application domains. However, these principles have to be tailored to the world of industrial automation and control systems to be effective in protecting these systems. From experience and application of engineering approaches, a number of standards and guidelines have been developed and documented for securing automation and control systems. These materials take into account the characteristics of industrial automation and control systems and present measures that should be taken to protect these systems.

Some of the important documents that can be used to address automation and control system security are:

- NIST SP 800-53, *Recommended Security Controls for Federal Information Systems*
- NIST SP 800-82, *Guide to Industrial Control Systems (ICS) Security*

- ANSI/ISA-62443-1-1 (99.01.01)-2007, *Security Technologies for Industrial Automation and Control Systems*
- North American Electric Reliability Corporation (NERC), *Critical Infrastructure Protection (CIP) Cyber Security Standards*
- Department of Homeland Security, *Catalog of Control Systems Security: Recommendations for Standards Developers*
- AMI-SEC Task Force, *AMI System Security Requirements*

This list is not meant to be all-inclusive, but it comprises some of the major standards and guidelines addressing industrial automation and control system security, and it covers the major protection approaches to securing these systems. These standards and guidelines will be explored in this chapter. Following this exposition, the controls described will be cross-correlated in a table to provide a concise summary of these protection mechanisms with references back to the authoritative document.

## NIST Special Publication 800-53, Revision 4, Recommended Security Controls for Federal Information Systems

NIST SP 800-53[1] provides a catalog of security controls that can be used to obtain compliance with requirements from governmental agencies, military services, and other sources. As stated in the document, "The security controls in the catalog facilitate the development of assessment methods and procedures that can be used to demonstrate control effectiveness in a consistent and repeatable manner—thus contributing to the organization's confidence that security requirements continue to be satisfied on an ongoing basis." In conjunction with NIST SP 800-82, NIST SP 800-53 provides the means to apply the appropriate controls to industrial automation and control systems.

NIST SP 800-53 includes a Privacy Control Appendix and also addresses areas such as:

- Application security
- Firmware integrity
- Distributed systems
- Advanced persistent threat
- Insider threat

- Mobile computing
- Cloud computing
- Application security
- Supply chain risk

The security controls in NIST SP 800-53 are divided into 18 families, each with a unique two-character identifier. Table 6-1 lists the control identifiers and summarizes the control families.

**Table 6-1. NIST SP 800-53 Control Families**

*Source: NIST SP 800-53 (2013)*

| Identifier | Family |
|---|---|
| AC | Access Control |
| AT | Awareness and Training |
| AU | Audit and Accountability |
| CA | Security Assessment and Authorization |
| CM | Configuration Management |
| CP | Contingency Planning |
| IA | Identification and Authentication |
| IR | Incident Response |
| MA | Maintenance |
| MP | Media Protection |
| PE | Physical and Environmental Protection |
| PL | Planning |
| PS | Personnel Security |
| RA | Risk Assessment |
| SA | System and Services Acquisition |
| SC | System and Communications Protection |
| SI | System and Information Integrity |
| PM | Program Management |

The format of a typical SP 800-53 control is as follows:

**SI-3 Malicious Code Protection**

Control – The organization:

- Employs malicious code protection mechanisms at information system entry and exit to detect and eradicate malicious code
- Updates malicious code protection mechanisms whenever new releases are available in accordance with organizational configuration management policy and procedures
- Configures malicious code protection mechanisms to:
  - Perform periodic scans of the information system [*Assignment: organization-defined frequency*] and real-time scans of files from external sources at [*Selection (one or more): endpoint; network entry/exit points*] as the files are downloaded, opened, or executed in accordance with organizational security policy.
  - [*Selection (one or more): block malicious code; quarantine malicious code; send alert to administrator*] [*Assignment: organization-defined action*] in response to malicious code detection.
- Addresses the receipt of false positives during malicious code detection and eradication and the resulting potential impact on the availability of the information system

Supplemental Guidance

Information system entry and exit points include, for example, firewalls, electronic mail servers, web servers, proxy servers, remote-access servers, workstations, notebook computers, and mobile devices. Malicious code includes, for example, viruses, worms, Trojan horses, and spyware. Malicious code can also be encoded in various formats (e.g., UUENCODE, Unicode), contained within compressed or hidden files, or hidden in files using steganography. Malicious code can be transported by different means including, for example, web accesses, electronic mail, electronic mail attachments, and portable storage devices. Malicious code insertions occur through the exploitation of information system vulnerabilities. Malicious code protection mechanisms include, for example, anti-virus signature definitions and reputation-based technologies. A variety of technologies and methods exist to limit or eliminate the effects of malicious code.

Pervasive configuration management and strong software integrity controls may be effective in preventing execution of unauthorized code. In addition to commercial off-the-shelf software, malicious code may also be present in custom-built software. This could include, for example, logic bombs, back doors, and other types of cyberattacks that could affect organizational missions and business functions. Traditional malicious code protection mechanisms are not built to detect such code. In these situations, organizations rely instead on other safeguards including, for example, secure coding practices, configuration management and control, trusted procurement processes, and monitoring practices to help ensure that software does not perform functions other than the functions intended. Organizations may determine that in response to the detection of malicious code, different actions may be warranted. For example, organizations can define actions in response to malicious code detection during periodic scans, actions in response to detection of malicious downloads, and/or actions in response to detection of maliciousness when attempting to open or execute files. Related controls are: CM-3, MP-2, SA-4, SA-8, SA-12, SA-13, SC-7, SC-26, SC-44, SI-2, SI-4, SI-7.

Control Enhancements

- The organization centrally manages malicious code protection mechanisms.
- The information system automatically updates malicious code protection mechanisms.
- Withdrawn from previous version. Incorporated into AC-6(10).
- The information system updates malicious code protection mechanisms only when directed by a privileged user.
- Withdrawn from previous version. Incorporated into MP-7.
- The organization:
  - Tests malicious code protection mechanisms [*Assignment: organization-defined frequency*] by introducing a known benign, nonspreading test case into the information system.

  - Verifies that both detection of the test case and associated incident reporting occur as required.
- The information system implements nonsignature-based malicious code detection mechanisms.
- The information system detects [*Assignment: organization-defined unauthorized operating system commands*] through the kernel application programming interface at [*Assignment: organization-defined information system hardware components*] and [*Selection (one or more): issues a warning; audits the command execution; prevents the execution of the command*].
- The information system implements [*Assignment: organization-defined security safeguards*] to authenticate [*Assignment: organization-defined remote commands*].
- The organization:
  - Employs [*Assignment: organization-defined tools and techniques*] to analyze the characteristics and behavior of malicious code.
  - Incorporates the results from malicious code analysis into organizational incident response and flaw remediation processes.

The NIST SP 800-53 controls are considered baseline controls, which can be tailored to more closely meet the needs of a particular organization and automation systems as discussed in NIST SP 800-82. NIST SP 800-53 describes the tailoring process as follows:

1. Identifying and designating common controls in initial security control baselines (common controls are controls that may be inherited by one or more organizational information systems)
2. Applying scoping considerations to the remaining baseline security controls
3. Selecting compensating security controls, if needed
4. Assigning specific values to organization-defined parameters via explicit assignment and selection statements
5. Supplementing baselines with additional security controls and control enhancements, if needed

6. Providing additional specification information for control implementation, if needed

The scoping guidance referred to in item 2 of the tailoring process is designed to ensure that only the controls appropriate to the particular mission and requirements of the organization are used. Some factors to consider in scoping include:

- Are there common controls specified by the organization?
- What are the primary security objectives?
- To what system components are the controls applicable?
- What technologies are employed in the system (e.g., cryptography, wireless)?
- What are the policy and regulatory requirements?
- What is the operational environment?
- Are there scalability requirements?
- Is public access permitted?

NIST SP 800-53 also provides for the use of compensating security controls, which are defined in NIST SP 800-53 as "The security controls employed in lieu of the recommended controls in the security control baselines described in NIST SP 800-53 and Committee on National Security Systems (CNSS) Instruction 1253 that provide equivalent or comparable protection for an information system or organization."

NIST SP 800-53 provides guidance to organizations in selecting security controls through the concept of baseline controls, which should be chosen based on the security category and associated impact level of information systems determined in accordance with FIPS Publication 199 and FIPS Publication 200, respectively. Appendix D of NIST SP 800-53 provides a listing of the security control baselines corresponding to the low-impact, moderate-impact, and high-impact information systems using the high water mark defined in FIPS Publication 200.

As a reference, Appendix F of NIST SP 800-53 provides a comprehensive catalog of security controls for information systems and organizations, arranged by control families.

Useful guidance is also provided in NIST SP 800-53 Appendix E. Appendix E describes the minimum assurance requirements for federal information systems and organizations and highlights the assurance-related controls in the security control baselines in Appendix D needed to ensure that the requirements are satisfied. Assurance is defined in NIST SP 800-53 as "the measure of confidence that the security functions, features, practices, policies, procedures, mechanisms, and architecture of organizational information systems accurately mediate and enforce established security policies." Appendix E specifies assurance-related controls in the low, moderate, and high baselines in tables to help organizations quickly define controls necessary to satisfy minimum assurance requirements, keeping in mind that achieving adequate security for organizational information systems requires the correct combination of both Appendix D functionality controls as well as security controls.

The low, moderate, and high assurance components of the baselines are given in Tables 6-2, 6-3, and 6-4 along with their associated assurance requirements.

## Minimum Assurance Requirements – Low-Impact Systems

Assurance Requirement: The organization, based on its security requirements, security policies, and needed security capabilities, has an expectation of: (1) a limited strength of security functionality; and (2) a limited degree of confidence supported by the depth and coverage of associated security evidence that the security functionality is complete, consistent, and correct.

**Table 6-2. Assurance-Related Controls for Low-Impact Systems**
*Source: NIST SP 800-53 (2013)*

| ID | Controls | ID | Controls |
|---|---|---|---|
| **AC** | AC-1 | **MP** | MP-1 |
| **AT** | AT-1, AT-2, AT-3, AT-4 | **PE** | PE-1, PE-6, PE-8 |
| **AU** | AU-1, AU-6 | **PL** | PL-1, PL-2, PL-4 |
| **CA** | CA-1, CA-2, CA-3, CA-5, CA-6, CA-7, CA-9 | **PS** | PS-1, PS-6, PS-7 |
| **CM** | CM-1, CM-2, CM-4, CM-8 | **RA** | RA-1, RA-3, RA-5 |
| **CP** | CP-1, CP-3, CP-4 | **SA** | SA-1, SA-2, SA-3, SA-4, SA-4(10), SA-5, SA-9 |
| **IA** | IA-1 | **SC** | SC-1, SC-39 |
| **IR** | IR-1, IR-2, IR-5 | **SI** | SI-1, SI-4, SI-5 |
| **MA** | MA-1 | | |

## Minimum Assurance Requirements – Moderate-Impact Systems

Assurance Requirement: The organization, based on its security requirements, security policies, and needed security capabilities, has an expectation of: (1) a moderate strength of security functionality; and (2) a moderate degree of confidence supported by the depth and coverage of associated security evidence that the security functionality is complete, consistent, and correct.

**Table 6-3. Assurance-Related Controls for Moderate-Impact Systems**
*Source: NIST SP 800-53 (2013)*

| ID | Controls | ID | Controls |
|---|---|---|---|
| **AC** | AC-1 | **MP** | MP-1 |
| **AT** | AT-1, AT-2, AT-2(2), AT-3, AT-4 | **PE** | PE-1, PE-6, PE-6(1), PE-8 |
| **AU** | AU-1, AU-6, AU-6(1), AU-6(3), AU-7, AU-7(1) | **PL** | PL-1, PL-2, PL-2(3), PL-4, PL-4(1), PL-8 |
| **CA** | CA-1, CA-2, CA-2(1), CA-3, CA-5, CA-6, CA-7, CA-7(1), CA-9 | **PS** | PS-1, PS-6, PS-7 |
| **CM** | CM-1, CM-2, CM-2(1), CM-2(3), CM-2(7), CM-3, CM-3(2), CM-4, CM-8, CM-8(1), CM-8(3), CM-8(5) | **RA** | RA-1, RA-3, RA-5, RA-5(1), RA-5(2), RA-5(5) |
| **CP** | CP-1, CP-3, CP-4, CP-4(1) | **SA** | SA-1, SA-2, SA-3, SA-4, SA-4(1), SA-4(2), SA-4(9), SA-4(10), SA-5, SA-8, SA-9, SA-9(2), SA-10, SA-11 |
| **IA** | IA-1 | **SC** | SC-1, SC-2, SC-39 |
| **IR** | IR-1, IR-2, IR-3, IR-3(2), IR-5 | **SI** | SI-1, SI-4, SI-4(2), SI-4(4), SI-4(5), SI-5, SI-7, SI-7(1), SI-7(7), SI-10, SI-16 |
| **MA** | MA-1 | | |

## Minimum Assurance Requirements – High-Impact Systems

Assurance Requirement: The organization, based on its security requirements, security policies, and needed security capabilities, has an expectation of: (1) a high strength of security functionality; and (2) a high degree of confidence supported by the depth and coverage of associated security evidence that the security functionality is complete, consistent, and correct.

**Table 6-4. Assurance-Related Controls for High-Impact Systems**

*Source: NIST SP 800-53 (2013)*

| ID | Controls | ID | Controls |
|---|---|---|---|
| **AC** | AC-1 | **MP** | MP-1 |
| **AT** | AT-1, AT-2, AT-2(2), AT-3, AT-4 | **PE** | PE-1, PE-6, PE-6(1), PE-6(4), PE-8 |
| **AU** | AU-1, AU-6, AU-6(1), AU-6(3), AU-6(5), AU-6(6), AU-7, AU-7(1), AU-10 | **PL** | PL-1, PL-2, PL-2(3), PL-4, PL-4(1), PL-8 |
| **CA** | CA-1, CA-2, CA-2(1), CA-2(2), CA-3, CA-5, CA-6, CA-7, CA-7(1), CA-8, CA-9 | **PS** | PS-1, PS-6, PS-7 |
| **CM** | CM-1, CM-2, CM-2(1), CM-2(2), CM-2(3), CM-2(7), CM-3, CM-3(1), CM-3(2), CM-4, CM-4(1), CM-8, CM-8(1), CM-8(2), CM-8(3), CM-8(4), CM-8(5) | **RA** | RA-1, RA-3, RA-5, RA-5(1), RA-5(2), RA-5(4), RA-5(5) |
| **CP** | CP-1, CP-3, CP-3(1), CP-4, CP-4(1), CP-4(2) | **SA** | SA-1, SA-2, SA-3, SA-4, SA-4(1), SA-4(2), SA-4(9), SA-4(10), SA-5, SA-8, SA-9, SA-9(2), SA-10, SA-11, SA-12, SA-15, SA-16, SA-17 |
| **IA** | IA-1 | **SC** | SC-1, SC-2, SC-3, SC-7(18), SC-7(21), SC-24, SC-39 |
| **IR** | IR-1, IR-2, IR-2(1), IR-2(2), IR-3, IR-3(2), IR-5, IR-5(1) | **SI** | SI-1, SI-4, SI-4(2), SI-4(4), SI-4(5), SI-5, SI-5(1), SI-6, SI-7, SI-7(1), SI-7(2), SI-7(5), SI-7(7), SI-7(14), SI-10, SI-16 |
| **MA** | MA-1 | | |

If, over time, organizations choose to raise the level of assurance in their information systems and increase the level of trustworthiness accordingly, they can add assurance-related controls to the controls in the baselines. These additions will make the functionality highly resistant to penetration, tamper, or bypass. Table 6-5 provides additional developmental and operational activities (e.g., in the SA, SI, and CM security control families) that organizations can select to achieve an enhanced level of assurance (up to and including high assurance).

**Table 6-5. Assurance-Related Controls for High-Impact Systems**
*Source: NIST SP 800-53 (2013)*

| ID | Controls | ID | Controls |
|---|---|---|---|
| **AC** | AC-25 | **MP** | No additional controls. |
| **AT** | AT-2(1), AT-3 (all enhancements) | **PE** | PE-6(2), PE-6(3) |
| **AU** | French (Congo (DRC)) AU-6(4), AU-6(7), AU-6(8), AU-6(9), AU-6(10), AU-10 (all enhancements), AU-11(1), AU-13 (plus enhancements), AU-14 (plus enhancements) | **PL** | PL-8 (all enhancements), PL-9 |
| **CA** | CA-2(3), CA-5(1), CA-7(3), CA-8 (all enhancements), CA-9(1) | **PS** | PS-6(2), PS-6(3) |
| **CM** | CM-2(6), CM-4(2), CM-8(6), CM-8(7), CM-8(8), CM-8(9), | **RA** | French (Congo (DRC)) RA-5(3), RA-5(6), RA-5(8), RA-5(10), RA-6 |
| **CP** | CP-3(2), CP-4(3), CP-4(4), CP-12 | **SA** | French (Congo (DRC)) SA-4(3), SA-4(5), SA-4(6), SA-4(7), SA-4(8), SA-9(1), SA-9(3), SA-9(4), SA-9(5), SA-10 (all enhancements), SA-11 (all enhancements), SA-12 (all enhancements), SA-13, SA-14, SA-15 (all enhancements), SA-17 (all enhancements), SA-18 (plus enhancements), SA-19 (plus enhancements), SA-20, SA-21 (plus enhancement), SA-22 (plus enhancement) |
| **IA** | No additional controls. | **SC** | French (Congo (DRC)) SC-2(1), SC-3 (all enhancements), SC-6, SC-7(22), SC-11 (plus enhancement), SC-29 (plus enhancement), SC-30 (plus enhancements), SC-31 (plus enhancements), SC-32, SC-34 (plus enhancements), SC-36 (plus enhancement), SC-37 (plus enhancement), SC-38, SC-39 (all enhancements) |
| **IR** | IR-3(1) | **SI** | French (Congo (DRC)) SI-4(1), SI-4(3), SI-4(7), SI-4(9), SI-4(10), SI-4(11), SI-4(12), SI-4(13), SI-4(14), SI-4(15), SI-4(16), SI-4(17), SI-4(18), SI-4(19), SI-4(20), SI-4(21), SI-4(22), SI-4(23), SI-4(24), SI-7(3), SI-7(6), SI-7(8), SI-7(9), SI-7(10), SI-7(11), SI-7(12), SI-7(13), SI-7(15), SI-7(16), SI-10 (all enhancements), SI-13 (plus enhancements), SI-14 (plus enhancement), SI15, SI-17 |
| **MA** | No additional controls. | | |

As stated previously in this chapter, used in conjunction with NIST SP 800-53, NIST SP 800-82 provides the means to apply effective and proven controls to industrial automation and control systems.

## NIST Special Publication 800-82, Guide to Industrial Control Systems Security

The purpose of this document is to "provide guidance for securing industrial control systems (ICS), including supervisory control and data acquisition (SCADA) systems, distributed control systems (DCS), and other systems performing control functions."

NIST SP 800-82[2] defines the major objectives for ICS security implementation as the following:

- **Restricting Logical Access to the ICS Network and Network Activity –** This includes using a demilitarized zone (DMZ) network architecture with firewalls to prevent network traffic from passing directly between the corporate and ICS networks, and having separate authentication mechanisms and credentials for users of the corporate and ICS networks. The ICS should also use a network topology that has multiple layers, with the most critical communications occurring in the most secure and reliable layer.
- **Restricting Physical Access to the ICS Network and Devices –** Unauthorized physical access to components could cause serious disruption of the ICS's functionality. A combination of physical access controls should be used, such as locks, card readers, and/or guards.
- **Protecting Individual ICS Components from Exploitation –** This includes deploying security patches in as expeditious a manner as possible, after testing them under field conditions; disabling all unused ports and services; restricting ICS user privileges to only those that are required for each person's role; tracking and monitoring audit trails; and using security controls, such as antivirus software and file integrity checking software, where technically feasible to prevent, deter, detect, and mitigate malware.
- **Restricting Unauthorized Modification of Data –** This includes data that is in transit (at least across the network boundaries) and at rest.

- **Maintaining Functionality During Adverse Conditions –** This consists of designing the ICS so that each critical component has a redundant counterpart. Additionally, if a component fails, it should fail in a manner that does not generate unnecessary traffic on the ICS or other networks, and does not cause another problem elsewhere, such as a cascading event.
- **Detecting Security Events and Incidents –** Detecting security events, which have not yet escalated into incidents, can help defenders break the attack chain before attackers attain their objectives. This includes the capability to detect failed ICS components, unavailable services, and exhausted resources that are important to provide proper and safe functioning of the ICS. The ICS should also allow for graceful degradation such as moving from *normal operation* with full automation to *emergency operation* with operators more involved and less automation to *manual operation* with no automation.
- **Restoring System After an Incident –** Incidents are inevitable and an incident response plan is essential. A major characteristic of a good security program is how quickly a system can be restored after an incident has occurred.

A risk management process should be employed throughout the organization, using a three-tiered approach to address risk at the (1) organization level; (2) mission/business process level; and (3) information system level (IT and ICS). Information on multi-tiered risk management and the risk management process can be found in NIST SP 800-39, *Managing Information Security Risk: Organization, Mission and Information System View*, and NIST SP 800-37 Revision 1, *Guide for Applying the Risk Management Framework to Federal Information Systems: A Security Life Cycle Approach*, as addressed in Chapter 5. In the discussion of risk, NIST SP 800-82 highlights ICS considerations as well as the impact that these considerations have on the risk management process.

## Network Segmentation and Segregation

NIST SP 800-82 promotes network segmentation and segregation to minimize access to sensitive information for unauthorized individuals, while ensuring that the organization can continue to operate effectively. Operational risk analysis should be performed to determine critical parts of each ICS network and operation and to help define what parts of the ICS need to be segmented. For example, one large ICS network can be partitioned into multiple ICS networks, where the partitioning is based on factors such as management authority, uniform policy and level of trust, functional criticality, and amount of communications traffic that crosses the domain boundary.

As detailed in the NIST SP 800-82 documentation, segmentation establishes security domains, or enclaves, that are typically defined as being managed by the same authority, enforcing the same policy, and having a uniform level of trust. Segmentation can minimize the method and level of access to sensitive information, ICS communication and equipment configuration; can make it significantly more difficult for a malicious cyber adversary; and can contain the effects of nonmalicious errors and accidents.

Traditionally, network segmentation and segregation is implemented at the gateway between domains. ICS environments often have multiple well-defined domains, such as operational LANs, control LANs, and operational DMZs, as well as gateways to non-ICS and less trustworthy domains, such as the Internet and the corporate LANs.

Network segmentation and segregation can be implemented through some of the following methods:

- Logical network separation is enforced by encryption or network device-enforced partitioning.
  - Virtual local area networks (VLANS) can be used.
  - Encrypted virtual private networks (VPNs) use cryptographic mechanisms to separate traffic combined in one network.
  - Unidirectional gateways restrict communications between connections to a single direction and therefore, segment the network.
- Physical network separation can completely prevent any interconnectivity of traffic between domains.
- Network traffic filtering can utilize a variety of technologies at various network layers to enforce security requirements and domains.
  - Network layer filtering restricts which systems are able to communicate with others on the network based on IP and route information.

- State-based filtering restricts which systems are able to communicate with others on the network based on their intended function or current state of operation.
- Port and/or protocol-level filtering restricts the number and type of services that each system can use to communicate with others on the network.
- Application filtering commonly filters the content of communications between systems at the application layer. This includes application-level firewalls, proxies, and content-based filter.

Another valuable security technique is boundary protection. In boundary protection, the flow of information between interconnected security domains is controlled. Boundary protection controls include gateways, routers, firewalls, guards, network-based malicious code analysis and virtualization systems, intrusion detection systems (networked and host-based), encrypted tunnels, managed interfaces, mail gateways, and unidirectional gateways (e.g., data diodes). Boundary protection devices determine whether data transfer is permitted, often by examining the data or associated metadata.

See Chapter 2 for a detailed discussion of firewalls used as boundary protection and network segregation.

## ICS Security Controls

In selecting and implementing security controls, the following steps are recommended:

1. **Execute the Risk Management Framework (RMF)** as discussed in Chapter 5, including categorization of the ICS as low-impact, moderate-impact, or high-impact for the security objectives of confidentiality, integrity, and availability. Examples of low, moderate, and high levels of security impact for ICS based on ISA99[3] are given in Table 6-6 and those based on product-produced, industry, and security concerns are shown in Table 6-7.

**Table 6-6. Security Impact Levels Based on ISA99**

*Source: NIST SP 800-82 (2015)*

| Impact Category | Low-Impact | Moderate-Impact | High-Impact |
|---|---|---|---|
| Injury | Cuts, bruises requiring first aid | Requires hospitalization | Loss of life or limb |
| Financial Loss | $1,000 | $100,000 | Millions |
| Environmental Release | Temporary damage | Lasting damage | Permanent damage, off-site damage |
| Interruption of Production | Minutes | Days | Weeks |
| Public Image | Temporary damage | Lasting damage | Permanent damage |

**Table 6-7. Security Impact Levels Based on Product, Industry, and Security Concerns**

*Source: NIST SP 800-82 (2015)*

| Category | Low-Impact | Moderate-Impact | High-Impact |
|---|---|---|---|
| Product Produced | • Nonhazardous materials or products<br>• Non-ingested consumer products | • Some hazardous products or steps during production<br>• High amount of proprietary information | • Critical infrastructure (e.g., electricity)<br>• Hazardous materials<br>• Ingested products |
| Industry Examples | • Plastic injection molding<br>• Warehouse applications | • Automotive metal industries<br>• Pulp and paper<br>• Semiconductors | • Utilities<br>• Petrochemical<br>• Food and beverage<br>• Pharmaceutical |
| Security Concerns | • Protection against minor injuries<br>• Ensuring uptime | • Protection against moderate injuries<br>• Ensuring uptime<br>• Capital investment | • Protection against major injuries/loss of life<br>• Ensuring uptime<br>• Capital investment<br>• Trade secrets<br>• Ensuring basic social services<br>• Regulatory compliance |

2. **Select Minimum Security Controls** planned or in place to protect the information system based on a set of requirements. These requirements can be determined through the following steps:

    a. Reference NIST FIPS 200[3] to determine a set of minimum security requirements from its 18 security-related areas with regard to protecting the confidentiality, integrity, and availability of information systems and the information processed, stored, and transmitted by those systems. Then, these requirements can be used to select baseline controls based on the impact levels of step 1.

b. Apply tailoring guidance through overlays to security control baselines described in NIST SP 800-53. An overlay is a fully specified set of security controls, control enhancements, and supplemental guidance that addresses the need for developing community-wide and specialized sets of security controls for organizations. Section 3.3 and Appendix I of NIST SP 800-53 provide additional information on the application of overlays.

c. Reference Appendix G of NIST SP 800-82, which includes an ICS-specific overlay of applicable controls that provide tailored baselines for low-impact, moderate-impact, and high-impact ICS. These tailored baselines can be utilized as starting specifications and recommendations that can be applied to specific ICS by responsible personnel. The use of an overlay does not in any way preclude organizations from performing further tailoring to add or remove controls and control enhancements (i.e., overlays can also be subject to tailoring) to reflect organization-specific needs, assumptions, or constraints.

3. **Implement Security Controls** for new systems or legacy systems as follows:

    a. For new developments, the security control selection process is applied from a requirements definition perspective since the systems do not yet exist and organizations are conducting initial security categorizations. The security controls included in the security plans for the information systems serve as a security specification and are expected to be incorporated into the systems during the development and implementation phases of the system development life cycle.

    b. For legacy systems, the security control selection process is applied from a gap analysis perspective when organizations are anticipating significant changes to the systems (e.g., during major upgrades, modifications, or outsourcing). Since the information systems already exist, organizations in all likelihood have completed the security categorization and security control selection processes resulting in the establishment of previously agreed-upon security controls in the respective security plans and the implementation of those controls within the systems.

4. **Assess Security Controls** by following the guidance in NIST SP 800-53A to determine the extent to which the security controls in the information system are effective in their application and operating as intended.

5. **Authorize System Operation** through a management decision to explicitly accept the risk to organization operations, assets, or individuals based on the implementation of an agreed-upon set of security controls.

6. **Monitor Security Controls** by continuously tracking changes to the system that may affect security controls and assess control effectiveness. NIST SP 800-137 provides guidance on information security continuous monitoring.

## NIST 800-53 Control Families

As discussed in NIST SP 800-53, its security controls are organized into 18 families; each family contains security controls related to the general security topic of the family. Security controls may involve aspects of policy, oversight, supervision, manual processes, actions by individuals, or automated mechanisms implemented by information systems/devices.

In summary, the 18 security-related areas are:

- **Access Control (AC)** – The process of granting or denying specific requests for obtaining and using information and related information processing services for physical access to areas within the information system environment.

- **Awareness and Training (AT)** – Policies and procedures to ensure that all information system users are given appropriate security training relative to their usage of the system and that accurate training records are maintained.

- **Audit and Accountability (AU)** – Independent review and examination of records and activities to assess the adequacy of system controls, to ensure compliance with established policies and operational procedures, and to recommend necessary changes in controls, policies, or procedures.

- **Security Assessment and Authorization (CA)** – Assurance that the specified controls are implemented correctly, operating as intended, and producing the desired outcome.
- **Contingency Planning (CP)** – Policies and procedures designed to maintain or restore business operations, including computer operations, possibly at an alternate location, in the event of emergencies, system failures, or disaster.
- **Configuration Management (CM)** – Policies and procedures for controlling modifications to hardware, firmware, software, and documentation to ensure the information system is protected against improper modifications prior to, during, and after system implementation.
- **Identification and Authentication (IA)** – The process of verifying the identity of a user, process, or device through the use of specific credentials (e.g., passwords, tokens, biometrics) as a prerequisite for granting access to resources in an IT system.
- **Incident Response (IR)** – Policies and procedures pertaining to incident response training, testing, handling, monitoring, reporting, and support services.
- **Maintenance (MA)** – Policies and procedures to manage all maintenance aspects of an information system.
- **Media Protection (MP)** – Policies and procedures to ensure secure handling of media. Controls cover access, labeling, storage, transport, sanitization, destruction, and disposal.
- **Physical and Environmental Protection (PE)** – Policies and procedures addressing physical, transmission, and display access control as well as environmental controls for conditioning (e.g., temperature, humidity) and emergency provisions (e.g., shutdown, power, lighting, fire protection).
- **Planning (PL)** – Development and maintenance of a plan to address information system security by performing assessments, specifying and implementing security controls, assigning security levels, and responding to incidents.

- **Personnel Security (PS)** – Policies and procedures for personnel position categorization, screening, transfer, penalty, and termination; this also addresses third-party personnel security.
- **Risk Assessment (RA)** – The process of identifying risks to operations, assets, or individuals by determining the probability of occurrence, the resulting impact, and additional security controls that would mitigate this impact.
- **System and Services Acquisition (SA)** – Allocation of resources for information system security to be maintained throughout the system's life cycle and the development of acquisition policies based on risk assessment results including requirements, design criteria, test procedures, and associated documentation.
- **System and Communications Protection (SC)** – Mechanisms for protecting both system and data transmission components.
- **System and Information Integrity (SI)** – Policies and procedures to protect information systems and their data from design flaws and data modification using functionality verification, data integrity checking, intrusion detection, malicious code detection, and security alert and advisory controls.
- **Program Management (PM)** – Provides security controls at the organizational level rather than the information-system level.

## Appendix G – ICS Overlay

NIST SP 800-82 defines an ICS overlay as a partial tailoring of the controls and control baselines in SP 800-53, Revision 4, and adds supplementary guidance specific to ICS. The overlay in NIST SP 800-82 provides guidance for securing ICS, including SCADA and DCS systems, PLCs, and other systems performing industrial control functions. Overlays provide a structured approach to help organizations tailor security control baselines and develop specialized security plans that can be applied to specific missions/business functions, environments of operation, and/or technologies. This specialization approach is important as the number of threat-driven controls and control enhancements in the catalog increases and organizations develop risk management strategies to address their specific protection needs within defined risk tolerances.

Table 6-8 (Table G-1 of NIST SP 800-82) provides a summary of the security controls and control enhancements from NIST SP 800-53 Appendix F that have been allocated to the initial security control baselines (i.e., Low, Moderate, and High) along with indications of ICS Supplemental Guidance and ICS tailoring. Controls and control enhancements for which there is ICS Supplemental Guidance are **bolded**. If the control baselines are supplemented by the addition of a control to the baseline, the control or control enhancement is underlined. If a control or control enhancement is removed from the baseline, the control or control enhancement is deleted.

Example:

| | | | | |
|---|---|---|---|---|
| AU-4 | Audit Storage Capacity | AU-4 **(1)** | AU-4 **(1)** | AU-4 **(1)** |

In this example from NIST SP 800-82, ICS Supplemental Guidance was added to Control Enhancement 1 of AU-4 (bolded). In addition, Control Enhancement 1 of AU-4 was added to the Low, Moderate (Mod), and High baselines (underlined).

**Table 6-8. Table G-1 Security Control Baselines**

| CNTL | Control Name | Low | Mod | High |
|---|---|---|---|---|
| AC-1 | Access Control Policy and Procedures | **AC-1** | **AC-1** | **AC-1** |
| AC-2 | Account Management | **AC-2** | **AC-2 (1) (2) (3) (4)** | **AC-2 (1) (2) (3) (4) (5)** (11) (12) (13) |
| AC-3 | Access Enforcement | **AC-3** | **AC-3** | **AC-3** |
| AC-4 | Information Flow Enforcement | Not Selected | AC-4 | AC-4 |
| AC-5 | Separation of Duties | Not Selected | **AC-5** | **AC-5** |
| AC-6 | Least Privilege | Not Selected | **AC-6 (1) (2) (5) (9) (10)** | **AC-6 (1) (2) (3) (5) (9) (10)** |
| AC-7 | Unsuccessful Logon Attempts | **AC-7** | **AC-7** | **AC-7** |
| AC-8 | System Use Notification | **AC-8** | **AC-8** | **AC-8** |
| AC-10 | Concurrent Session Control | Not Selected | Not Selected | **AC-10** |
| AC-11 | Session Lock | Not Selected | **AC-11 (1)** | **AC-11 (1)** |
| AC-12 | Session Termination | Not Selected | **AC-12** | **AC-12** |
| AC-14 | Permitted Actions without Identification or Authentication | AC-14 | AC-14 | AC-14 |

**Table 6-8. Table G-1 Security Control Baselines**

| | | | | |
|---|---|---|---|---|
| AC-17 | Remote Access | **AC-17** | **AC-17 (1) (2) (3) (4)** | **AC-17 (1) (2) (3) (4)** |
| AC-18 | Wireless Access | **AC-18** | **AC-18 (1)** | **AC-18 (1) (4)** (5) |
| AC-19 | Access Control for Mobile Devices | **AC-19** | **AC-19** (5) | **AC-19** (5) |
| AC-20 | Use of External Information Systems | **AC-20** | **AC-20** (1) (2) | **AC-20** (1) (2) |
| AC-21 | Collaboration and Information Sharing | **AC-21** | **AC-21** | **AC-21** |
| AC-22 | Publicly Accessible Content | **AC-22** | **AC-22** | **AC-22** |
| AT-1 | Security Awareness and Training Policy and Procedures | **AT-1** | **AT-1** | **AT-1** |
| AT-2 | Security Awareness Training | **AT-2** | **AT-2** (2) | **AT-2** (2) |
| AT-3 | Role-Based Security Training | **AT-3** | **AT-3** | **AT-3** |
| AT-4 | Security Training Records | AT-4 | AT-4 | AT-4 |
| AU-1 | Audit and Accountability Policy and Procedures | **AU-1** | **AU-1** | **AU-1** |
| AU-2 | Audit Events | **AU-2** | **AU-2** (3) | **AU-2** (3) |
| AU-3 | Content of Audit Records | **AU-3** | **AU-3** (1) | **AU-3** (1) (2) |
| AU-4 | Audit Storage Capacity | AU-4 **(1)** | AU-4 **(1)** | AU-4 **(1)** |
| AU-5 | Response to Audit Processing Failures | AU-5 | AU-5 | AU-5 (1) (2) |
| AU-6 | Audit Review, Analysis, and Reporting | AU-6 | AU-6 **(1)** (3) | AU-6 **(1)** (3) (5) (6) |
| AU-7 | Audit Reduction and Report Generation | Not Selected | AU-7 (1) | AU-7 (1) |
| AU-8 | Time Stamps | **AU-8** | **AU-8 (1)** | **AU-8 (1)** |
| AU-9 | Protection of Audit Information | AU-9 | AU-9 (4) | AU-9 (2) (3) (4) |
| AU-10 | Nonrepudiation | Not Selected | Not Selected | **AU-10** |
| AU-11 | Audit Record Retention | AU-11 | AU-11 | AU-11 |
| AU-12 | Audit Generation | AU-12 | AU-12 | AU-12 **(1) (3)** |
| CA-1 | Security Assessment and Authorization Policies and Procedures | **CA-1** | **CA-1** | **CA-1** |
| CA-2 | Security Assessments | **CA-2** | **CA-2** (1) | **CA-2** (1) **(2)** |
| CA-3 | System Interconnections | **CA-3** | **CA-3** (5) | **CA-3** (5) |
| CA-5 | Plan of Action and Milestones | CA-5 | CA-5 | CA-5 |
| CA-6 | Security Authorization | CA-6 | CA-6 | CA-6 |
| CA-7 | Continuous Monitoring | **CA-7** | **CA-7** (1) | **CA-7** (1) |
| CA-8 | Penetration Testing | Not Selected | Not Selected | **CA-8** |
| CA-9 | Internal System Connections | **CA-9** | **CA-9** | **CA-9** |

**Table 6-8. Table G-1 Security Control Baselines**

| | | | | |
|---|---|---|---|---|
| CM-1 | Configuration Management Policy and Procedures | **CM-1** | **CM-1** | **CM-1** |
| CM-2 | Baseline Configuration | CM-2 | CM-2 (1) (3) (7) | CM-2 (1) (2) (3) (7) |
| CM-3 | Configuration Change Control | Not Selected | CM-3 (2) | CM-3 (1) (2) |
| CM-4 | Security Impact Analysis | **CM-4** | **CM-4** | **CM-4** (1) |
| CM-5 | Access Restrictions for Change | Not Selected | CM-5 | CM-5 **(1) (2) (3)** |
| CM-6 | Configuration Settings | CM-6 | CM-6 | CM-6 (1) (2) |
| CM-7 | Least Functionality | **CM-7** (1) | **CM-7** (1) (2) ~~(4)~~ (5) | **CM-7** (1) (2) (5) |
| CM-8 | Information System Component Inventory | CM-8 | CM-8 (1) (3) (5) | CM-8 (1) (2) (3) (4) (5) |
| CM-9 | Configuration Management Plan | Not Selected | CM-9 | CM-9 |
| CM-10 | Software Usage Restrictions | **CM-10** | **CM-10** | **CM-10** |
| CM-11 | User-Installed Software | CM-11 | CM-11 | CM-11 |
| CP-1 | Contingency Planning Policy and Procedures | **CP-1** | **CP-1** | **CP-1** |
| CP-2 | Contingency Plan | **CP-2** | **CP-2 (1) (3) (8)** | **CP-2 (1) (2) (3) (4) (5) (8)** |
| CP-3 | Contingency Training | CP-3 | CP-3 | CP-3 (1) |
| CP-4 | Contingency Plan Testing | CP-4 | CP-4 (1) | CP-4 (1) (2) |
| CP-6 | Alternate Storage Site | Not Selected | CP-6 (1) (3) | CP-6 (1) (2) (3) |
| CP-7 | Alternate Processing Site | Not Selected | CP-7 (1) (2) (3) | CP-7 (1) (2) (3) (4) |
| CP-8 | Telecommunications Services | Not Selected | **CP-8** (1) (2) | **CP-8** (1) (2) (3) (4) |
| CP-9 | Information System Backup | CP-9 | CP-9 (1) | CP-9 (1) (2) (3) (5) |
| CP-10 | Information System Recovery and Reconstitution | **CP-10** | **CP-10** (2) | **CP-10** (2) (4) |
| CP-12 | Safe Mode | **CP-12** | **CP-12** | **CP-12** |
| IA-1 | Identification and Authentication Policy and Procedures | **IA-1** | **IA-1** | **IA-1** |
| IA-2 | Identification and Authentication (Organizational Users) | **IA-2 (1) (12)** | **IA-2 (1) (2) (3) (8)** (11) **(12)** | **IA-2 (1) (2) (3) (4) (8) (9)** (11) **(12)** |
| IA-3 | Device Identification and Authentication | **IA-3** | **IA-3** (1) (4) | **IA-3** (1) (4) |
| IA-4 | Identifier Management | IA-4 | IA-4 | IA-4 |

**Table 6-8. Table G-1 Security Control Baselines**

| | | | | |
|---|---|---|---|---|
| IA-5 | Authenticator Management | **IA-5** (1) (11) | **IA-5** (1) (2) (3) (11) | **IA-5** (1) (2) (3) (11) |
| IA-6 | Authenticator Feedback | **IA-6** | **IA-6** | **IA-6** |
| IA-7 | Cryptographic Module Authentication | IA-7 | IA-7 | IA-7 |
| IA-8 | Identification and Authentication (Non- Organizational Users) | **IA-8 (1) (2) (3) (4)** | **IA-8 (1) (2) (3) (4)** | **IA-8 (1) (2) (3) (4)** |
| IR-1 | Incident Response Policy and Procedures | **IR-1** | **IR-1** | **IR-1** |
| IR-2 | Incident Response Training | IR-2 | IR-2 | IR-2 (1) (2) |
| IR-3 | Incident Response Testing | Not Selected | IR-3 (2) | IR-3 (2) |
| IR-4 | Incident Handling | IR-4 | IR-4 (1) | IR-4 (1) (4) |
| IR-5 | Incident Monitoring | IR-5 | IR-5 | IR-5 (1) |
| IR-6 | Incident Reporting | **IR-6** | **IR-6 (1)** | **IR-6 (1)** |
| IR-7 | Incident Response Assistance | IR-7 | IR-7 (1) | IR-7 (1) |
| IR-8 | Incident Response Plan | IR-8 | IR-8 | IR-8 |
| MA-1 | System Maintenance Policy and Procedures | **MA-1** | **MA-1** | **MA-1** |
| MA-2 | Controlled Maintenance | MA-2 | MA-2 | MA-2 (2) |
| MA-3 | Maintenance Tools | Not Selected | MA-3 (1) (2) | MA-3 (1) (2) (3) |
| MA-4 | Nonlocal Maintenance | MA-4 | MA-4 (2) | MA-4 (2) **(3)** |
| MA-5 | Maintenance Personnel | MA-5 | MA-5 | MA-5 (1) |
| MA-6 | Timely Maintenance | Not Selected | MA-6 | MA-6 |
| MP-1 | Media Protection Policy and Procedures | **MP-1** | **MP-1** | **MP-1** |
| MP-2 | Media Access | MP-2 | MP-2 | MP-2 |
| MP-3 | Media Marking | Not Selected | MP-3 | MP-3 |
| MP-4 | Media Storage | Not Selected | MP-4 | MP-4 |
| MP-5 | Media Transport | Not Selected | MP-5 (4) | MP-5 (4) |
| MP-6 | Media Sanitization | MP-6 | MP-6 | MP-6 (1) (2) (3) |
| MP-7 | Media Use | MP-7 | MP-7 (1) | MP-7 (1) |
| PE-1 | Physical and Environmental Protection Policy and Procedures | **PE-1** | **PE-1** | **PE-1** |
| PE-2 | Physical Access Authorizations | PE-2 | PE-2 | PE-2 |
| PE-3 | Physical Access Control | PE-3 | PE-3 | PE-3 (1) |
| PE-4 | Access Control for Transmission Medium | Not Selected | PE-4 | PE-4 |
| PE-5 | Access Control for Output Devices | Not Selected | PE-5 | PE-5 |
| PE-6 | Monitoring Physical Access | **PE-6** | **PE-6** (1) **(4)** | **PE-6** (1) **(4)** |

## Table 6-8. Table G-1 Security Control Baselines

| | | | | |
|---|---|---|---|---|
| PE-8 | Visitor Access Records | PE-8 | PE-8 | PE-8 (1) |
| PE-9 | Power Equipment and Cabling | Not Selected | PE-9 (1) | PE-9 (1) |
| PE-10 | Emergency Shutoff | Not Selected | PE-10 | PE-10 |
| PE-11 | Emergency Power | PE-11 (1) | PE-11 (1) | PE-11 (1) (2) |
| PE-12 | Emergency Lighting | PE-12 | PE-12 | PE-12 |
| PE-13 | Fire Protection | **PE-13** | **PE-13** (3) | **PE-13** (1) (2) (3) |
| PE-14 | Temperature and Humidity Controls | PE-14 | PE-14 | PE-14 |
| PE-15 | Water Damage Protection | **PE-15** | **PE-15** | **PE-15** (1) |
| PE-16 | Delivery and Removal | PE-16 | PE-16 | PE-16 |
| PE-17 | Alternate Work Site | Not Selected | PE-17 | PE-17 |
| PE-18 | Location of Information System Components | Not Selected | Not Selected | PE-18 |
| PL-1 | Security Planning Policy and Procedures | **PL-1** | **PL-1** | **PL-1** |
| PL-2 | System Security Plan | PL-2 (3) | PL-2 (3) | PL-2 (3) |
| PL-4 | Rules of Behavior | PL-4 | PL-4 (1) | PL-4 (1) |
| PL-7 | Security Concept of Operations | | PL-7 | PL-7 |
| PL-8 | Information Security Architecture | Not Selected | PL-8 | PL-8 |
| PS-1 | Personnel Security Policy and Procedures | **PS-1** | **PS-1** | **PS-1** |
| PS-2 | Position Risk Designation | PS-2 | PS-2 | PS-2 |
| PS-3 | Personnel Screening | PS-3 | PS-3 | PS-3 |
| PS-4 | Personnel Termination | PS-4 | PS-4 | PS-4 (2) |
| PS-5 | Personnel Transfer | PS-5 | PS-5 | PS-5 |
| PS-6 | Access Agreements | PS-6 | PS-6 | PS-6 |
| PS-7 | Third-Party Personnel Security | PS-7 | PS-7 | PS-7 |
| PS-8 | Personnel Sanctions | PS-8 | PS-8 | PS-8 |
| RA-1 | Risk Assessment Policy and Procedures | **RA-1** | **RA-1** | **RA-1** |
| RA-2 | Security Categorization | RA-2 | RA-2 | RA-2 |
| RA-3 | Risk Assessment | RA-3 | RA-3 | RA-3 |
| RA-5 | Vulnerability Scanning | **RA-5** | **RA-5** (1) (2) (5) | **RA-5** (1) (2) (4) (5) |
| SA-1 | System and Services Acquisition Policy and Procedures | **SA-1** | **SA-1** | **SA-1** |
| SA-2 | Allocation of Resources | SA-2 | SA-2 | SA-2 |
| SA-3 | System Development Life Cycle | SA-3 | SA-3 | SA-3 |

**Table 6-8. Table G-1 Security Control Baselines**

| SA-4 | Acquisition Process | **SA-4 (10)** | **SA-4 (1) (2) (9) (10)** | **SA-4 (1) (2) (9) (10)** |
|---|---|---|---|---|
| SA-5 | Information System Documentation | SA-5 | SA-5 | SA-5 |
| SA-8 | Security Engineering Principles | Not Selected | SA-8 | SA-8 |
| SA-9 | External Information System Services | SA-9 | SA-9 (2) | SA-9 (2) |
| SA-10 | Developer Configuration Management | Not Selected | SA-10 | SA-10 |
| SA-11 | Developer Security Testing and Evaluation | Not Selected | SA-11 | SA-11 |
| SA-12 | Supply Chain Protection | Not Selected | Not Selected | SA-12 |
| SA-15 | Development Process, Standards, and Tools | Not Selected | Not Selected | SA-15 |
| SA-16 | Developer-Provided Training | Not Selected | Not Selected | SA-16 |
| SA-17 | Developer Security Architecture and Design | Not Selected | Not Selected | SA-17 |
| SC-1 | System and Communications Protection Policy and Procedures | **SC-1** | **SC-1** | **SC-1** |
| SC-2 | Application Partitioning | Not Selected | **SC-2** | **SC-2** |
| SC-3 | Security Function Isolation | Not Selected | Not Selected | **SC-3** |
| SC-4 | Information in Shared Resources | Not Selected | **SC-4** | **SC-4** |
| SC-5 | Denial of Service Protection | **SC-5** | **SC-5** | **SC-5** |
| SC-7 | Boundary Protection | SC-7 | SC-7 (3) (4) (5) (7) **(18)** | SC-7 (3) (4) (5) (7) (8) **(18)** (21) |
| SC-8 | Transmission Confidentiality and Integrity | Not Selected | SC-8 **(1)** | SC-8 **(1)** |
| SC-10 | Network Disconnect | Not Selected | **SC-10** | **SC-10** |
| SC-12 | Cryptographic Key Establishment and Management | **SC-12** | **SC-12** | **SC-12** (1) |
| SC-13 | Cryptographic Protection | SC-13 | SC-13 | SC-13 |
| SC-15 | Collaborative Computing Devices | SC-15 | SC-15 | SC-15 |
| SC-17 | Public Key Infrastructure Certificates | Not Selected | SC-17 | SC-17 |
| SC-18 | Mobile Code | Not Selected | SC-18 | SC-18 |
| SC-19 | Voice over Internet Protocol | Not Selected | **SC-19** | **SC-19** |
| SC-20 | Secure Name/Address Resolution Service (Authoritative Source) | **SC-20** | **SC-20** | **SC-20** |
| SC-21 | Secure Name/Address Resolution Service (Recursive or Caching Resolver) | SC-21 | SC-21 | SC-21 |

### Table 6-8. Table G-1 Security Control Baselines

| SC-22 | Architecture and Provisioning for Name/Address Resolution Service | **SC-22** | **SC-22** | **SC-22** |
|---|---|---|---|---|
| SC-23 | Session Authenticity | Not Selected | **SC-23** | **SC-23** |
| SC-24 | Fail in Known State | Not Selected | **SC-24** | **SC-24** |
| SC-28 | Protection of Information at Rest | Not Selected | **SC-28** | **SC-28** |
| SC-39 | Process Isolation | **SC-39** | **SC-39** | **SC-39** |
| SC-41 | Port and I/O Device Access | **SC-41** | **SC-41** | **SC-41** |
| SI-1 | System and Information Integrity Policy and Procedures | **SI-1** | **SI-1** | **SI-1** |
| SI-2 | Flaw Remediation | **SI-2** | **SI-2 (2)** | **SI-2** (1) **(2)** |
| SI-3 | Malicious Code Protection | **SI-3** | **SI-3 (1) (2)** | **SI-3 (1) (2)** |
| SI-4 | Information System Monitoring | **SI-4** | **SI-4 (2) (4) (5)** | **SI-4 (2) (4) (5)** |
| SI-5 | Security Alerts, Advisories, and Directives | **SI-5** | **SI-5** | **SI-5** (1) |
| SI-6 | Security Function Verification | Not Selected | Not Selected | **SI-6** |
| SI-7 | Software, Firmware, and Information Integrity | Not Selected | **SI-7 (1) (7)** | **SI-7 (1) (2) (5) (7)** (14) |
| SI-8 | Spam Protection | Not Selected | **SI-8 (1) (2)** | **SI-8 (1) (2)** |
| SI-10 | Information Input Validation | Not Selected | SI-10 | SI-10 |
| SI-11 | Error Handling | Not Selected | SI-11 | SI-11 |
| SI-12 | Information Handling and Retention | SI-12 | SI-12 | SI-12 |
| SI-13 | Predictable Failure Prevention | Not Selected | Not Selected | **SI-13** |
| SI-14 | Non-Persistence | Not Selected | Not Selected | Not Selected |
| SI-15 | Information Output Filtering | Not Selected | Not Selected | Not Selected |
| SI-16 | Memory Protection | Not Selected | SI-16 | SI-16 |
| SI-17 | Fail-Safe Procedures | SI-17 | SI-17 | SI-17 |

The Program Management (PM) family is deployed organization-wide, supporting the information security program. It is not associated with security control baselines and is independent of any system impactlevel. The PM family from NIST 800-82 is given as follows:

**Table 6-9. PM Family**

| | | |
|---|---|---|
| PM-1 | Information Security Program Plan | PM-1 |
| PM-2 | Senior Information Security Officer | PM-2 |
| PM-3 | Information Security Resources | PM-3 |
| PM-4 | Plan of Action and Milestones Process | PM-4 |
| PM-5 | Information System Inventory | PM-5 |
| PM-6 | Information Security Measures of Performance | PM-6 |
| PM-7 | Enterprise Architecture | PM-7 |
| PM-8 | Critical Infrastructure Plan | PM-8 |
| PM-9 | Risk Management Strategy | PM-9 |
| PM-10 | Security Authorization Process | PM-10 |
| PM-11 | Mission/Business Process Definition | PM-11 |
| PM-12 | Insider Threat Program | PM-12 |
| PM-13 | Information Security Workforce | PM-13 |
| PM-14 | Testing, Training, and Monitoring | PM-14 |
| PM-15 | Contacts with Security Groups and Associations | PM-15 |
| PM-16 | Threat Awareness Program | PM-16 |

Appendix B of this text provides a detailed listing of the ICS Supplemental Guidance for the NIST SP 800-53 security controls.

The application of these security controls will reduce the risk to industrial automation and control systems and will mitigate vulnerabilities. Another perspective, which reinforces the controls discussed NIST 800-82, includes the controls recommended by ANSI/ISA-62443-1-1 (99.01.01)-2007.

## ANSI/ISA-62443-1-1 (99.01.01)-2007, Security Technologies for Industrial Automation and Control Systems

ANSI/ISA-62443-1-1 (99.01.01)-2007[4] provides information on a variety of cybersecurity technologies for industrial automation and control systems (IACS). This data can be used to mitigate risk and improve the security posture of industrial automation and control systems. The controls of

ANSI/ISA-62443-1-1 (99.01.01)-2007 are organized into the following categories:

- Authentication and authorization technologies
- Filtering/blocking/access control technologies
- Encryption technologies and data validation
- Management, audit, measurement, monitoring, and detection tools
- Industrial automation and control systems computer software
- Physical security controls

## Authentication and Authorization

ANSI/ISA-62443-1-1 (99.01.01)-2007 describes authorization as "the initial step in protecting an IACS system and its critical assets from unwanted breaches. It is the process of determining who and what should be allowed into or out of a system." Authentication is defined as "the process of positively identifying potential network users, hosts, applications, services, and resources, using a combination of identification factors or credentials. The result of this authentication process then becomes the basis for permitting or denying further actions." In addition, authentication is further divided into the following two areas:

- **User Authentication** – Traditional computer authentication, such as "logging into a computer" or activating a human-machine interface (HMI) to adjust a process
- **Network Service Authentication** – The ability for networked devices to distinguish between authorized and unauthorized remote requests for IACS data or to perform actions on the IACS

The major components of authentication and authorization technologies spelled out in ANSI/ISA-62443-1-1 (99.01.01)-2007 are summarized as follows:

- **Role-Based Authorization** – Role-based access control (RBAC) uses roles, hierarchies, and constraints to organize user access levels and is attractive in situations where employees change positions frequently in an organization.
- **Password Authentication** – These technologies authenticate an individual, using a secret known by the user, such as a password or PIN.

ANSI/ISA-62443-1-1 (99.01.01)-2007 points to an issue to be considered in a control environment where, "during a major crisis when human intervention is critically required, an operator may panic and have difficulty remembering the password and either be locked out completely or delayed in being able to respond to the event. If the password has been entered wrong and the system has a limit on allowed wrong password entries, the operator may be locked out permanently until an authorized employee can reset the account."

- **Challenge/Response Authentication** – This authentication method uses a challenge/response, which requires a requestor and service provider both to share a secret. A challenge is sent to the service provider, and it answers with a unique, expected response.
- **Physical/Token Authentication** – The person requesting access uses a token or smart card, which might use multifactor authentication, to gain access to a system.
- **Biometric Authentication** – Biometric authentication uses the biological characteristics of an individual to confirm his or her identity. Characteristics include fingerprints, retina signatures, facial recognition, and signature dynamics.
- **Location-Based Authentication** – The authentication of a requestor is based on his or her physical location, as determined by a GPS or IP address.

## Filtering/Blocking/Access Control

This control uses filtering/blocking/access control technologies, such as firewalls. A firewall provides for tracking successful and unsuccessful transmissions through the firewall, restricting data entering or leaving a network, and supporting interactions among networks. A firewall can filter by applications, ports, or IP addresses.

Specifically, ANSI/ISA-62443-1-1 (99.01.01)-2007 states:

> Firewalls enforce access control policies using mechanisms that either block or permit certain types of traffic, thus regulating the flow of information. Firewalls typically block traffic from the outside of a protected area to the inside of a protected area, while permitting users on the inside to communicate with outside services. While it is important

to have a firewall separating a company's enterprise network from the Internet, it is even more important to have firewalls between the enterprise network and industrial automation and control systems LANs.

The three primary types of firewalls are summarized as follows:

- **Packet Filtering** – Evaluates the address information in each packet of data relative to certain criteria and, as a result, can either drop the packet, forward the packet, or send a message to the originator. It looks at the data packet to get information about the source and destination addresses, the session's communications protocol (TCP, UDP, or ICMP), and the source and destination application port for the desired service. Because the processing of packets in this type of firewall requires a small amount of overhead, it has minimal impact on system performance.
- **Stateful Inspection** – Filters packets at the network layer, determines whether session packets are valid, and evaluates the packet at the application layer. A stateful inspection firewall uses an inspection engine to extract state-related information. This information includes packet sequence numbers, IP addresses, and ports engaged in the connection. It maintains this information in a dynamic state table and evaluates subsequent connection attempts. This type of firewall offers a high level of security, good performance, and transparency to end users, but is relatively expensive.
- **Application Proxy** – Is commonly a host computer that is running proxy server software, making it a proxy server. It operates by transferring a copy of each accepted data packet from one network to another, thereby masking the data's origin. A proxy server can control which services a workstation uses on the Internet, and it aids in protecting the network from outsiders who may be trying to get information about the network's design. This type of firewall is also called an *application layer gateway*. It offers a high level of security but has a significant impact on network performance.

Firewalls are discussed in more detail in Chapter 2.

## Encryption Technologies Data Validation

Symmetric and asymmetric encryption technologies (detailed in Chapter 2) are used to protect sensitive process-related information. Data validation technologies safeguard the accuracy and completeness of information used in the industrial process. As stated in ANSI/ISA-62443-1-1 (99.01.01)-2007, "Symmetric key encryption can be used to distinguish communication by devices that are not part of the desired network. This feature is generally attractive for SCADA and other process control systems that wish to allow little or no access to the control network, but it is difficult to deploy in systems requiring unrestricted Internet access." Encryption can also be employed in a VPN, which provides for authentication, integrity, and protection from unauthorized disclosure. The most common VPN implementations are Internet Protocol Security (IPSec), Secure Shell (SSH), and Secure Sockets Layer (SSL).

## Management, Audit, Measurement, Monitoring, and Detection

Audit and related mechanisms are used to evaluate the security posture of an industrial automation and control system, and forensic methods are used to detect and analyze incidents. Some of the events that can be tracked in typical audit and logging environments for automation systems are:

- Access to files, folders, or other resources
- Account logon events to automation system elements
- Changes to specific accounts
- Directory service accesses
- Instances of elevated privileges to specific users or accounts
- Process accesses
- Security policy modifications
- System events

Regarding management issues, ANSI/ISA-62443-1-1 (99.01.01)-2007 discusses host configuration management (HCM) tools that systems administrators use to manage resources centrally, control access to systems, and set a general level of security for each host on a network. However, it notes, "industrial automation and control systems do not typically use HCM tools because of the policy-driven nature of the tools. Most of the configuration is done through the IACS application and is established because of functional needs rather than administrative or security policies."

Other management aids, automated software management (ASM) tools, are applications used to distribute software across a network to specified hosts or groups of hosts. They are employed, to some degree, in IACS applications and are used to manage updates to antivirus and firewall software, for application version control, and for patch management.

## Industrial Automation and Control Systems Computer Software

Automation and control system software might contain vulnerabilities due to programming errors and should be tested and evaluated to eliminate any such risk factors. ANSI/ISA-62443-1-1 (99.01.01)-2007 identifies the three main types of software that have to be considered in industrial automation and control systems as server and workstation operating systems, real-time and embedded operating systems, and web servers and Internet technologies. Many automation and control system hosts and HMI computers use the Windows, Unix, and Linux operating systems and are therefore vulnerable to attacks designed for these systems. In many cases, real-time devices and controllers use specific vendor-developed operating systems.

## Physical Security Controls

These controls are defined by ANSI/ISA-62443-1-1 (99.01.01)-2007 as "any physical measures, either active or passive, that limit physical access to any information assets in the IACS environment." Physical security devices are designed to protect computer media assets and documents, physical equipment assets, personnel, and facilities. The three main categories of physical security control devices are:

- **Active** – Active devices are devices that can be engaged or disengaged as a function of policy, time, control algorithms, or other input. Typical examples are gates, locks, movable barriers, and doors.
- **Identification and Monitoring Devices** – These devices include various types of cameras, recording devices, sensors, and biometric devices. They usually require monitoring by a human to make security decisions.
- **Passive** – Passive devices are hard installations, such as concrete barriers, fences, and walls.

### Personnel Security Controls

In ANSI/ISA-62443-1-1 (99.01.01)-2007, personnel security falls under the category of physical security, and its elements are hiring policies, terms and conditions of employment, and related company policies and practices.

## North American Electric Reliability Corporation, Critical Infrastructure Protection Cyber Security Standards

Since 2007, the North American Electric Reliability Corporation (NERC) has developed a number of versions of Critical Infrastructure Protection (CIP) standards as a means of ensuring protection of critical infrastructure and critical cyber assets. The latest version is Version 5, which the Federal Energy Regulatory Commission (FERC) approved in November 2013. The CIP Version 5 standards represent an improvement over the currently effective CIP Version 3 Reliability Standards. The NERC CIP requirements evolved from Version 1 in 2008 to Version 3 in 2010 and then, to Version 4. However, before Version 4 was finally approved, it was superseded by Version 5, which was adopted in 2013. Key provisions of Version 5 became effective on April 1, 2016, with the remainder becoming effective 1 year later. In the meantime, responsible entities must continue to comply with Version 3, while preparing to comply with Version 5.

A major change in Version 5 was a focus on the risk to a Bulk Electric System (BES), rather than the risk to a registered entity's operations. Thus, registrants are required to identify all BES Cyber Assets and BES Cyber Systems and categorize them as having a high-, medium-, or low-impact on the reliable operation of the BES.

The highlights of the NERC CIP Cyber Security Standards, Version 3, will be summarized in the following paragraphs, with the major changes resulting from Version 5 reviewed thereafter.

The NERC CIP Cyber Security Standards, Version 3,[5] are divided into standards CIP-002-4 through CIP-009-4 and address "the identification and protection of critical cyber assets to support reliable operation of the Bulk Electric System (BES)."

These eight standards were designed for the electric utility industry, but the security principles listed are sound and applicable to the manufacturing and process industries. The following list identifies the major requirements of Version 3:

- CIP-002-4 - R1, R2, R3
- CIP-003-4 - R1, R2, R3, R4, R5, R6
- CIP-004-4 - R1, R2, R3, R4
- CIP-005-4 - R1, R2, R3, R4, R5
- CIP-006-4 - R1, R2, R3, R4, R5, R6, R7, R8
- CIP-007-4 - R1, R2, R3, R4, R5, R6, R7, R8
- CIP-008-4 - R1, R2
- CIP-009-4 - R1, R2, R3, R4, R5

The purpose of each standard is summarized as follows:

- **CIP-002-4, Cyber Security—Critical Cyber Asset Identification**
  This standard requires the responsible entity to identify and document the critical cyber assets associated with the respective bulk electric system.
  - R1. Critical asset identification
  - R2. Critical cyber asset identification
  - R3. Annual approval by senior manager or delegates of the list of critical assets and the list of critical cyber assets
- **CIP-003-4, Cyber Security—Security Management Controls Standard**
  This standard requires that "responsible entities have minimum security management controls in place to protect critical cyber assets."
  - R1. Cybersecurity policy implementation
  - R2. Leadership assignment
  - R3. Exceptions documentation
  - R4. Information protection
  - R5. Access control
  - R6. Change control and configuration management

- **CIP-004-4, Cyber Security—Personnel and Training**
  This standard requires that "personnel having authorized cyber or authorized unescorted physical access to critical cyber assets, including contractors and service vendors, have an appropriate level of personnel risk assessment, training, and security awareness."
  - R1. Awareness program
  - R2. Cybersecurity training
  - R3. Personnel risk assessment
  - R4. Access control
- **CIP-005-4, Cyber Security—Electronic Security Perimeter(s)**
  This standard addresses "the identification and protection of the electronic security perimeter(s) inside which all critical cyber assets reside, as well as all access points on the perimeter."
  - R1. Electronic security perimeter for critical assets
  - R2. Electronic access controls
  - R3. Monitoring electronic access
  - R4. Cyber vulnerability assessment
  - R5. Documentation review and maintenance
- **CIP-006-4, Cyber Security—Physical Security of Critical Cyber Assets**
  This standard "is intended to ensure the implementation of a physical security program for the protection of critical cyber assets."
  - R1. Physical security plan
  - R2. Protection of physical access control systems
  - R3. Protection of electronic access control systems
  - R4. Physical access controls
  - R5. Monitoring physical access
  - R6. Logging physical access
  - R7. Access log retention
  - R8. Maintenance and testing

- **CIP-007-4, Cyber Security—Systems Security Management Standard**
This standard requires "responsible entities to define methods, processes, and procedures for securing those systems determined to be critical cyber assets, as well as the other (noncritical) cyber assets within the electronic security perimeter(s)." Because this standard describes the security controls for protecting the critical infrastructure, it is one of the most important in the CIP list.
    - R1. Test procedures
    - R2. Ports and services
    - R3. Security patch management
    - R4. Malicious software prevention
    - R5. Account management
    - R6. Security status monitoring
    - R7. Disposal or redeployment
    - R8. Cyber vulnerability assessment
- **CIP-008-4, Cyber Security—Incident Reporting and Response Planning**
This standard CIP-008-4 "ensures the identification, classification, response, and reporting of cybersecurity incidents related to critical cyber assets."
    - R1. Cybersecurity incident response plan
    - R2. Cybersecurity incident documentation
- **CIP-009-4, Cyber Security—Recovery Plans for Critical Cyber Assets**
Standard CIP-009-4 "ensures that recovery plan(s) are put in place for critical cyber assets and that these plans follow established business continuity and disaster recovery techniques and practices."
    - R1. Recovery plans
    - R2. Exercises of recovery plans
    - R3. Change control
    - R4. Backup and restore
    - R5. Testing backup media

In Version 5 of the NERC CIP Cyber Security Standard, the existing versions of CIP-002 through CIP-009 have been significantly revised, and two new standards, CIP-010 and CIP-011, have been added. While the FERC approved these standards in its Order No. 791, it also directed NERC to develop and file modifications to these standards.[1]

Version 5 uses new terms to define the assets subject to CIP protections—the BES Cyber System and a cyber asset. The BES Cyber System is "one or more BES Cyber Assets logically grouped by a responsible entity to perform one or more reliability tasks for a functional entity." A BES Cyber Asset is an asset that if rendered unavailable, degraded, or misused would, within 15 minutes of its required operation, misoperation, or nonoperation, adversely impact one or more facilities, systems, or equipment, which, if destroyed, degraded, or otherwise rendered unavailable when needed, would affect the reliable operation of the Bulk Electric System. Redundancy of affected facilities, systems, and equipment shall not be considered when determining adverse impact. Each BES Cyber Asset is included in one or more BES Cyber Systems. (A cyber asset is not a BES Cyber Asset if, for 30 consecutive calendar days or less, it is directly connected to a network within an Electronic Security Perimeter (ESP), it is a cyber asset within an ESP or to a BES Cyber Asset, and it is used for data transfer, vulnerability assessment, maintenance, or troubleshooting purposes.)

The modifications to the standards are reviewed as follows:[6]

- **CIP-002-5, Cyber Security—BES Cyber System Categorization**
  This standard requires responsible entities to implement a process that identifies all BES Cyber Systems impacting the Bulk Electric System as having a high-, medium-, or low-impact. The assets to be considered include control centers and backup control centers; transmission stations and substations; generation resources; systems and facilities critical to system restoration, including blackstart resources and cranking paths and initial switching requirements; special protection systems that support the reliable operation of the Bulk Electric System; and specified protection systems for distribution providers.
- **CIP-003-5, Cyber Security—Security Management Controls**
  This standard requires that each responsible entity review and obtain CIP senior manager approval at least once every 15 months of

documented cybersecurity policies for its high-impact and medium-impact BES Cyber Systems that address the following specified topics:

- Personnel and training
- Electronic security perimeters, including interactive remote access
- Physical security of BES Cyber Systems
- System security management
- Incident reporting and response planning
- Recovery plans for BES Cyber Systems
- Configuration change management and vulnerability assessments
- Information protection
- Declaring and responding to CIP exceptional circumstances

For low-impact BES Cyber Systems, the standard requires that the responsible entity implement in a manner that identifies, assesses, and corrects deficiencies, one or more documented cybersecurity policies that address cybersecurity awareness, physical security controls, electronic access controls for external routable protocol connections and dial-up connectivity, and cybersecurity incident response. These policies must be reviewed and approved by the CIP senior manager at least once every 15 calendar months.

- **CIP-004-5, Cyber Security—Personnel and Training**
  This standard requires documented processes or programs for security awareness, cybersecurity training, personnel risk assessment, and access management.
  - **Security Awareness –** For security awareness, a responsible entity must have a program related to high-impact and medium-impact BES Cyber Systems that at least once each calendar quarter reinforces cybersecurity practices for personnel who have authorized electronic or authorized unescorted physical access to BES Cyber Systems.
  - **Cyber Security Training –** This standard requires that a responsible entity implement—in a manner that identifies, assesses, and corrects deficiencies—a cybersecurity training

program that contains the training content listed in the standard and that is appropriate to individual roles, functions, or responsibilities. This portion of the standard applies to high-impact BES Cyber Systems and medium-impact BES Cyber Systems with external routable connectivity. The training program must require completion of the training prior to granting authorized electronic access and authorized unescorted physical access to applicable cyber assets, except during CIP exceptional circumstances, and must require completion of the training at least once every 15 calendar months.

- **Personnel Risk Assessments** – This standard requires that a responsible entity implement—in a manner that identifies, assesses, and corrects deficiencies—documented personnel risk assessment programs to attain and retain authorized electronic or authorized unescorted physical access to BES Cyber Systems that include each of the following:
    - A process to confirm identity
    - A process to perform a 7-year criminal history records check as part of each personnel risk assessment that includes current residence and other locations where, during the 7-year period, the subject has resided for 6 consecutive months or more
    - Criteria or a process to evaluate criminal history records checks for authorizing access
    - Criteria or a process for verifying personnel risk assessments performed for contractors or service vendors conducted according to requirements listed above
    - A process to ensure that individuals with authorized electronic or authorized unescorted physical access have completed a personnel risk assessment within the last 7 years

This portion of the standard applies to high-impact BES Cyber Systems and medium-impact BES Cyber Systems with external routable connectivity.

- **Access Management Program** – This standard also requires implementation—in a manner that identifies, assesses, and corrects deficiencies—of access management programs for high-impact and medium-impact BES Cyber Systems with external routable connectivity that include the following specified requirements:

  - A process to authorize electronic access, unescorted physical access into a physical security perimeter, and access to designated storage locations for BES Cyber System information based on need, except for CIP exceptional circumstances
  - Verification at least once each calendar quarter that individuals with active electronic access or unescorted physical access have authorization records
  - For electronic access, verification at least once every 15 calendar months that all user accounts, user account groups, or user role categories, and their specific, associated privileges are correct and are those that the responsible entity determines are necessary
  - Verification at least once every 15 calendar months that access to the designated storage locations for BES Cyber System information are correct and are those that the responsible entity determines are necessary for performing assigned work functions

- **Access Revocation Program** – This standard requires a responsible entity to implement—in a manner that identifies, assesses, and corrects deficiencies—one or more documented access revocation programs that include the following specified requirements for high-impact and medium-impact BES Cyber Systems with external routable connectivity:
  - A process to initiate removal of an individual's ability for unescorted physical access and interactive remote access upon a termination action, and to complete removals within 24 hours of the termination action
  - For reassignments or transfers, revocation of the individual's access that the responsible entity determines is not necessary by the end of the next calendar day following the date that

the responsible entity determines that the individual no longer requires that access

- For terminations, revocation of the individual's access to the designated storage locations for BES Cyber System information by the end of the next calendar day following the effective date of the termination
- For high-impact BES Cyber Systems, a process for terminations; revocation of the individual's nonshared user accounts within 30 calendar days of the effective date of the termination
- For high-impact BES Cyber Systems, a process for terminations; change passwords for shared accounts known to the user within 30 calendar days of the termination and for reassignments or transfers, change passwords for shared accounts known to the user within 30 calendar days following the date that the responsible entity determines that the individual no longer requires that access

- **CIP-005-5 – Cyber Security – Electronic Security Perimeter(s)**
  This standard requires responsible entities to implement one or more documented processes that include the requirements specified in the standard and listed below:
  - For high-impact and medium-impact BES Cyber Systems, the processes must require all applicable cyber assets connected to a network via a routable protocol must reside with a defined ESP.
  - For high-impact and medium-impact BES Cyber Systems with external routable connectivity, the processes must require that all external routable connectivity be through an identified electronic access point. (The nonroutable blanket exemption has been removed from CIP-002-5.)
  - For electronic access points for high-impact and medium-impact BES Cyber Systems, the processes must require inbound and outbound access permissions, including the reason for granting access, and deny all other access by default.
  - For high-impact and medium-impact BES Cyber Systems with dial-up connectivity, where technically feasible, the processes

must require that authentication be performed when establishing dial-up connectivity with applicable cyber assets.

  - For electronic access points for high-impact and medium-impact BES Cyber Systems, the processes must have one or more methods for detecting known or suspected malicious communications for both inbound and outbound communications.

- **CIP-006-5, Cyber Security—Physical Security of BES Cyber Systems** This standard requires each responsible entity to implement—in a manner that identifies, assesses, and corrects deficiencies—one or more documented physical security plans that include the applicable specified requirements.
  - Physical security plan
  - Monitoring for unauthorized access
  - Visitor control
  - Physical access control system maintenance and testing
- **CIP-007-5, Cyber Security—Systems Security Management** This standard addresses system security by specifying technical, operational, and procedural requirements in support of protecting high-impact and medium-impact BES Cyber Systems against compromise that could lead to misoperation or instability of the Bulk Electric System.
  - Address each of the specified requirements related to ports and services
  - Security patch management
  - Malicious code prevention
  - Security event monitoring
  - System access control
- **CIP-008-5, Cyber Security—Incident Reporting and Response Planning** This standard requires each responsible entity to document one or more cybersecurity incident response plans for high-impact and medium-impact BES Cyber Systems that include the following:

- A process to identify, classify, and respond to cybersecurity incidents
- A process to determine if the incident is a reportable cybersecurity incident and notify the Electricity Sector Information Sharing & Analysis Center (E-ISAC) within 1 hour from the time that the determination of a reportable cybersecurity incident is made. (E-ISAC is operated on behalf of the Electricity Subsector by the North American Electric Reliability Corporation and provides security services to electricity service owners and operators in the United States, Canada, and portions of Mexico.)
- The roles and responsibilities of cybersecurity incident response groups or individuals
- Incident handling procedures for cybersecurity incidents

- **CIP-009-5, Cyber Security—Recovery Plans for BES Cyber Systems** This standard specifies the controls needed to protect data that would be useful in the investigation of an event requiring execution of a BES Cyber System recovery plan and requires operational testing to support the recovery of BES Cyber Systems. This standard also establishes a timeline for a responsible entity to determine lessons learned and update recovery plans. More specifically, this standard requires each responsible entity to have one or more documented recovery plans for high-impact and certain medium-impact BES Cyber Systems that include the following:
  - Conditions for activation of the recovery plan
  - Roles and responsibilities of responders
  - One or more processes for the backup and storage of information required to recover BES Cyber System functionality
  - One or more processes to verify the successful completion of the backup processes and to address any backup failures
  - One or more processes to preserve data, per cyber asset capability, for determining the cause of a cybersecurity incident that triggers activation of the recovery plan

- **CIP-010-1, Cyber Security—Configuration Change Management and Vulnerability Assessments**
  This new standard consolidates the configuration change management and vulnerability assessment from previous versions of CIP-003, CIP-005, and CIP-007. Each responsible entity is required to implement—in a manner that identifies, assesses, and corrects deficiencies—one or more documented processes for high-impact and medium-impact BES Cyber Systems that meet the following specified requirements:

  - Develop a baseline configuration that includes operating systems or firmware where no independent operation system exists, any commercially available or open-source application software intentionally installed, any custom software installed, any logical network accessible ports, and any security patches applied.

  - Authorize and document changes that deviate from the existing baseline configuration.

  - Update the baseline configuration within 30 calendar days of completing a change that deviates from the existing baseline configuration.

  - Prior to a change that deviates from the existing baseline configuration, determine the required cybersecurity controls in CIP-005 and CIP-007 that could be impacted by the change, verify following the change that required cybersecurity controls are not adversely affected, and document the results of the verification.

  - For high-impact BES Cyber Systems, where technically feasible, for each change that deviates from the existing baseline configuration, prior to implementing any change in the production environment, test changes in a test environment or in a production environment where the test is performed in a manner that minimizes adverse effects, that models the baseline configuration to ensure that required cybersecurity controls in CIP-005 and CIP-007 are not adversely affected, and document the results of the testing.

- **CIP-011-1, Cyber Security—Information Protection**
  This new standard consolidates information protection requirements from previous versions of CIP-003 and CIP-007. It requires each responsible entity to implement—in a manner that identifies, assesses,

and corrects deficiencies—one or more documented information protection programs for high-impact and medium-impact BES Cyber Systems that must include methods to identify information that meets the definition of BES Cyber System information and procedures for protecting and securely handling BES Cyber System information (including storage, transit, and use).

This standard also requires a responsible entity to implement one or more documented processes that meet the specified requirements for BES Cyber Asset reuse and disposal. Those processes must require that the responsible entity take action to prevent the unauthorized retrieval of BES Cyber System information from the cyber asset data storage media prior to the release for reuse of the applicable cyber assets that contain BES Cyber System information. Those processes must also require that the responsible entity must take action to prevent the unauthorized retrieval of BES Cyber System information from the cyber asset or destroy the data storage media prior to the disposal of the applicable cyber assets that contain BES Cyber System information.

## Department of Homeland Security, Catalog of Control Systems Security: Recommendations for Standards Developers

The Department of Homeland Security (DHS) catalog[7] is based on NIST SP 800-53 and is described in the document as presenting "a compilation of practices that various industry bodies have recommended to increase the security of control systems from both physical and cyberattacks." The controls in the catalog are divided into 19 categories or families that are taken primarily from NIST SP 800-53, Revision 3, with some controls from ANSI/ISA–62443-2-1 (99.02.01)–2009.[8] There are 250 controls in the DHS control systems security catalog as instead of 347 in NIST SP 800-53.

The control families from the catalog and their relationship to NIST SP 800-53 categories are shown in Table 6-10.

**Table 6-10. DHS Security Controls Catalog and Relationship to NIST SP 800-53**

*Source: Department of Homeland Security (2011)*

| Catalog of Recommendations Subsections | | NIST SP 800-53 Revision 3 Families |
|---|---|---|
| 2.1 | Security Policy | Each of the 18 control family initial elements relates to its own policy and procedures (e.g., AC-1 through PM-1) |
| 2.2 | Organizational Security | Access Control—AC<br>Program Management—PM |
| 2.3 | Personnel Security | Access Control—AC<br>Personnel Security—PS |
| 2.4 | Physical and Environmental Security | Access Control—AC<br>Physical and Environmental Security—PE |
| 2.5 | System and Services Acquisition | System and Services Acquisition—SA |
| 2.6 | Configuration Management | Configuration Management—CM |
| 2.7 | Strategic Planning | Planning—PL |
| 2.8 | System and Communication Protection | Access Control—AC<br>System and Communication Protection—SC |
| 2.9 | Information and Document Management | Contingency Planning—CP<br>Media Protection—MP |
| 2.10 | System Development and Maintenance | Maintenance—MA |
| 2.11 | Security Awareness and Training | Awareness and Training—AT |
| 2.12 | Incident Response | Incident Response—IR |
| 2.13 | Media Protection | Media Protection—MP |
| 2.14 | System and Information Integrity | System and Information Integrity—SI |
| 2.15 | Access Control | Access Control—AC<br>Identification Authentication—IA |
| 2.16 | Audit and Accountability | Audit and Accountability—AU |
| 2.17 | Monitoring and Reviewing Control System Security Policy | Security Assessment and Authorization—CA |
| 2.18 | Risk Management and Assessment | Assessment and Authorization—CA<br>Risk Assessment—RA |
| 2.19 | Security Program Management | Program Management—PM |

## AMI System Security Requirements

Security requirements for the Advanced Metering Infrastructure (AMI) have been developed by the AMI Security (AMI-SEC) Task Force, under the title of *AMI System Security Requirements*.[9] The purpose of these requirements, as stated in the task force document, is:

> To provide the utility industry along with supporting vendor communities and other stakeholders a set of security requirements that should be applied to AMI implementations to ensure the high level of information assurance, availability, and security necessary to maintain a reliable system and consumer confidence. While this specification focuses on AMI, the security requirements contained in the document may be extended to other network-centric, Smart Grid solutions.

A three-letter identifier is assigned to each AMI security control to indicate the security requirements of that control. The letters in the identifiers indicate:

- F = functional requirements
- S = supporting services to functional requirements
- A = assurance requirements
- Remaining letters = help associate the requirement with its requirement class

Table 6-11 summarizes the AMI system security control requirements.

**Table 6-11. AMI System Security Controls**

*Source: (AMI-SEC) AMI System Security Requirements (2008)*

| Control | Description |
|---|---|
| Confidentiality and Privacy (FCP) | These requirements provide user, service, or object protection, against discovery and misuse of identity by other users/ subjects. |
| Integrity (FIN) | Information integrity increases assurance that sensitive data have neither been modified nor deleted in an unauthorized or undetected manner. |
| Availability (FAV) | This requirement addresses the ability of the system to continue to operate and satisfy business/mission needs under diverse operating conditions, including but not limited to peak load conditions, attacks, maintenance operations, and normal operating conditions. |

**Table 6-11. AMI System Security Controls**

*Source: (AMI-SEC) AMI System Security Requirements (2008)*

| | |
|---|---|
| Identification (FID) | This area addresses controls associated with who an actor claims to be. |
| Authentication (FAT) | This section covers requirements around the proof of identity of an actor. |
| Authorization (FAZ) | This requirement covers the approval of an actor to perform an action. |
| Nonrepudiation (FNR) | These requirements address the ability to tie an actor to an action. |
| Accounting (FAC) | This area addresses the recording of activity by actors/ elements throughout the system and providing the means to perform a successful audit of events that occur on the system. |
| Anomaly Detection Services (FAS) | These requirements address services applied to detect events outside of the bounds of normally anticipated or desired behavior, such as attacks, intrusions, or errors. |
| Boundary Services (FBS) | This section provides requirements for boundary services that provide isolation between system elements or between the system and external entities. |
| Cryptographic Services (FCS) | This section covers cryptographic services, which include encryption, signing, key management and key revocation with the security objectives of identification and authentication, nonrepudiation, trusted path, trusted channel and data separation. |
| Notification and Signaling Services (FNS) | These requirements cover notification and signaling services, which are oriented towards providing system activity information and command result logging. |
| Resource Management Services (FRS) | This section covers resource management services requirements, including management of runtime resources such as network/communication paths, processors, memory or disk space (e.g., for audit log capacity), and other limited system resources. |
| Trust and Certificate Services (FTS) | This section addresses the description of relationships between entities and the faith placed on the relationship certificates that have uses outside of cryptography; for example, material relating to creation, storage, and revocation of certificates. |
| Development Rigor (ADR) | This section contains AMI System Security Specification requirements regarding the activities involved in developing Smart Grid system solutions. Topics include acquisition issues, configuration management, and development practices. |
| Organizational Rigor (AOR) | This section contains requirements regarding the policies employed by the organization(s) with access to assets of a deployed Smart Grid system. These requirements reflect on an organization's ability to continue to operate a Smart Grid system reliably over time. Topics include training procedures, personnel security, strategic planning, and monitoring and reviewing security policies. |

**Table 6-11. AMI System Security Controls**

*Source: (AMI-SEC) AMI System Security Requirements (2008)*

| | |
|---|---|
| Handling/Operating Rigor (AHR) | This section contains requirements regarding the activities involved in the day-to-day operation of deployed Smart Grid systems. Topics include information and document management policies, incident response procedures, maintenance procedures, physical and environmental security, and media protection. |
| Accountability (AAY) | The corresponding requirements are based on security auditing and recognizing, recording, storing, and analyzing information related to security relevant activities. The resulting audit records can be examined to determine which security relevant activities took place and who (which user) is responsible for them. |
| Access Control (AAC) | The focus of this section is ensuring that resources are only accessed by the appropriate personnel and that those personnel are correctly identified. Mechanisms also need to be put into place to monitor access activities for inappropriate activity. |

The format of a typical AMI security function is as follows:

## Identification (FID)

This section covers requirements around verifying who a user claims to be.

**FID.1** – The security function shall require each user to be successfully identified before allowing any other system's security function-mediated actions on behalf of that user unless it is one of the following that may be allowed before the user is identified:

[*Assignment: list of system's security function-mediated actions*]

**FID.2** – The security function shall associate the following user security attributes with subjects acting on behalf of that user:

[*Assignment: list of user security attributes*]

**FID.3** – The security function shall enforce the following rules on the initial association of user security attributes with subjects acting on behalf of users:

[*Assignment: rules for the initial association of attributes*]

**FID.4** – The security function shall enforce the following rules governing changes to the user security attributes associated with subjects acting on the behalf of users:

[*Assignment: rules for the changing of attributes*]

**FID.5** – The security function shall uniquely identify (and authenticate) [*Assignment: users, processes acting on behalf of users, devices, etc.*] before establishing a connection.

**FID.6** – The organization shall manage user identifiers by:

1. Uniquely identifying each user
2. Verifying the identity of each user
3. Receiving authorization to issue a user identifier from an appropriate organization official
4. Issuing the user identifier to the intended party
5. Disabling the user identifier after [*Assignment: organization-defined time period*] of inactivity
6. Archiving user identifiers

**FID.7** – The security function shall have mechanisms to uniquely identify (and authenticate) [*Assignment: users, processes acting on behalf of users, etc.*].

**FID.8** – The security function shall appropriately label information in storage, in process, and in transmission.

## Consolidation of Best Practices Controls for Industrial Automation and Control Systems

The preceding sections presented some of the most effective controls that can be applied to automation and control systems, and provided excellent guidance for securing these systems. A consolidation of the best practices from all the reviewed controls would be useful to the practitioners and management in the industrial control and automation field for use in everyday applications.

Table 6-12 provides this information with references to the source documents to give a solid basis for the justification and origin of these controls. This table is intended to provide a quick, useful reference to implementing the best security practices for the control engineer, technician, supervisor, and management of the key elements of manufacturing operations, process plants, and the critical infrastructure.

**Table 6-12. Controls Summary Table and Source References**

| Control # | Industrial Automation and Control System Security Control | Comments | Source |
|---|---|---|---|
| 1. | Implement Access Control | Implement processes for arbitrating requests for access to critical system information. | SP 800-82<br>TR99.00.0<br>SP 800-53<br>DHS catalog<br>AMI |
| 2. | Conduct Risk Assessment | Identify risks to systems including potential impacts, probabilities, and mitigation options. | SP 800-53<br>DHS catalog<br>AMI |
| 3. | Remove Default Accounts | Many automation system manufacturers have default accounts and passwords in place for maintenance or system access. These accounts should be deleted and any default passwords changed or eliminated. | SP 800-82<br>TR99.00.01<br>SP 800-53<br>DHS catalog<br>AMI |
| 4. | Develop Security Policies | Have policies in place that express management intent, address the security mission, define roles and responsibilities, and address the use of computational resources. | SP 800-82<br>TR99.00.01<br>SP 800-53<br>DHS catalog<br>AMI |
| 5. | Develop Security Plan | Generate a plan to define and implement security controls, conduct incident response, and evaluate security throughout the system's life cycle. | SP 800-82<br>TR99.00.01<br>SP 800-53<br>DHS catalog<br>AMI |
| 6. | Develop Personnel Screening Policies and Procedures | Implement personnel-related policies and procedures, including screening, transfers, and termination. | SP 800-82<br>TR99.00.01<br>SP 800-53<br>DHS catalog<br>AMI |
| 7. | Manage Maintenance | Develop and implement policies and procedures to address all parts of system maintenance. | SP 800-82<br>TR99.00.01<br>SP 800-53<br>DHS catalog<br>AMI |

**Table 6-12. Controls Summary Table and Source References**

| | | | |
|---|---|---|---|
| 8. | Protect System Information Integrity | Implement policies and procedures to ensure the integrity of data. | SP 800-82<br>TR99.00.01<br>SP 800-53<br>DHS catalog<br>AMI |
| 9. | Develop Hiring Policies | Implement policies for employee background checks, employment terms, responsibilities, and so on. | SP 800-82<br>TR99.00.01<br>SP 800-53<br>DHS catalog<br>AMI |
| 10. | Develop Acquisition Policies | Apply risk assessment–based acquisition policies throughout the system's life cycle. | SP 800-82<br>SP 800-53<br>DHS catalog<br>AMI |
| 11. | Provide for Strong Authentication | Employ strong authentication mechanisms at interfaces to the Internet and any other public networks. | SP 800-82<br>TR99.00.01<br>SP 800-53<br>DHS catalog<br>AMI |
| 12. | Test and evaluate critical software | Implement mechanisms to evaluate software elements for vulnerabilities. | SP 800-82<br>TR99.00.01<br>DHS catalog<br>AMI |
| 13. | Implement Audit Procedures | Provide for independent evaluation of audit records to determine adequacy of security controls and compliance. | SP 800-82<br>TR99.00.01<br>SP 800-53<br>DHS catalog<br>AMI |
| 14. | Implement an Effective Password Policy | Establish a strong password policy that specifies characteristics, such as length, types of characters, audit frequency, and change periods. The password policy should take into account the balance required between the types of password protected access mechanisms used and the need for an automation system operator to quickly access system components under stress in the event of an emergency. | SP 800-82<br>TR99.00.01<br>SP 800-53<br>DHS catalog<br>AMI |
| 15. | Identify Critical System Elements | Determine the critical system elements that have to be protected and included in disaster recovery planning. | SP 800-82<br>TR99.00.01<br>SP 800-53<br>DHS catalog<br>AMI |

**Table 6-12. Controls Summary Table and Source References**

| | | | |
|---|---|---|---|
| 16. | Apply Defense in Depth | Apply security in layers, beginning with perimeter defense and employing SCADA-aware firewalls, and intrusion detection and prevention systems. | SP 800-82<br>TR99.00.01<br>SP 800-53<br>DHS catalog |
| 17. | Protect Critical Data and Media | Protect sensitive information in all stages of its life cycle including its use, storage, transit, and disposal. Encryption should be applied according to good security principles. | SP 800-82<br>TR99.00.01<br>SP 800-53<br>DHS catalog<br>AMI |
| 18. | Provide for Physical Security | Implement physical security controls including back-up power, fire detection and prevention, physical access controls, cameras, badges, and tokens. | SP 800-82<br>TR99.00.01<br>SP 800-53<br>DHS catalog<br>AMI |
| 19. | Develop an Up-to-Date System Diagram | Prepare and maintain a diagram of the automation system, including topologies, devices (such as PLCs), software in use, protection mechanisms, and locations of devices, transducers, and sensors. | SP 800-82<br>SP 800-53<br>DHS catalog<br>AMI |
| 20. | Protect Equipment | When possible, place equipment in locations that prevent unauthorized access. | SP 800-82<br>TR99.00.01<br>SP 800-53<br>DHS catalog<br>AMI |
| 21. | Develop and Implement a Patching Policy | When compatible with real-time and production requirements, test and install patches to critical software that address vulnerabilities. Patches should not be installed on production systems without evaluation on test systems or off-line to ensure that no faults are introduced into production equipment. | SP 800-82<br>TR99.00.01<br>SP 800-53<br>DHS catalog |
| 22. | Conduct Vulnerability Assessments | When compatible with real-time and production requirements, perform vulnerability assessments in a manner that does not interfere with system operation. | SP 800-82<br>TR99.00.01<br>SP 800-53<br>DHS catalog<br>AMI |
| 23. | Implement System Logging | Log and review critical system parameters in a manner compatible with the production requirements of automation and control systems. | SP 800-82<br>TR99.00.01<br>SP 800-53<br>DHS catalog<br>AMI |

**Table 6-12. Controls Summary Table and Source References**

| | | | |
|---|---|---|---|
| 24. | Manage and Control Remote Access | It is important to control and manage remote access using strong authentication and encrypted links, such as VPNs. Give special security consideration to wireless devices and desktop modems. | SP 800-82<br>TR99.00.01<br>SP 800-53<br>DHS catalog<br>AMI |
| 25. | Protect Communications | Implement methods to protect system transmission elements. | SP 800-82<br>TR99.00.01<br>SP 800-53<br>DHS catalog<br>AMI |
| 26. | Implement Filtering and Screening | Use devices such as firewalls to protect critical data and support access control. | SP 800-82<br>TR99.00.01<br>SP 800-53<br>DHS catalog<br>AMI |
| 27. | Provide Security Awareness Training | Provide personnel with security awareness training to alert them to issues such as social engineering and phishing. They should also be informed on how to determine if an incident has occurred, whom to notify if an incident has occurred, and any actions to be taken in that event. | SP 800-82<br>TR99.00.01<br>SP 800-53<br>DHS catalog<br>AMI |
| 28. | Maintain Only Necessary Services | Disable any unnecessary systems services and unused open ports. | SP 800-82<br>TR99.00.01<br>SP 800-53<br>DHS catalog |
| 29. | Undergo Certification and Accreditation | Obtain certification and accreditation and accept residual risk. | SP 800-82<br>SP 800-53<br>DHS catalog<br>AMI<br>(accreditation) |
| 30. | Employ Configuration Management | Implement configuration management practices that document any system hardware and software changes. | SP 800-82<br>TR99.00.01<br>SP 800-53<br>DHS catalog<br>AMI |
| 31. | Implement Back-Ups | Arrange for backing up system hardware, software, and data in the event of a disaster or other failure of critical systems. | SP 800-82<br>TR99.00.01<br>DHS catalog<br>AMI |
| 32. | Develop Business Continuity and Disaster Recovery Plans | Develop, test, and implement disaster recovery and business continuity plans. | NIST 800-82<br>SP 800-53<br>DHS catalog<br>AMI |

**Table 6-12. Controls Summary Table and Source References**

| | | | |
|---|---|---|---|
| 33. | Secure Extranets | Minimize and secure partner networks or extranets if they are necessary. | SP 800-82<br>TR99.00.01<br>SP 800-53,<br>DHS catalog<br>AMI |
| 34. | Use Antivirus Software Intelligently | Implement antivirus software, taking into account the limitations and requirements of automation systems, including its effect on response time and memory capacity. | SP 800-82<br>TR99.00.01<br>SP 800-53,<br>DHS catalog<br>AMI |
| 35. | Implement Program Management | The Program Management plan:<br>Specifies the individuals within the organization responsible for the security program management controls and the appointment of a senior information security officer<br>Ensures that all capital planning and investment requests include the resources needed to implement the information security program<br>Develops an enterprise architecture with consideration for information security<br>Defines mission/business processes with consideration for information security and the resulting risk to the organization, personnel, and the critical infrastructure | SP 800-53<br>DHS catalog<br>AMI |
| 36. | Predictable Failure Prevention | Control that protects the information system from harm by considering mean time to failure for critical components in specific environments of operation. | SP 800-53<br>DHS catalog<br>AMI |

## Summary

Chapter 6 presents the most popular and effective controls specifically designed to secure industrial automation systems as well as controls that could be adapted to such systems. These standards and guidelines form a comprehensive, structured methodology for protecting manufacturing facilities, process plants, and the critical infrastructure. A useful output of this chapter is the consolidation of these best practices to provide a single reference source as a baseline for securing industrial automation and control systems.

## Review Questions for Chapter 6

6.1 NIST SP 800-82, *Guide to Industrial Control Systems (ICS) Security*, partitions security controls into which one of the following groups?

A. Management controls, operational controls, and technical controls

B. Management controls, administrative controls, and technical controls

C. Management controls, operational controls, and physical controls

D. Logical controls, operational controls, and technical controls

6.2 NIST SP 800-82 defines five major security objectives for an industrial control system (ICS) implementation. Which of the following is NOT one of those objectives?

A. Restricting logical access to the ICS network and network activity

B. Restricting physical access to the ICS network and devices

C. Protecting individual ICS components from exploitation

D. Shutting down the ICS after an incident

6.3 In NIST SP 800-82, policies and procedures for controlling modifications to hardware, firmware, software, and documentation to ensure the information system is protected against improper modifications prior to, during, and after system implementation is known as which of the following?

A. Configuration management

B. Risk management

C. System management

D. System maintenance

6.4 In the NIST SP 800-82 controls, the process of granting or denying specific requests for obtaining and using information and related information processing services for physical access to areas within the information system environment is known as which of the following?

A. Authentication

B. Audit and accountability

C. Access control

D. Identification

6.5 The controls of ANSI/ISA-62443-1-1 (99.01.01)-2007 are organized into six categories. Which of the following is NOT one of those categories?

A. Encryption technologies and data validation

B. Risk mitigation technologies

C. Authentication and authorization technologies

D. Filtering/blocking/access control technologies

6.6 ANSI/ISA-62443-1-1 (99.01.01)-2007 describes which of the following as "the initial step in protecting an industrial automation and control system (IACS) and its critical assets from unwanted breaches. It is the process of determining who and what should be allowed into or out of a system"?

A. Authorization

B. Authentication

C. Identification

D. Confirmation

6.7 Which of the following are the major components of authentication and authorization technologies spelled out in ANSI/ISA-62443-1-1 (99.01.01)-2007?

A. Role-based, password, and challenge response

B. Rule-based, user ID, and challenge response

C. Role-based, password, and call-back

D. Rule-based, password, and call-back

6.8 Which of the following does ANSI/ISA-62443-1-1 (99.01.01)-2007 identify as the three main types of software that have to be considered in industrial automation and control system software?

A. Mobile operating systems, real-time and embedded operating systems, and web servers and Internet technologies

B. Server and workstation operating systems, real-time and embedded operating systems, and wireless technologies

C. Server and workstation operating systems, real-time and embedded operating systems, and web servers and Internet technologies

D. Server and workstation operating systems, real-time and embedded operating systems, and mobile technologies

6.9 In ANSI/ISA-62443-1-1 (99.01.01)-2007, what are the three main categories of physical security elements?

A. Active, identification and monitoring devices, and passive

B. Active, real time, and passive

C. Active, identification and monitoring devices, and real time

D. Reactive, identification and monitoring devices, and passive

6.10 The North American Electric Reliability Corporation (NERC) Critical Infrastructure Protection (CIP) Cyber Security Standards are divided into standards CIP-002-4 through CIP-009-4 and address "the identification and protection of critical cyber assets to support reliable operation of the Bulk Electric System (BES)." Which of the following is NOT one of the eight standards?

A. Risk assessment

B. Critical cyber asset identification

C. Security management controls

D. Electronic security perimeters

6.11 Which NERC CIP standard "ensures the identification, classification, response, and reporting of cybersecurity incidents related to critical cyber assets?"

A. Physical security of critical cyber assets

B. Systems security management

C. Recovery plans for critical cyber assets

D. Incident reporting and response planning

6.12 In NIST SP 800-53, the security controls are divided into 18 families, each with a unique, two-character identifier. The control families are themselves partitioned into three classes. Which of the three classes does the control family titled CP (contingency planning) fall under?

A. Technical

B. Management

C. Operational

D. Administrative

6.13 The NIST SP 800-53 controls are considered baseline controls, which can be tailored to more closely meet the needs of a particular organization and automation system. Which of the following is NOT one of the elements of the NIST SP 800-53 tailoring process?

A. Selecting (or specifying) compensating security controls, if needed, to adjust the preliminary set of controls to obtain an equivalent set deemed to be more feasible to implement

B. Applying scoping guidance to the initial baseline security controls to obtain a preliminary set of applicable controls for the tailored baseline

C. Specifying organization-defined parameters in the security controls via explicit assignment and selection statements to complete the definition of the tailored baseline

D. Developing custom parameters to meet the needs of the control system

6.14 The NIST SP 800-53 tailoring process refers to guidance "designed to ensure that only the controls appropriate to the particular mission and requirements of the organization are used." This guidance is referred to as which of the following?

A. Correlating

B. Assessing

C. Evaluating

D. Scoping

6.15 NIST SP 800-53 provides for the use of security controls, which the publication defines as "a management, operational, or technical control (i.e., safeguard or countermeasure) employed by an organization in lieu of a recommended security control … that provides an equivalent or comparable level of protection for an information system and the information processed, stored, or transmitted by that system." This type of additional control is known as which of the following?

A. Supporting

B. Supplemental

C. Compensating

D. Custom

6.16 Recognizing some of the special requirements of industrial automation and control systems, NIST SP 800-53 defines supplements to the baseline control systems to address automation and control issues. Which one of the following is NOT one of these supplements?

A. Risk evaluation

B. Predictable failure prevention

C. Emergency power

D. Access enforcement

6.17 The Department of Homeland Security's *Catalog of Control Systems Security* is based on which of the following security controls documents?

A. NIST SP 800-82

B. NIST SP 800-53

C. NIST SP 800-39

D. The NERC CIP Cyber Security Standards

6.18 The Advanced Metering Infrastructure (AMI) Security Requirements are characterized by letters. Which of the following is NOT one of the AMI security requirements categories?

A. "R" are risk requirements

B. "F" are functional requirements

C. "S" are supporting services to functional requirements

D. "A" are assurance requirements

6.19 The AMI security requirement FIN refers to which of the following security principles?

A. Inventory

B. Final evaluation

C. Integrity

D. Internal

6.20 The recording of activity by actors/elements throughout the system and providing the means to perform a successful audit of events that occur on the system refers to which of the following AMI security controls?

A. Accounting

B. Anomaly detection

C. Authentication

D. Authorization

6.21 A major change in Version 5 of the NERC CIP standard was a focus on which of the following?

A. The risk to a Bulk Electric System (BES)

B. The risk to a registered entity's operations

C. The risk to physical and environmental factors

D. The risk to personnel

6.22 The North American Electric Reliability Corporation (NERC) Critical Infrastructure Protection (CIP) Cyber Security Standards, Version 3, are divided into standards CIP-002-4 through CIP-009-4 and address which of the following?

A. Availability of technologies to support reliable operation of the Bulk Electric System (BES)

B. The identification and protection of critical cyber assets to support reliable operation of the Bulk Electric System (BES)

C. Integrity of software to support reliable operation of the Bulk Electric System (BES)

D. Accountability of methodologies for the protection of critical cyber assets to support reliable operation of the Bulk Electric System (BES)

6.23 In NIST SP 800-82, *Guide to Industrial Control Systems (ICS) Security*, risk assessment, planning, systems and services acquisition, certification, accreditation, and security assessments fall under what category of controls?

A. Technical

B. Logical

C. Management

D. Operational

6.24 In NIST SP 800-82, *Guide to Industrial Control Systems (ICS) Security*, controls such as personnel security and contingency planning fall under which of the following control categories?

A. Management

B. Operational

C. Technical

D. Physical

6.25 NIST SP 800-53 provides guidance to organizations in selecting security controls through the concept of baseline controls, which should be chosen based on the security category and associated impact level of information systems determined in accordance with which of the following publications?

A. NIST 800-37 and FIPS 200

B. NIST 800-37 and NIST 800-39

C. FIPS 109 and FIPS 200

D. FIPS 199 and FIPS 200

6.26 ANSI/ISA-62443-1-1 (99.01.01)-2007, *Security Technologies for Industrial Automation and Control Systems*, defines which of the following principles as "the process of positively identifying potential network users, hosts, applications, services, and resources, using a combination of identification factors or credentials. The result of this process then becomes the basis for permitting or denying further actions"?

A. Authentication

B. Authorization

C. Accountability

D. Accessibility

6.27 ANSI/ISA-62443-1-1 (99.01.01)-2007, *Security Technologies for Industrial Automation and Control Systems*, divides authentication into which of the following two areas?

A. User authentication and group authentication

B. Group authentication and network service authentication

C. User authentication and network service authentication

D. Platform authentication and user authentication

6.28 ANSI/ISA-62443-1-1 (99.01.01)-2007 defines which of the following as something that "can be used to distinguish communication by devices that are not part of the desired network. This feature is generally attractive for SCADA and other process control systems that wish to allow little or no access to the control network, but is difficult to deploy in systems requiring unrestricted Internet access"?

A. Asymmetric key encryption

B. Digital signature

C. Biometrics

D. Symmetric key encryption

6.29 ANSI/ISA-62443-1-1 (99.01.01)-2007 states that encryption can also be employed in a virtual private network (VPN), which provides for authentication, integrity, and protection from unauthorized disclosure. Which of the following is NOT one of the most common VPN implementations?

A. Hypertext Transfer Protocol (HTTP)

B. Internet Protocol Security (IPSec)

C. Secure Shell (SSH)

D. Secure Sockets Layer (SSL)

6.30 What standards address "the identification and protection of critical cyber assets to support reliable operation of the Bulk Electric System (BES)"?

A. ANSI/ISA-62443-1-1 (99.01.01)-2007

B. NIST SP 800-53

C. NIST SP 800-82

D. NERC CIP Cyber Security Standards

6.31 Which of the following is NOT one of the NERC CIP Cyber Security Standards?

A. Risk Management

B. Critical Cyber Asset Identification

C. Security Management Controls

D. Personnel and Training

6.32 The following are characterized as what type of guidance in NIST SP 800- 53, which includes determining the primary security objectives, which system components the controls are applicable to, and what technologies are employed in the system?

A. Custom

B. Complementary

C. Scoping

D. Supplemental

6.33 NIST SP 800-53 security controls are organized into how many families?

A. 15

B. 17

C. 18

D. 20

6.34 The Organizational Security controls in the Department of Homeland Security *Catalog of Control Systems Security* relate to which of the following control categories in NIST SP 800-53?

A. Personnel Security (PS)

B. System and Services Acquisition (SA)

C. Access Control (AC) and Program Management (PM)

D. Contingency Planning (CP) and Media Protection (MP)

6.35 The Monitoring and Reviewing Control System Security Policy controlsin the *Catalog of Control Systems Security* relate to which of the following control categories in NIST SP 800-53?

A. Audit and Accountability (AU)

B. Program Management (PM)

C. Security Assessment and Authorization (CA)

D. System and Information Integrity (SI)

6.36 The AMI System Security Requirement that addresses "the proof of identity of an actor" is which of the following?

A. Authentication

B. Identification

C. Authorization

D. Nonrepudiation

6.37 The AMI System Security Requirement that is "based on security auditing and recognizing, recording, storing, and analyzing information related to security relevant activities" is which of the following?

A. Authorization

B. Authentication

C. Accountability

D. Integrity

## References

1. NIST SP 800-53. *Recommended Security Controls for Federal Information Systems.* Revision 4. Washington, DC: NIST (National Institute of Standards and Technology), 2013.

2. NIST SP 800-82. *Guide to Industrial Control Systems (ICS) Security.* Revision 2. Washington, DC: NIST (National Institute of Standards and Technology), 2015.

3. U. S. Department of Commerce. "Minimum Security Requirements for Federal Information and Information Systems." Federal Information Processing Standards (FIPS) Publication 200. March 2006. http://csrc.nist.gov/publications/fips/fips200/FIPS-200-final-march.pdf.

4. ANSI/ISA-62443-1-1 (99.01.01)-2007. *Security for Industrial Automation and Control Systems – Part 1-1: Terminology, Concepts, and Models.* Research Triangle Park, NC: ISA (International Society of Automation).

5. NERC. *Critical Infrastructure Protection (CIP) Cyber Security Standards.* Atlanta, GA: NERC (North American Electric Reliability Corporation), 2011.

6. V&E Law (Vincent & Elkins LLP). *Summary of CIP Version 5 Standards.* Houston, TX. 2014.

7. U.S. Department of Homeland Security. *Catalog of Control Systems Security: Recommendations for Standards Developers.* Washington, DC: U.S. Department of Homeland Security, April 2011.

8. ANSI/ISA-162443-2-1 (99.02.01)-2009. *Security for Industrial Automation and Control Systems: Establishing an Industrial Automation and Control Systems Security Program.* Washington, DC: NIST (National Institute of Standards and Technology), 2009.

9. Advanced Metering Infrastructure Security Task Force (AMI-SEC). *AMI System Security Requirements.* V1.01, 2008.

# 7 Industrial Automation and Control System Security Training

*Chapter 6 presented the application of standards and guidelines to industrial automation and control system security and subsequently developed a foundation of security controls that are effective in reducing risk and protecting these systems. Chapter 7 reinforces these concepts by developing automation and control system security training concepts, training sources, and templates for designing and generating such programs.*

## Background

Training for industrial automation and control system security is addressed in many different formats, philosophies, levels, and degrees of formality. Some automation and control system security issues are covered peripherally in IT-based information system security courses; others are covered in depth in courses aimed at control system engineers and technicians; and still others are somewhere in between.

Until recently, there has been a dearth of high-quality courses focused primarily on industrial automation and control system security. Most courses offered by various bodies were heavily focused on information system security and paid little attention to real-time systems used to control the nation's manufacturing facilities, process plants, and critical infrastructure. Because of the threat of terrorism, increased concerns of natural disasters, and the demands of a growing population, focus is shifting from purely information system security to the security of automation and control systems within governmental agencies, educational institutions, and private training organizations.

Training in the field of securing industrial automation and control systems is critical to supporting the goals of protecting the nation's manufacturing and process industry base and its critical infrastructure by building a cadre of trained personnel, maintaining technological leadership, developing advanced security-related hardware and software, and raising awareness of threats to national security.

With these goals in mind, key issues are who should be responsible for such training and for what level the training should be designed. In organizations, specific individuals should be assigned the responsibility of training employees involved with the security of automation and control systems as well as vendors and outside contractors. Some of the major sources and methods of training for automation and control systems security are presented in the following sections of this chapter.

## Training Sources and Approaches

Some of the useful industrial automation and control system training programs and related initiatives that are available include the following:

- Idaho National Laboratory (http://energy.gov/oe/downloads/introduction-scada-security-managers-and-operators), (https://www.sgiclearinghouse.org/CyberSecurity?q=node/1275&lb=1
- Sandia National Laboratories (http://energy.sandia.gov/?page_id=6912)
- ISA (www.isa.org/training)
- US-CERT (http://www.us-cert.gov/control_systems)
- SANS (http://www.sans.org/security-resources/)
- National Initiative for Cybersecurity Education (NICE) (http://nist.gov/nice)
- National Security Agency (NSA) and the Department of Homeland Security (DHS) National Centers of Academic Excellence (CAE)

## Idaho National Laboratory

The Idaho National Laboratory, along with the Sandia National Laboratories, is home to the National SCADA Test Bed (NSTB) Program. This program was established by the U.S. Department of Energy and is focused on ensuring the "secure, reliable, and efficient distribution of power."

As described on its website, the NSTB program is chartered to perform the following functions:

- Assess selected control systems and control system components to identify cyber vulnerabilities. Testing is planned and conducted in collaboration with the interested industry partner and may be performed in the laboratory (test bed) environment or at the partner's site.
- Provide control system security training through workshops that describe common cyber vulnerabilities found in control systems and effective methods for mitigation.
- Share with appropriate standards organizations information that can be used to support the development of improved industry standards applicable to control system security.
- Participate in conferences that include control system security topics to share information obtained through system assessments and analyses.

Relative to SCADA, the Idaho National Laboratory training is aimed at "IT/control system managers, IT/control system security personnel, network and control system support engineers, and control system designers and developers who are involved in or responsible for control system cybersecurity." The course offerings are:

- **Introductory SCADA Security** (4 hours) comprised of:
  - SCADA Network Communications Overview
  - Common Vulnerabilities of Control Systems
  - Inadequate Policies and Procedures
  - Poorly Designed Control System Networks
  - Misconfigured or Unpatched Operating Systems and Embedded Devices

- Inappropriate Use of Wireless Communication
- Inadequate Authentication of Control System Communications
- Inadequate Identification and Control of Access to Control System
- Lack of Detection and Logging of Intrusion
- Dual Use of Control System Networks
- Lack of Security Checking of Control System Software/Applications
- Lack of Change Management/Change Control Procedures and Agreements
- Potential Mitigation Strategies Based on Multiple Levels of Implementation

- **Intermediate SCADA Security** (8 hours), covering the following material:
  - General Security Observations and Pitfalls
  - SCADA Network Communications Overview
  - Potential SCADA Network Entry Points and Defenses
  - SCADA Network Scanning and Vulnerability Identification (in a safe manner)
  - Network Monitoring and Simple Intrusion Detection
  - Dissecting SCADA Protocols
  - Common Programming Pitfalls
  - Modern Hardware and OS Mitigation Strategies
  - Incident Response Essentials for the SCADA Community
- **Advanced SCADA Security Red/Blue Team** (5 days), with the following topics, which are similar to those covered in Intermediate SCADA Security but include hands-on exercises using SCADA equipment:
  - General Security Observations and Pitfalls
  - Control System Network Communications Overview

- Potential Control System Network Entry Points and Defenses
- Control System Network Scanning and Vulnerability Identification
- Network Monitoring and Simple Intrusion Detection
- Dissecting Control System Protocols
- Common Programming Pitfalls
- Modern Hardware and OS Mitigation Strategies
- Incident Response Essentials for the Control System Community

## Sandia National Laboratories

The Sandia National Laboratories offer a course on SCADA assessment titled "Methodologies for Assessing SCADA Systems and an Overview of Related Security and Vulnerability Concerns." This course is offered "at Sandia's discretion to individuals with need-to-know and by invitation." The course is customizable and includes the following areas:

- SCADA and Other Digital Control System Uses in Infrastructures and Industry
- Vulnerabilities in SCADA Components and Systems
- Methodologies and Tools to Assess SCADA Systems in a Successful, Measurable, and Reproducible Manner

## International Society of Automation

The International Society of Automation (ISA) offers courses on securing industrial automation and control systems, including the following:

- **Industrial Networking and Security,** covering the following topics:
  - TCP/IP Networking Over Ethernet, Over Serial Links (PPP) and Through Other Networks
  - Making Networks Reliable: Redundancy/Name Services/Fault Tolerance/Spanning Tree
  - Secure Architectures: Layering Based on Function, Firewall Placements, Use of DMZs, Patch/Virus Update Management

  - Understanding Packets and Protocols: Understanding IP, TCP, UDP, and Application Protocols
  - Building a Plant Floor Web Server Using HTML, XML, Client/Server Side Scripting
  - Network Security Issues: Risks and Vulnerabilities | Attack Methods and Technologies
  - Applying: Virus Protection | Firewall Basics | Encryption Basics
  - Advanced Security: VPN Technology and Application, VLAN Technology and Application, Static/Dynamic Routing
  - Firewall Technology: Basic/Advanced, ACL Definitions, Stateful Inspection, HIDS/NIDS
  - User Authentication: Strong Authentication, Password Strategies, Multifactor, Centralized Policy Management
  - Practical Industrial Intranet Applications Using ISA-99 Recommendations and Standards
- **Using ANSI/ISA-99 to Secure Your Control System**, with the following topics:
  - Understanding the Current Industrial Security Environment: What Is Electronic Security for Industrial Automation and Control Systems? | How IT and the Plant Floor Are Different and How They Are the Same
  - How Cyberattacks Happen: Understanding the Threat Sources | The Steps to Successful Cyberattacks
  - Creating A Security Program: Critical Factors for Success | Understanding ISA-99 Part 2: Establishing an Industrial Automation and Control Systems Security Program
  - Using ISA-99: Risk Analysis: Business Rationale | Risk Identification, Classification, and Assessment | The DNSAM Methodology
  - Using ISA-99: Addressing Risk with Security Policy, Organization, and Awareness: CSMS Scope | Organizational Security | Staff Training and Security Awareness

- Using ISA-99: Addressing Risk with Selected Security Countermeasures: Personnel Security | Physical and Environmental Security | Network Segmentation | Access Control
- Using ISA-99: Addressing Risk with Implementation Measures: Risk Management and Implementation | System Development and Maintenance | Information and Document Management
- Using ISA-99: Monitoring and Improving the CSMS: Compliance and Review | Improve and Maintain the CSMS

- **Advanced Industrial Cybersecurity**, focusing on the following subjects:
  - TCP/IPV6 Networking: Over Ethernet | Over Serial Links (PPTP) | Using other Industrial Protocols
  - Converting from IPV4 to IPV6 without Causing Security Problems
  - Making Networks Secure: Local, Global, and Organizational Policies
  - COTS OS: Vulnerabilities and Hardening Techniques
  - Security/Performance Problems Associated with Patch/Virus Update Management
  - Secure Architectures: Layering Based on Function and on Firewall/VPN Placements
  - Building a Secure Plant Floor Web Server: HTML | XML | Methods to Enhance Security
  - Network Security Issues: Risks and Vulnerabilities | Attack Methods and Technologies
  - Applying: IPS/Virus Protection/Access-ACLs | Identification/Authentication/Nonrepudiation
  - Network Security: VPN Technology and Application | VLAN Technology and Application | Managed Switch/Router Configuration
  - Security Management: Access, Strong Authentication, Password Strategies, Multifactor, Centralized Policy Management

  - Practical Cybersecurity Applications: ANSI/ISA-99 | NIST | NCSDOCIPEP | DHS | DOE | CPNI
- **Cybersecurity for Automation, Control, and SCADA Systems**, which includes the following subjects:
  - Defining Industrial Cybersecurity
  - Risk Assessment
  - Threats and Vulnerabilities
  - Security Policies, Programs, and Procedures
  - Understanding TCP/IP, Hackers, and Malware
  - Technical Countermeasures
  - Architectural and Operational Strategies
- **Introduction to Industrial Automation Security and the ANSI/ISA-99 Standards**, covering the following areas:
  - Understanding the Current Industrial Security Environment: What Is Electronic Security for Industrial Automation and Control Systems? | Trends in Security Incidents
  - How IT and the Plant Floor Are Different and How They Are the Same
  - Current Security Standards and Practices
  - Creating a Security Program: Critical Factors for Success | Understanding ISA-99 Part 2: Establishing an Industrial Automation and Control Systems Security Program
  - Using ISA-99 – Risk Analysis: Business Rationale | Risk Identification, Classification, and Assessment
  - Using ISA-99 – Addressing Risk with Security Policy, Organization, and Awareness: CSMS Scope | Organizational Security | Staff Training and Security Awareness | Business Continuity Plan | Security Policies and Procedures
  - Using ISA-99 – Addressing Risk with Selected Security Countermeasures: Personnel Security | Physical and Environmental Security | Network Segmentation | Access

Control: Account Administration, Authentication, and Authorization

– Using ISA-99 – Addressing Risk with Implementation Measures: Risk Management and Implementation | System Development and Maintenance | Information and Document Management | Incident Planning and Response

– Using ISA-99 – Monitoring and Improving the CSMS: Compliance and Review | Improve and Maintain the CSMS

## U.S. Computer Emergency Readiness Team

The U.S. Computer Emergency Readiness Team (US-CERT) offers the following instructor-led courses:

- **Introduction to Control Systems Cybersecurity** – Introduces participants to industrial control system fundamentals and comparisons of IT and control system architectures. The elements of the course are:
  - Cybersecurity Landscape: Understanding the Risks
  - Industrial Control Systems Applications
  - Current State of Cybersecurity in Industrial Control Systems
  - Practical Applications of Cybersecurity
- **Industrial Control System Security (ICS) for Management** – Provides background and basic understanding of the current industrial control system cybersecurity landscape for managers. The course includes the following topics:
  - Overview of the Elements of the Risk Equation and How It Applies to Cybersecurity of an ICS
  - Threats to an ICS and Its Components
  - Tools That Can Be Used to Help Mitigate the Cybersecurity Risk
- **Intermediate Cybersecurity for Industrial Control Systems Part 1** – Consists of technical instruction on the protection of industrial control systems using offensive and defensive methods. The material includes:

  - Current Security in ICS
  - Strategies Used Against ICS
  - Defending the ICS
  - Preparation and Further Reading for Part 2

- **Intermediate Cybersecurity for Industrial Control Systems Part 2** – Hands-on training with laboratory exercises to impart knowledge of how attacks against control systems are launched and succeed. The course topics include:
  - Supervisory Control and Data Acquisition (SCADA) and Control System Overview
  - Risk to Industrial Control Systems
  - Exploit Demonstration
  - Basic Control Security Considerations
  - Network: Security, Identification, and Remediation
  - Network: Defense, Detection, and Analysis

- **ICS Advanced Cybersecurity** – Includes hands-on training on protecting and securing industrial control systems from cyberattacks, including a Red Team/Blue Team exercise that will be conducted within an actual control systems environment. The course comprises the following subjects:
  - Overview of the DHS Control Systems Security Program
  - Review of Cybersecurity for Industrial Control Systems
  - Demonstration Showing How a Control System Can Be Attacked from the Internet
  - Hands-on Classroom Training on Network Discovery Techniques and Practices
  - Hands-on Classroom Training on Network Exploitation, Network Defense Techniques and Practices
  - A 12-Hour Exercise where Participants are Either Attacking (Red Team) or Defending (Blue Team)

  - Red Team/Blue Team Exercise Lessons Learned and Round-Table Discussion

## SANS

The SANS organization offers an advanced SCADA security hands-on training course designed for administrators and automation system personnel. The material addresses vulnerabilities and risks to SCADA systems, effects on enterprise security, the use of open source security tools, and testing methods. The course subject areas include the following:

- Review of Industrial Control Systems (ICS), Operating System Kernels, and Network Security SCADA Penetration Testing
- SCADA Vulnerability Assessment Methodology
- SCADA Vulnerability Analysis
- Embedded Device Fuzzing (an attack using random, malformed, or bad data)
- Protocol Analysis
- Methods for Compromising and Dissecting Common Security Controls Found in ICS Environments

## National Initiative for Cybersecurity Education

The National Initiative for Cybersecurity Education (NICE) is a partnership of government, industrial, educational, and professional organizations with the mission "to enhance the overall cybersecurity posture of the United States by accelerating the availability of educational and training resources designed to improve the cyber behavior, skills, and knowledge of every segment of the population." The specific goals of NICE are to:

1. Raise awareness among the American public about the risks of online activities
2. Broaden the pool of skilled workers capable of supporting a cyber-secure nation
3. Develop and maintain an unrivaled, globally competitive cybersecurity workforce

The National Institute of Standards and Technology (NIST) has been assigned the lead in the NICE initiative and will coordinate the government activities involving the Department of Homeland Security (DHS), the Department of Defense (DoD), the Department of Education (ED), the National Science Foundation (NSF), the National Security Agency (NSA), and the Office of Personnel Management (OPM).

The objectives[1] defined by NICE to achieve the desired strategic goals include the following:

1. Improve citizens' knowledge to allow them to make smart choices as they manage online risk
2. Improve knowledge of cybersecurity within organizations so that resources are well applied to meet the most obvious and serious threats
3. Enable access to cybersecurity resources
4. Improve K-12 science, technology, engineering, and mathematics (STEM) education, emphasizing the important role of mathematics and computational thinking
5. Increase the quantity and quality of academic computer science courses in high schools
6. Increase the quantity and quality of undergraduate and graduate cybersecurity curricula for students in computer science and, more broadly, IT and security-related degree programs
7. Incentivize, support, and recognize excellence in graduate-level cybersecurity research and development
8. Develop a usable cybersecurity competency framework (human resources and curriculum focus)
9. Provide a framework for focusing on cybersecurity training to meet evolving needs
10. Study the application of professionalization, certification, and licensing standards on cybersecurity career fields

In order to advance its objectives, NICE has published a draft document, the *NICE Cybersecurity Workforce Framework*, which classifies the typical duties and skill requirements of cybersecurity workers. This framework was developed because cybersecurity professionals often do not fit into standard job descriptions and titles (http://csrc.nist.gov/nice/framework/).[2]

The *NICE Cybersecurity Workforce Framework* develops a vocabulary that can be used by any type of organization. It organizes cybersecurity work into the following categories:

- Design
- Operation
- Maintenance
- Incident response
- Information gathering
- Analysis

In another initiative, NIST, the U.S. Department of Education, NICE, and the National Cybersecurity Education Council (NCEC) are developing a strategic public-private partnership to promote formal cybersecurity education. The effort is designed to support the NICE goal of broadening the pool of skilled workers capable of supporting a cyber-secure nation. In order to accomplish this goal, the partnership is collaborating with state and local governments to establish and support formal cybersecurity education programs for kindergarten through 12th grade and for higher education and vocational programs.

## National Security Agency and the Department of Homeland Security National Centers of Academic Excellence

National Security Agency (NSA) and the Department of Homeland Security (DHS) are sponsors of the National Centers of Academic Excellence (CAE) initiative. This program has the goal of reducing vulnerabilities in the national information infrastructure. In order to accomplish this goal, CAE are established in educational institutions to promote research and education in information assurance (IA). The students graduating from these institutions with training and research experience in information assurance will be able to apply their skills to securing the nation's infrastructure and information systems.

Institutions can be designed as CAE in IA Education (CAE/IAE) and Research (CAE-R) programs. The designations have terms of 5 years, after which the institution must successfully reapply in order to retain its CAE designation.

Applicants for a National Center of Academic Excellence in IA Education designation are 4-year colleges and graduate-level universities and must pass a stringent evaluation and review. To apply for designation of a National Center of Academic Excellence in research, the college or university must meet the Carnegie Foundation's classifications of Research University/Very High (RU/VH), Research University/High (RU/H), and Doctoral Research University (DRU) or the equivalent.

## Training Support Guidelines

There are a number of documents that can provide guidance in setting up and conducting training applicable to industrial automation and control systems. Two of the important ones are reviewed here to provide a general understanding of their content and capabilities. These publications are:

- NIST SP 800-50, *Building an Information Technology Security Awareness and Training Program,* http://csrc.nist.gov/publications/nistpubs/800-50/NIST-SP800-50.pdf
- NIST SP 800-16, *Revision 1, Information Security Training Requirements: A Role- and Performance-Based Model (Draft),* http://csrc.nist.gov/publications/PubsDrafts.html#800?16?rev1

### NIST Special Publication 800-50

NIST SP 800-50,[3] *Building an Information Technology Security Awareness and Training Program,* provides guidance in developing and conducting information system security training. The document is aimed at IT security, but it is useful in supporting automation and control system security awareness and training in that it provides useful principles that can be used in setting up such programs.

The publication describes the following four critical steps in the life cycle of a security awareness and training program:

1. **Awareness and Training Program Design** – Needs assessment is conducted and a training strategy is developed and approved. This

strategic planning document identifies implementation tasks to be performed in support of established security training goals.

2. **Awareness and Training Material Development** – Available training sources, scope, content, and development of training material, including solicitation of contractor assistance if needed, are addressed.

3. **Program Implementation** – Effective communication and rollout of the awareness and training program are conducted. It also addresses options for delivery of awareness and training material (web-based, distance learning, video, on-site, etc.).

4. **Post-Implementation** – Guidance on keeping the program current and monitoring its effectiveness is provided. Effective feedback methods are described (surveys, focus groups, benchmarking, etc.).

The document also defines the following three common models used in managing a security training function:

- **Centralized** – All responsibility resides with a central authority.
- **Partially Decentralized** – Policy and strategy lie with a central authority, but implementation responsibilities are distributed.
- **Fully Decentralized** – Only policy development resides with a central authority, and all other responsibilities are delegated to individual organizational components.

NIST SP 800-50 also provides a useful perspective of learning as a continuum, going from awareness to training and then to education, as shown in Figure 7-1.

NIST SP 800-50 discusses the differences among awareness, training, and education as:

- **Awareness** – Focuses attention on security
- **Training** – Strives to produce relevant and needed security skills and competencies
- **Education** – Integrates all the security skills and competencies of the various functional specialties into a common body of knowledge and strives to produce IT security specialists

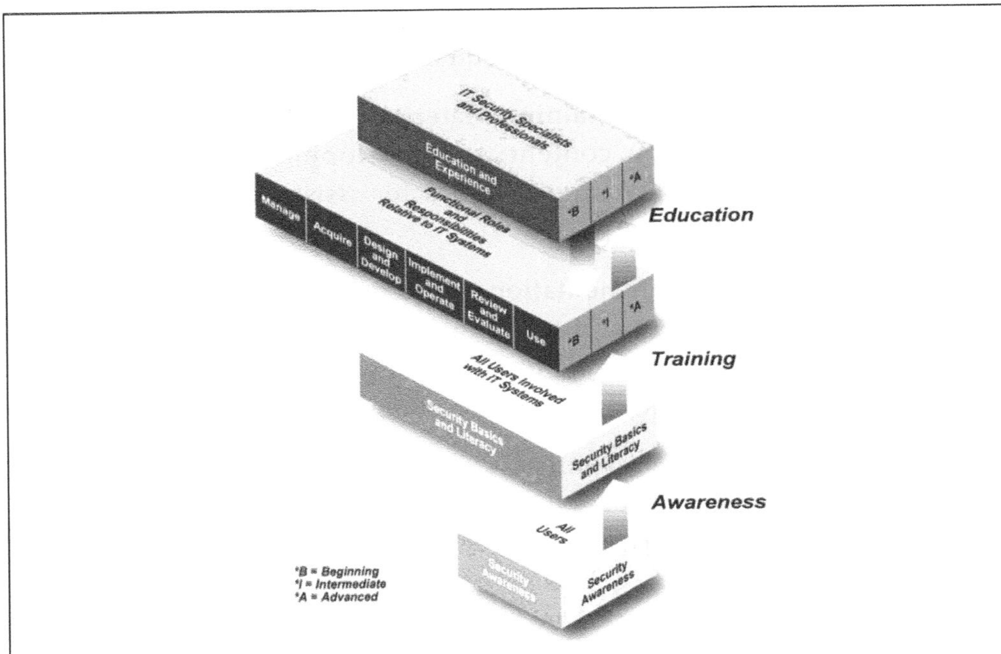


**Figure 7-1. Security Learning Continuum**
*Source: NIST SP 800-50 (2003)*

This publication provides additional details on implementing awareness and training that can be applied to industrial automation and control systems.

## NIST Special Publication 800-16

NIST SP 800-16, *Information Security Training Requirements: A Role- and Performance-Based Model, Draft,*[4] complements NIST SP 800-50. As shown in Figure 7-1, it defines learning as a continuum, beginning with awareness, transitioning to training, and then moving into education. The model is role-based in that it "defines the information security learning needed as a person assumes different roles within an organization and different responsibilities in relation to information systems." NIST SP 800-16 expands on the continuum as follows:

1. **Basic Security Awareness** – Required for employees, including contractor employees who are involved in any way with information systems.

2. **Awareness Training (Basics and Literacy)** – A transitional stage between basic awareness and role-based training. It provides the

foundation for subsequent specialized or role-based training by providing a universal baseline of key security terms and concepts.

3. **Role-Based Training** – Focuses on providing the knowledge, skills, and abilities specific to an individual's roles and responsibilities relative to information systems. At this level, training recognizes the differences between beginning, intermediate, and advanced skill requirements.

4. **The Education Level** – Focuses on developing the ability and vision to perform complex multidisciplinary activities and the skills needed to further the information security profession and to keep pace with threats and technology changes.

5. **Professional Development** – Intended to ensure that users, from the beginner to the career security professional, possess a required level of knowledge and competence necessary for their roles. Professional development validates skills through certification and advanced education.

To implement instructional design, NIST SP 800-16 presents the five-phase ADDIE Instructional Design Model, which comprises the phases of analysis, design, development, implementation, and evaluation. The model is illustrated in Figure 7-2.

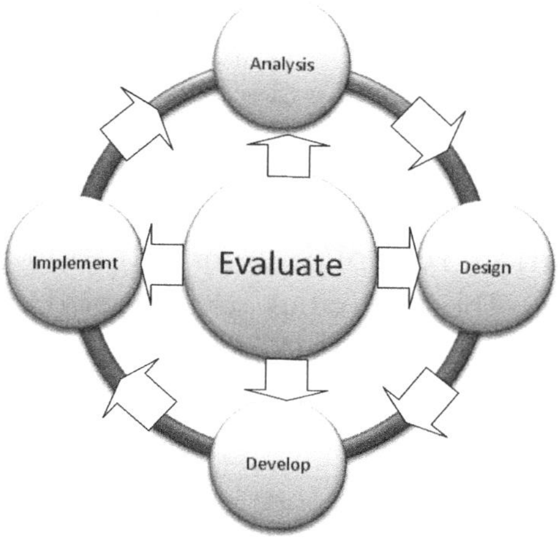


**Figure 7-2. ADDIE Instructional Design Model**
*Source: NIST SP 800-16 (2009)*

The five phases of the instructional design model consist of the following activities:

1. **The Analysis Phase** – Compiles an inventory of all tasks associated with each role and conducts a needs assessment to select those tasks for which training needs to be developed. Task descriptions can usually be found in position descriptions, performance plans, assignment letters, system security plans, and directives, or can be determined through interviews with current practitioners. These tasks should yield:

   – The knowledge level required

   – The desired behavioral outcomes

   – Performance measures for each task

   In addition, prerequisite knowledge should be determined.

2. **The Design Phase** – Develops the learning objectives for each of the tasks defined during the analysis phase. Then, the steps required to accomplish each objective have to be identified. The design phase consists of the following steps:

   – Develop learning objectives (enabling and terminal)

   – Identify and list steps required

   – Develop performance/knowledge tests

   – Develop required entry behaviors (optional)

   – Sequence learning objectives

3. **The Development Phase** – Identifies and lists the activities that will help the training recipients achieve each objective (e.g., lecture, simulation, role-play, demonstration, lab exercise, table-top exercise). These activities are then used to select the appropriate delivery method(s). The steps of the development phase are summarized as follows:

   – List activities required for each objective

   – Select appropriate delivery method(s)

   – Gather reference material

   - Review existing course material
   - Develop course material
   - Develop lesson plans and instructor guide
   - Conduct expert review of course materials

4. **The Implementation Phase** – Tests a draft version of the course or module with a selected audience for feedback. Adjustments are then made as required. The steps in this phase are:
   - Pilot course to hand-picked audience
   - Make course adjustments
   - Train the trainer (if required)
   - Full-scale delivery

5. **The Evaluation Phase** – Conducts formative and summative evaluations, described as follows:
   - *Formative* – During all phases, analysis, design, development, and implementation, a formative analysis evaluates if what is being taught is what was intended to be taught.
   - *Summative* – Evaluates if the student was satisfied with the course and instructor, if the student achieved his or her objectives, if supervisors of the students felt the student was a more effective employee as a result of the course, and if the training program is providing the types of courses that are most needed.

NIST SP 800-16 also provides templates for training modules that provide a roadmap for teaching the module. Figure 7-3 illustrates a typical module.

INFORMATION SECURITY TRAINING
Curriculum Module
**Security Program (2.2D)**

Training Area: **Management**
Responsibility: **Implement & Operate**

**Definition** — The implementation and use of organizational structures and processes for information security program goal-setting, prioritizing, and related decision-making activities; these encompass such elements as organization-specific policies, guidelines, requirements, roles, responsibilities, and resource allocation.

**Behavioral Outcome** — Individuals who are responsible for the implementation and daily operations of an information security program have a sufficient understanding of the appropriate program elements and requirements to be able to apply them in a manner which provides adequate and appropriate levels of protection for the organization's IT resources.

**Knowledge Levels** —

1. Beginning — Identify, Understand, Apply
2. Intermediate — Interpret, Analyze, Decide
3. Advanced — Investigate, Approve, Direct

**Terminal Learning Objectives** —

At the conclusion of this module, individuals will be able to:

1. Beginning — Apply organization-specific information security program elements to the implementation of the program and identify areas of weakness.

2. Intermediate — Analyze patterns of non-compliance and take appropriate administrative or programmatic actions to minimize security risks.

3. Advanced — Direct the implementation of appropriate operational structures and processes to ensure an effective information security program.

**Figure 7-3. Example Instructional Module**
*Source: NIST SP 800-16 (2009)*

**Applicable Roles (minimum)** —

| Senior Information Resources Management Official | | |
|---|---|---|
| Chief Information Officer | Program and Functional Managers | |
| System Administrator | Internal Auditor | System Owner |
| Information Resources Manager | Information System Security Officer | |
| Network Administrator | Security Administrator | |
| Senior Agency Information Security Officer/Manager and Staff | | |
| Assessor | Incident Response Coordinator | |

## INFORMATION SECURITY TRAINING TOPICS

1. Laws and Regulations
2. Information Security Program
3. System Environment
4. System Interconnection
5. Information Sharing
6. Security Objectives
7. Risk Management
8. Management Controls
9. Acquisition/Development/Installation/Implementation Controls
10. Operational Controls
11. Awareness, Training, and Education Controls
12. Technical Controls

## RESOURCE DOCUMENTS

NIST SP 800-12
NIST SP 800-100

**Figure 7-3. Example Instructional Module (continued)**

## Common Training Subjects

In reviewing the courses and training programs discussed in this chapter, a common set of topics can be extracted to use as a basis for training in the security of industrial automation and control systems. These proposed core topics are summarized in Table 7-1.

**Table 7-1. Industrial Automation and Control System Core Topics**

| Topic | Description |
|---|---|
| Security Policies and Procedures | Guiding document |
| Incorporating Applicable Standards and Guidelines | ANSI/ISA-99, NIST |
| Risk Management and Countermeasures | Risk analysis and mitigation |
| SCADA Network Communications | TCIP/IP, Ethernet, wireless, etc. |
| Common Control System Vulnerabilities and Threat Sources | Vulnerability identification, network scanning |
| Access Control and Authentication for Control Systems | Multifactor, passwords |
| Logging and Intrusion Detection | Logging of system accesses, intrusions, and abnormal activity |
| Change Management Procedures | Document and confirm system changes |
| Typical SCADA Entry Points | Typical targets of choice |
| Application and Control System Software | Vulnerabilities and security weaknesses, COTS OS |
| Security Awareness Training | Awareness training actively conducted in accordance with security policy |
| Incident Response | Responding to incidents in an automation environment |
| Encryption Methods | Encryption in an automation/control environment |
| Virtual Private Networks and VLANs | Secure remote access methods and issues |
| Firewalls and Antivirus Software | Screening devices and methods in time-critical automation/control environments |

## Summary

Training in the area of industrial automation and control system security is now offered by a variety of sources and types of institutions, including secondary schools, universities, community colleges, technical schools, industry, technical societies, and governmental bodies. This type of training was not always as available as it is now, which is the result of increased awareness of threats against automation and control systems that are critical to the nation.

This chapter reviewed some of the more prominent training programs and their content. In addition, governmental and NIST recommendations and guidelines applicable to conducting automation and control system-related training were presented to provide a basis for developing effective programs in this area. The chapter concluded with a proposed listing of fundamental cybersecurity topics for automation and control system security that can be used as a foundation for automation system-based training. Chapter 8 will explore what the future holds for security in the realm of industrial automation and control systems.

## Review Questions for Chapter 7

7.1 What organizations host the National SCADA Test Bed (NSTB) program?

A. NIST and the NSA

B. The Idaho National Laboratory and the Sandia National Laboratories

C. ISA and SANS

D. US-CERT and NIST

7.2 Which of the following is NOT a function of the National SCADA Test Bed (NSTB)?

A. Assess selected control systems and control system components to identify cyber vulnerabilities

B. Develop new standards and guidelines for governmental agencies

C. Provide control system security training through workshops that describe common cyber vulnerabilities found in control systems

D. Share information that can be used to support the development of improved industry standards applicable to control system security with the appropriate standard's organizations

7.3 What organization developed and teaches the application of 62443-2-1 (99.02.01)-2009, *Security for Industrial Automation and Control Systems: Establishing an Industrial Automation and Control Systems Security Program*?

A. SANS

B. NIST

C. ISA

D. NSA

7.4 In a training program hands-on exercise, what is the function of the Red Team?

A. Mitigating

B. Scanning

C. Defending

D. Attacking

7.5 In a training program hands-on exercise, what is the function of the Blue Team?

A. Defending

B. Attacking

C. Scanning

D. Mitigating

7.6 What is fuzzing?

A. Modifying memory locations

B. An attack using random or malformed bad data

C. Scanning to determine vulnerabilities

D. Responding to an attack by counterattacking

7.7 The National Initiative for Cybersecurity Education (NICE) is a partnership of government, industrial, educational, and professional organizations. Which of the following is NOT one of the goals of NICE?

A. Raise awareness among the American public about the risks of online activities

B. Broaden the pool of skilled workers capable of supporting a cyber-secure nation

C. Develop and maintain an unrivaled, globally competitive cybersecurity workforce

D. Develop secure digital control systems

7.8 Which of the following organizations is NOT a participant in the NICE initiative?

A. NIST

B. NASA

C. Department of Homeland Security (DHS)

D. NSA

7.9 The *NICE Cybersecurity Workforce Framework* develops a vocabulary that can be used by any type of organization in categorizing cybersecurity work. Which of the following is NOT a category in this framework?

A. Penetration testing

B. Design

C. Operation

D. Maintenance

7.10 The U.S. National Centers of Academic Excellence initiative is sponsored by which of the following organizations?

A. NIST and the Department of Energy

B. SANS and the Department of Education

C. NSA and the Department of Homeland Security

D. The Department of Defense and the US-CERT

7.11 The designation of an institution as a National Center of Academic Excellence has a duration of how many years before the institution has to reapply?

A. Two years

B. Ten years

C. Five years

D. Seven years

7.12 Which of the following is NOT one of the critical steps in the life cycle of a security awareness and training program as described in NIST SP 800-50, *Building an Information Technology Security Awareness and Training Program*?

A. Awareness and training program design

B. Program implementation

C. Post-implementation

D. Pre-implementation needs review

7.13 Which of the following is NOT one of the three common models used in managing a security training function as described in NIST SP 800-50?

A. Centralized

B. Partially Decentralized

C. Partitioned

D. Fully Decentralized

7.14 According to NIST SP 800-50, learning is a continuum comprising which of the following elements?

A. Awareness, skill, knowledge

B. Awareness, training, education

C. Training, education, skill

D. Education, awareness, skill

7.15 NIST SP 800-16, *Information Security Training Requirements: A Role- and Performance-Based Model (Draft)*, defines the following as what type of training?

Training that "focuses on providing the knowledge, skills, and abilities specific to an individual's roles and responsibilities relative to information systems. At this level, training recognizes the differences between beginning, intermediate, and advanced skill requirements."

A. Role-based training

B. Task-oriented training

C. Responsibility training

D. Knowledge and skills training

7.16 Which one of the following is NOT one of the five phases of the industrial design model described in NIST SP 800-16?

A. Analysis

B. Training

C. Design

D. Evaluation

7.17 What are the two types of evaluations in the evaluation phase of the industrial design model described in NIST SP 800-16?

A. Formative and summative

B. Formative and intuitive

C. Summative and descriptive

D. Descriptive and formative

## References

1. NICE (National Initiative for Cybersecurity Education). "Building a Digital Nation." Strategic Plan (Draft). NICE, 2011.

2. NICE (Cybersecurity Workforce Framework, (http://csrc.nist.gov/nice/framework/), NICE, 2014.

3. NIST SP 800-50. *Building an Information Technology Security Awareness and Training Program*. Washington, DC: NIST (National Institute of Standards and Technology), 2003.

4. NIST SP 800-16. *Information Security Training Requirements: A Role- and Performance-Based Model*. Revision 1 (Draft). Washington, DC: NIST (National Institute of Standards and Technology), 2009.

# 8

# Industrial Automation and Control System Trends, Approaches, and Issues

*Chapter 7 emphasized the importance of training to the cybersecurity of industrial automation and control systems. It pointed out sources of that training, identified government and industrial initiatives in promoting security awareness and training, and reviewed effective methodologies for implementing a cybersecurity training program applicable to industrial automation and control systems. Chapter 8 looks into related trends that will affect automation and control systems and provides guidelines for taking advantage of these developments.*

## Automation and Control System Trends

Because industrial automation and control systems are complex, their future characteristics will be affected by advances in a number of domains. These domains include hardware, software, virtualization, storage, mobile devices, cloud computing, communications, HMI, networking, encryption, security, artificial intelligence, standards, and robotics. Advances in these technologies will influence capabilities, performance, reliability, trustworthiness, cost, maintenance, training, and rates of adoption. In addition, increased and more stringent regulatory requirements will require increasing effort and result in higher costs to meet compliance, traceability, environmental, and legal obligations.

Software, hardware, and the control systems that use them will be more adaptable, secure, efficient, and self-healing as improvements in these areas are made. Some of the important elements that will characterize future software, hardware, and control systems are summarized in Figures 8-1, 8-2, and 8-3, respectively.

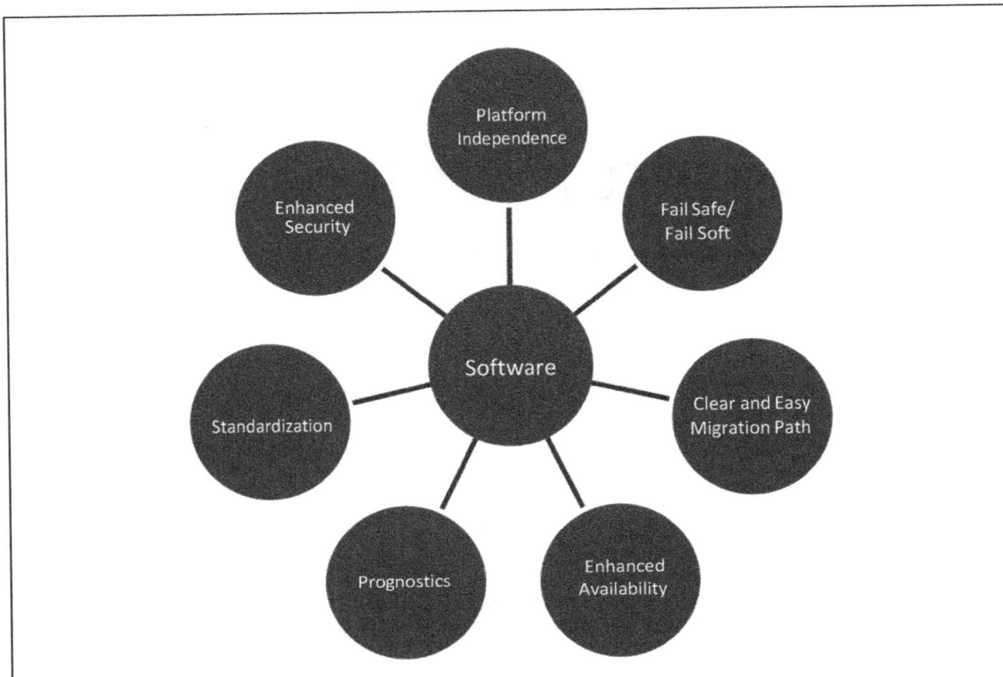


**Figure 8-1. Future Software Characteristics**

It is interesting to note that there are some common threads implicit and explicit among these three areas. These commonalities include the following characteristics:

- Adaptive and self-organizing
- Self-diagnosis
- Fail soft
- Resilience
- Prognostics
- Enhanced availability
- Intrinsic security
- Availability of and adherence to standards

**Figure 8-2. Future Hardware Characteristics**

**Figure 8-3. Future Control System Characteristics**

- Wide use of wireless technology
- Incorporation of web-based and mobile devices
- Use of self-powered nanotechnology sensors and devices
- Higher-speed processing
- Agent-based (A software agent is a computer program that performs tasks for a user using assumptions based on past behavior or defined preferences.)
- Reconfigurable

In the hardware realm, advances in chip technology will yield higher performance multi-core computing engines using less power than conventional single-core designs. Processor workloads can be assigned to those cores with the highest power efficiency and those that are best able to complete the required processing. In addition, cores not needed can be powered down or run at diminished capacity to save energy. In order to take advantage of high-performance, multi-core designs, software must be developed to support efficient partitioning of tasks among the cores and not waste the available capabilities of such systems. These capabilities will result in automation systems with faster response times, lower costs, increased computational power, and support for semi-autonomous operation.

The continuous improvement in tablet computers and intuitive HMIs (human-machine interfaces) will provide remote platforms from which to monitor and control processes as well as to conduct maintenance functions. Many of these functions will be performed wirelessly. Correspondingly, wireless security will have to keep pace to reduce or eliminate exposure to threats against mobile devices.

Another technology that will become increasingly important in industrial automation and control systems is virtualization. Virtualization is based on the concept of a virtual machine (VM) running on a physical computing platform under the control of a virtual machine monitor (VMM), known as a *hypervisor*. Virtualization reduces the effort required to maintain hardware and software systems by eliminating the binding of software to specific hardware platforms through the logical partitioning of physical computing resources into multiple execution environments, including servers, applications, and operating systems. Thus, applications can be centrally managed, and cybersecurity mechanisms can be distributed among virtualization servers, using virtual instances. The control and automation software applications and operating system no longer have to be individually installed and maintained on every client. A typical virtual server can support a variety of client instances on a single hardware platform through protocols, such as the Remote Desktop Protocol (RDP). Additional advantages of virtualization are summarized as follows:

- Alarm management capabilities can be easily implemented to identify critical system situations or system errors.
- Clients can be deleted or added rapidly without affecting plant operation.
- Existing systems can be ported to different hardware platforms with minimum effort.
- Software updates can be applied during system operation.

For cloud computing, virtualization provides a level of freedom of choice for the customer and efficiencies and cost savings for cloud providers. The migration to virtualized clouds will, in many instances, be initially through hybrid clouds while customers test and evaluate the cloud paradigm in the context of their own business requirements. As users become accustomed to employing cloud computing, they will embrace the ability to have large numbers of servers available on demand and, conversely, to rapidly reducing the number of servers in use when they are not required. Virtualization, if applied properly, can also result in minimal ramp-up times, reduction of time to develop and market applications, and rapid acquisition or reduction of resources, as requirements vary. Cloud computing was covered in detail in Chapter 4.

Regarding virtualization security, conventional attacks such as denial of service can be used against VMs while other attacks exploit specific VM vulnerabilities. Because multiple VMs share the same physical hardware, a vulnerability in one VM can be exploited to attack other VMs or the host systems. Some of the vulnerabilities of the virtual environment include the following:

- **Keystroke Logging** – Logging of keystrokes and screen updates allows information to be passed across virtual terminals in the VM and written to host files, thus permitting the monitoring of encrypted terminal connections inside the VM.

- **One VM Accessing Another VM** – Monitoring of one VM by another VM using hacker techniques through a virtual hub or switch.
- **Shared Clipboard Technology** – Allows data to be transferred between VMs and the host, providing a means of moving data between malicious programs in VMs of different security realms.
- **Host VM Monitoring** – Monitoring of network packets coming from or going to a VM by the host enables the host to affect the VM by the following actions and, if the host is compromised, in a negative way:
  - Adjusting the number of CPUs, amount of memory, amount and number of virtual disks, and number of virtual network interfaces available to a VM
  - Monitoring and configuring resources available to the VMs, including CPU, memory, disk, and network usage of VMs
  - Monitoring the applications running inside the VM
  - Starting, stopping, pausing, and restarting VMs
  - Viewing, copying, and modifying data stored on the VM's virtual disks

## Penetration Testing of Industrial Automation and Control Systems

With more assessment tools and methods for evaluating computer systems and networks being available in general, there will be increased pressure to apply these approaches to industrial automation and control systems in the future. One particularly effective vulnerability assessment method that is widely used and specified in many evaluation paradigms is penetration testing. However, penetration testing of industrial automation and control systems, if not properly conducted, can cause damage to these systems and disrupt operations. The commonly used penetration testing tools create additional traffic on automation and control systems and, if applied to such systems that have limited memory and processing capability, can severely impact their safety and performance. However, if done properly, penetration testing can provide valuable insight into the security posture of industrial automation and control systems.

Some useful guidelines for conducting penetration testing on automation and control systems are:

- Document and understand the automation/control system configuration to identify areas of high risk.
- Identify all routes of access to the automation/control system from outside elements, including the corporate network.
- Where possible, conduct testing on independent test systems in a controlled laboratory setting.
- In laboratory test systems, ensure that intrusion detecting systems and firewall configurations are duplicates of those in the production system.
- Evaluate testing procedures and approaches in a laboratory test setting before applying them to online production systems.
- Understand the proper operation of the automation/control system to be tested in order to more effectively identify the results of attack scenarios.
- Conduct testing on redundant automation/control system components.
- Use passive data collection techniques when possible.
- Be cognizant of risks when conducting testing on active, operational systems, and have recovery procedures in place in the event of a system failure caused by the testing.
- Conduct penetration testing from computers external to the network to simulate an attacker's approach.

A penetration test is usually conducted with no prior knowledge of the automation/control system, that is, with a "black box" approach. Penetration testing begins with a reconnaissance phase comprised of the following steps:

- **Footprinting** – Gathering information about the target automation/control network in order to develop a mapping of the network and associated systems
- **Scanning** – Conducting a scan of the network to discover applications and open ports, services running on ports, and information about the operating system

- **Enumerating** – Using the results of the scanning phase to develop a complete map of the network, discover vulnerable resources, and identify valid user accounts

Following reconnaissance, the network can be assessed for vulnerabilities by acquiring access to the automation/control system. If an attack was launched on the system, the acquiring access phase is where the actual attack is implemented. In this phase, an attacker could access the operating system and network, obtain higher-level system privileges, launch buffer overflow and denial-of-service attacks, and insert viruses and Trojan horses.

Some of the common assessment tools used in penetration testing include Nmap, Metasploit, Nessus, STAT Scanner, Wireshark, Ettercap, and software debuggers.

## Formal Methods Used to Quantify and Standardize Important Concepts and Applications

In many fields, as advances occur, formal methods exist to quantify and standardize important concepts and applications. Three extant methods that can be applied to industrial automation and control systems are Information Security Continuous Monitoring (ISCM), the Smart Grid Maturity Model (SGMM), and the Automation Maturity Model (AMM).

### ISCM Strategy

NIST SP 800-137,[1] *Information Security Continuous Monitoring (ISCM) for Federal Information Systems and Organizations*, defines the process for developing an ISCM strategy and implementing an ISCM program. ISCM is defined as "maintaining ongoing awareness of information security, vulnerabilities, and threats to support organizational risk management decisions."

The ISCM process comprises the following steps:

1. **Define** an ISCM strategy based on risk tolerance that maintains clear visibility into assets, awareness of vulnerabilities, up-to-date threat information, and mission/business impacts.
2. **Establish** an ISCM program, determining metrics, status monitoring frequencies, control assessment frequencies, and an ISCM technical architecture.

3. **Implement** an ISCM program and collect the security-related information required for metrics, assessments, and reporting. Automate collection, analysis, and reporting of data where possible.
4. **Analyze** the data collected and report findings, determining the appropriate response. It may be necessary to collect additional information to clarify or supplement existing monitoring data.
5. **Respond** to findings with technical, management, and operational mitigating activities or with acceptance, transference/sharing, or avoidance/rejection.
6. **Review and Update** the monitoring program, adjusting the ISCM strategy and maturing measurement capabilities to increase visibility into assets and awareness of vulnerabilities, further enable data-driven control of the security of an organization's information infrastructure, and increase organizational resilience.

The ISCM process is depicted in Figure 8-4.

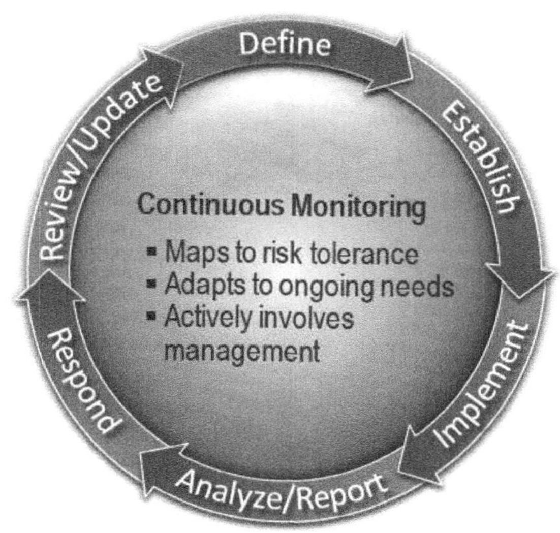


**Figure 8-4. The ISCM Process**
*Source: NIST SP 800-137 (2011)*

In the ISCM process, the status of automation security in a facility can be determined by establishing appropriate metrics. The metrics are defined through the following activities:

- Maintaining an understanding of threats and threat activities
- Assessing all security controls
- Collecting, correlating, and analyzing security-related information
- Providing actionable communication of security status across all tiers of the organization
- Active management of risk by organizational officials

NIST SP 800-137 defines the process for managing information security and information system-related risk as the Risk Management Framework (RMF). The RMF integrates information system security and risk management activities as illustrated in Figure 8-5. It applies equally well to both automation and control systems and information systems.

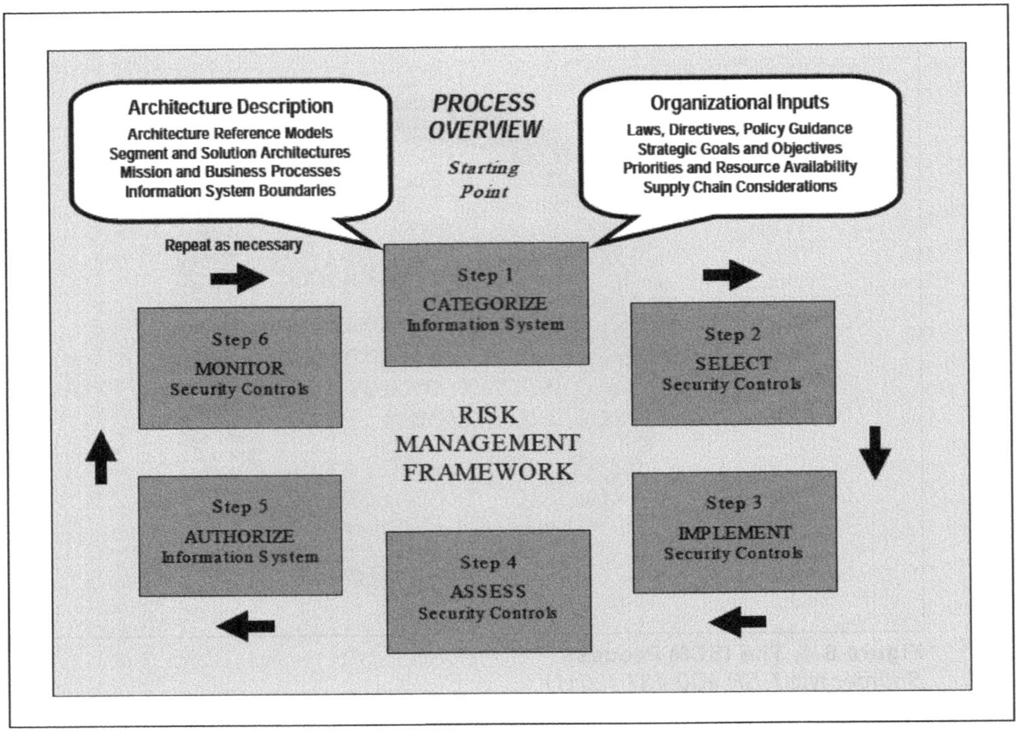


**Figure 8-5. The Risk Management Framework**
*Source: NIST SP 800-137 (2011)*

The frequency of monitoring security controls for effectiveness and relevance is a function of the following criteria as presented in NIST SP 800-137:

- **Security Control Volatility** – Security control volatility is a measure of how frequently a control is likely to change over time subsequent to its implementation. Volatile security controls such as configuration management are assessed more frequently than, for example, personnel screening controls that tend to remain static over long periods. The latter would, therefore, typically require less frequent assessment, whether the objective is establishing security control effectiveness or supporting calculation of a metric.
- **System Categorizations/Impact Levels** – In general, security controls implemented on systems that are categorized as high-impact are monitored more frequently than controls implemented on moderate-impact systems, which are in turn monitored more frequently than controls implemented on low-impact systems.
- **Security Controls or Specific Assessment Objects Providing Critical Functions** – Security controls or assessment objects that provide critical security functions (e.g., log management servers, firewalls) are candidates for more frequent monitoring.
- **Security Controls with Identified Weaknesses** – Existing risks documented in security assessment reports are considered for more frequent monitoring to ensure that the risks stay within tolerance.
- **Organizational Risk Tolerance** – Organizations with a low tolerance for risk (e.g., organizations that process, store, or transmit large amounts of proprietary and/or personally identifiable information [PII]), organizations with numerous high-impact systems, and organizations facing specific persistent threats) monitor more frequently than organizations with a higher tolerance for risk.
- **Threat Information** – Organizations consider current credible threat information, including known exploits and attack patterns when establishing monitoring frequencies. For instance, if a specific attack is developed that exploits a vulnerability of an implemented technology, temporary or permanent increases to the monitoring frequencies for related controls or metrics may help provide protection from the threat.
- **Vulnerability Information** – Organizations consider current vulnerability information with respect to technology products when

establishing monitoring frequencies. For instance, if a specific product manufacturer provides software patches monthly, an organization might consider conducting vulnerability scans on that product at least that often.

- **Risk Assessment Results** – Results from organizational and/or system-specific assessments of risk (either formal or informal) are examined and taken into consideration when establishing monitoring frequencies. For instance, if a system-specific risk assessment identifies potential threats and vulnerabilities related to nonlocal maintenance (NIST SP 800-53, MA-4), the organization considers more frequent monitoring of the records kept on nonlocal maintenance and diagnostic activities.
- **Reporting Requirements** – Reporting requirements do not drive the ISCM strategy but may play a role in the frequency of monitoring.

The ISCM process Respond step conducts a response to findings with technical, management, and operational mitigating activities or acceptance, transference/sharing, or avoidance/rejection. In accordance with NIST SP 800-137, the respond activity mitigation recommendations are developed using the *security automation domain* concepts presented in Appendix D of NIST SP 800-137. Note that the usage of the term "security automation domain" in the ISCM process is not the same as "automation and control system security" as conventionally used. Appendix D defines a security automation domain as "an information security area that includes a grouping of tools, technologies, and data. Data within the domains is captured, correlated, analyzed, and reported to present the security status of the organization that is represented by the domains monitored." The 11 security automation domains that support continuous monitoring and that are used to categorize the remediation recommendations are illustrated in Figure 8-6.

Appendix D recommends sets of NIST SP 800-53 security controls that are applicable to the security automation domains. These controls are given in Table 8-1.

**Figure 8-6. The Security Automation Domains**
*Source: NIST SP 800-137 (2011)*

**Table 8-1. NIST SP 800-53 Security Controls Applicable to Security Automation Domains**
*Source: NIST SP 800-137 (2011)*

| Security Automation Domain | Controls |
|---|---|
| Vulnerability Management and Patch Management | SI-2, Flaw Remediation; CA-2, Security Assessments; CA-7, Continuous Monitoring; CM-3, Configuration Change Control; IR-4, Incident Handling; IR-5, Incident Monitoring; MA-2, Controlled Maintenance; RA-5, Vulnerability Scanning; SA-11, Developer Security Testing; and SI-11, Error Handling. |
| Event Management and Incident Management | AU-2, Auditable Events; AU-3, Content of Audit Records; AU-4, Audit Storage Capacity; AU-5, Response to Audit Processing Failures; AU-6, Audit Review, Analysis, and Reporting; AU-7, Audit Reduction and Report Generation; AU-8, Time Stamps; AU-12, Audit Generation; CA-2, Security Assessments; CA-7, Continuous Monitoring; IR-5, Incident Monitoring; RA-3, Risk Assessment; and SI-4, Information System Monitoring. |
| Malware Detection | CA-2, Security Assessments; CA-7, Continuous Monitoring; IR-5, Incident Monitoring; RA-3, Risk Assessment; SA-12, Supply Chain Protection; SA-13, Trustworthiness; SI-3, Malicious Code Protection; SI-4, Information System Monitoring; SI-7, Software and Information Integrity; and SI-8, Spam Protection. |

**Table 8-1. NIST SP 800-53 Security Controls Applicable to Security Automation Domains**

*Source: NIST SP 800-137 (2011)*

| | |
|---|---|
| Asset Management | CA-7, Continuous Monitoring; CM-2, Baseline Configuration; CM-3, Configuration Change Control; CM-4, Security Impact Analysis; CM-8, Information System Component Inventory; and SA-10, Developer Configuration Management. |
| Configuration Management | AC-2, Account Management; AC-3, Access Enforcement; AC-5, Separation of Duties; AC-7, Unsuccessful Login Attempts; AC-9, Previous Logon (Access) Notification; AC-10, Concurrent Session Control; AC-11, Session Lock; AC-19, Access Control for Mobile Devices; AC-20, Use of External Information Systems; AC-22, Publicly Accessible Content; CA-2, Security Assessments; CA-7, Continuous Monitoring; CM-2, Baseline Configuration; CM-3, Configuration Change Control; CM-5, Access Restrictions for Change; CM-6, Configuration Settings; CM-7, Least Functionality; IA-2, Identification and Authentication (Organizational Users); IA-3, Device Identification and Authentication; IA-4, Identifier Management; IA-5, Authenticator Management; IA-8, Identification and Authentication (Non-Organizational Users); IR-5, Incident Monitoring; MA-5, Maintenance Personnel; PE-3, Physical Access Control; RA-3, Risk Assessment; SA-7, User Installed Software; SA-10, Developer Configuration Management; and SI-2, Flaw Remediation. |
| Network Management | AC-4, Information Flow Enforcement; AC-17, Remote Access; AC-18, Wireless Access; CA-7, Continuous Monitoring; CM-2, Baseline Configuration; CM-3, Configuration Change Control; CM-4, Security Impact Analysis; CM-6, Configuration Settings; CM-8, Information System Component Inventory; SC-2, Application Partitioning; SC-5, Denial-of-Service Protection; SC-7, Boundary Protection; SC-10, Network Disconnect; SC-32, Information System Partitioning; and SI-4, Information System Monitoring. |
| License Management | CA-7, Continuous Monitoring; CM-8, Information System Component Inventory; and SA-6, Software Usage Restrictions. |
| Information Management | AC-4, Information Flow Enforcement; AC-17, Remote Access; CA-3, Information System Connections; CA-7, Continuous Monitoring; CM-7, Least Functionality; SC-9, Transmission Confidentiality; and SI-12, Information Output Handling and Retention. |
| Software Assurance | CA-7, Continuous Monitoring; SA-4, Acquisitions; SA-8, Security Engineering Principles; SA-11, Developer Security Testing; SA-12, Supply Chain Protection; SA-13, Trustworthiness; SA-14, Critical Information System Components; and SI-13, Predictable Failure Prevention. |

## The Smart Grid Maturity Model (SGMM)

The SGMM2 was initially developed by electric power utilities and is now maintained by the Carnegie Mellon University Software Engineering Institute (SEI). As stated in the SGMM definition document, "The model provides a framework for understanding the current state ofSmart Grid deployment and capability within an electric utility and provides a context for establishing future strategies and work plans as they pertain to Smart Grid implementations." Even though the model is focused on the Smart Grid, its fundamental principles and concepts are applicable and extendable to automation systems.

The SEI provides a suite of SGMM product components to support the model as follows:

1. **Smart Grid Maturity Model Definition** – This refers to the formal description of the model that includes an overview of the model architecture and is the basis for the other product suite elements.
2. **Smart Grid Maturity Model Matrix** – Expected characteristics contained in the model are summarized in a tabular format.
3. **Smart Grid Maturity Model Compass Assessment** – The assessment is a survey that is to be completed by or on behalf of utilities. Compass questions are designed to collect demographic and performance data and to characterize the status of the responding utility's Smart Grid implementation in the context of the SGMM.
4. **Smart Grid Maturity Model Navigator Process** – This refers to a process developed to help organizations chart a technical, organizational, and operational path through their grid modernization effort.
5. **Smart Grid Maturity Model Training** – This consists of two courses that support the transition and use of the SGMM.

The scope of the SGMM is illustrated in Figure 8-7:

The model comprises eight domains with each domain having six defined levels of maturity, ranging from Level 0 (lowest) to Level 5 (highest). The model defines a domain as "a logical grouping of Smart Grid characteristics" with each characteristic comprising "a concise, declarative statement to support the consistent evaluation of its implementation and its independence from other

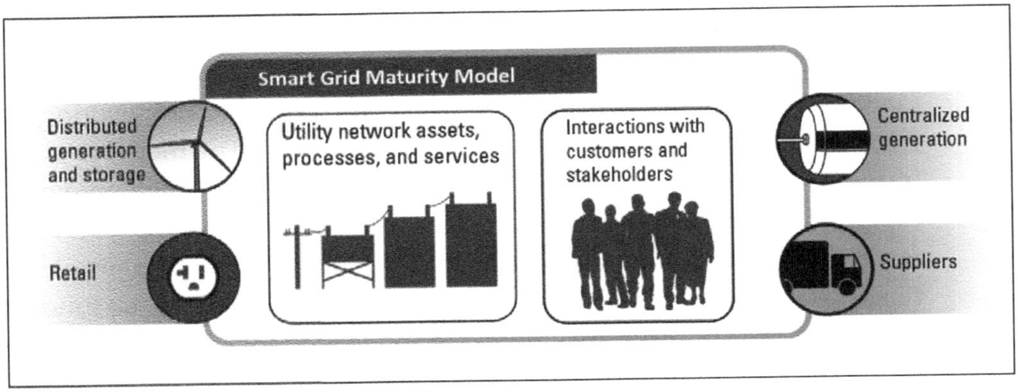


**Figure 8-7. The Scope of the SGMM**
*Source: The Smart Grid Maturity Model (2011)*

characteristics at the same level within the model. The model also requires that for a given level of maturity to be achieved, the expected characteristics for that level and all lower levels in that domain must be sufficiently implemented." The eight domains of the SGMM are listed as follows:

1. **Strategy, Management, and Regulatory (SMR)** – This represents the capabilities and characteristics that enable an organization to successfully develop a Smart Grid vision and strategy; establish internal governance and management processes; and promote collaborative relationships with stakeholders to implement that strategy and vision.
2. **Organization and Structure (OS)** – This represents the organizational capabilities and characteristics that enable an organization to align and operate as required to achieve its desired Smart Grid transformation.
3. **Grid Operations (GO)** – This represents the organizational capabilities and characteristics that support the reliable, secure, safe, and efficient operation of the electrical grid.
4. **Work and Asset Management (WAM)** – This represents the organizational capabilities and characteristics that support the optimal management of assets and workforce resources (i.e., people and equipment) that are central to meeting Smart Grid goals.
5. **Technology (TECH)** – This represents the organizational capabilities and characteristics that enable effective strategic technology

planning for Smart Grid capabilities and the establishment of rigorous engineering and business processes for the evaluation, acquisition, integration, and testing of new Smart Grid technology.

6. **Customer (CUST)** – This represents the organizational capabilities and characteristics that enable customer participation toward achieving the benefits of the Smart Grid transformation.
7. **Value Chain Integration (VCI)** – This represents the organizational capabilities and characteristics that underlie an electric utility's ability to achieve its Smart Grid goals by successfully managing the utility's organizational interdependencies with both the supply chain for the production of electricity and the demand chain for its delivery.
8. **Societal and Environmental (SE)** – This represents the organizational capabilities and characteristics that enable an organization to contribute to achieving societal goals regarding the reliability, safety, and security of our electric power infrastructure; the quantity and sources of the energy we use; and the impact of the infrastructure and our energy use on the environment and our quality of life.

An SGMM Compass assessment provides an organization with a maturity rating for each of the model's eight domains. The model states that "the six levels of maturity represent defined stages of an organization's progress toward achieving its Smart Grid vision in terms of automation, efficiency, reliability, energy and cost savings, integration of alternative energy sources, improved customer interaction, and access to new business opportunities and markets." The maturity levels are hierarchical in that an organization must reach a Level 1 before progressing to the next levels in sequence. The levels of maturity within each domain are described by a group of expected characteristics and a group of optional informative characteristics.

Expected characteristics are defined as "the capabilities and characteristics that an organization must implement or exhibit to achieve the corresponding maturity level within a domain. Each expected characteristic is presented as a concise, declarative statement to support the consistent evaluation of its implementation and its independence from other characteristics at the same level within the model." Informative characteristics give an additional perspective that provides information concerning whether an organization has achieved a given level of maturity within a domain.

The six maturity levels defined in the SGMM, which are stages of an organization's progress toward achieving its Smart Grid vision, are summarized in Figure 8-8.

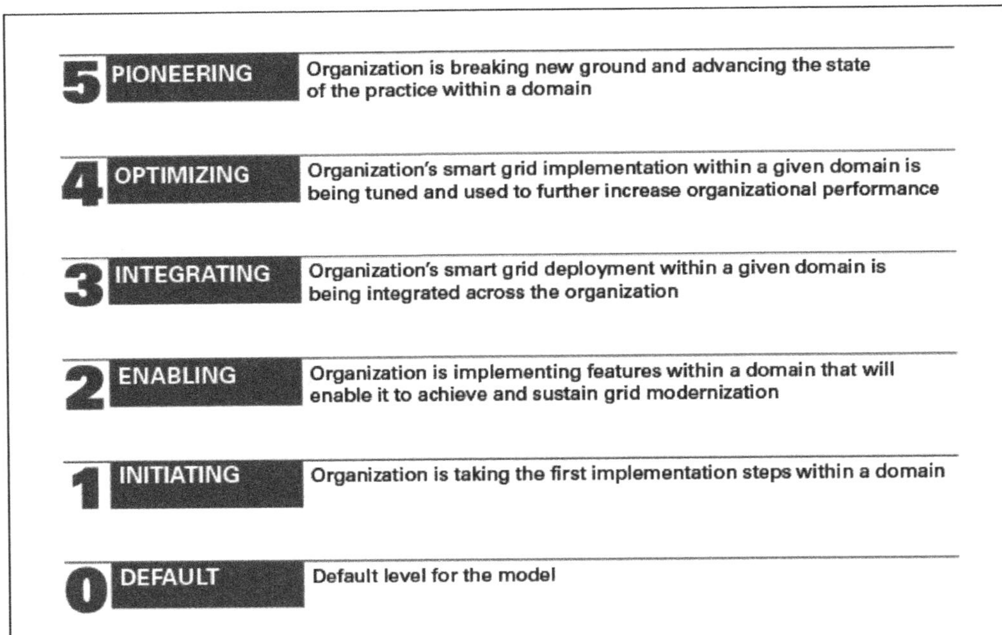


**Figure 8-8. The SGMM Maturity Levels**
*Source: The Smart Grid Maturity Model (2011)*

The characteristics of each maturity level are summarized as follows:

- **Level 0: Default** – The default or entry level of the model across the eight domains of utility. This level precedes an organization taking significant initial steps toward the adoption of Smart Grid technology.
- **Level 1: Initiating** – This level denotes that an organization is taking the first implementation steps toward a Smart Grid within a domain.
- **Level 2: Enabling** – Level 2 focuses on implementing features that will enable an organization to achieve and sustain grid modernization.
- **Level 3: Integrating** – Level 3 Smart Grid deployments are being integrated across the organization.

- **Level 4: Optimizing** – At Level 4, Smart Grid implementations within a given domain are being fine-tuned and used to further increase organizational performance.
- **Level 5: Pioneering** – Organizations at Level 5 are achieving new results with a given domain and advancing the state of the practice.

The maturity rating for each domain in an organization is determined by the SGMM Compass Assessment, which comprises both domain-specific and organizational attribute questions. A typical domain maturity profile determined by an SGMM assessment is shown in Figure 8-9.

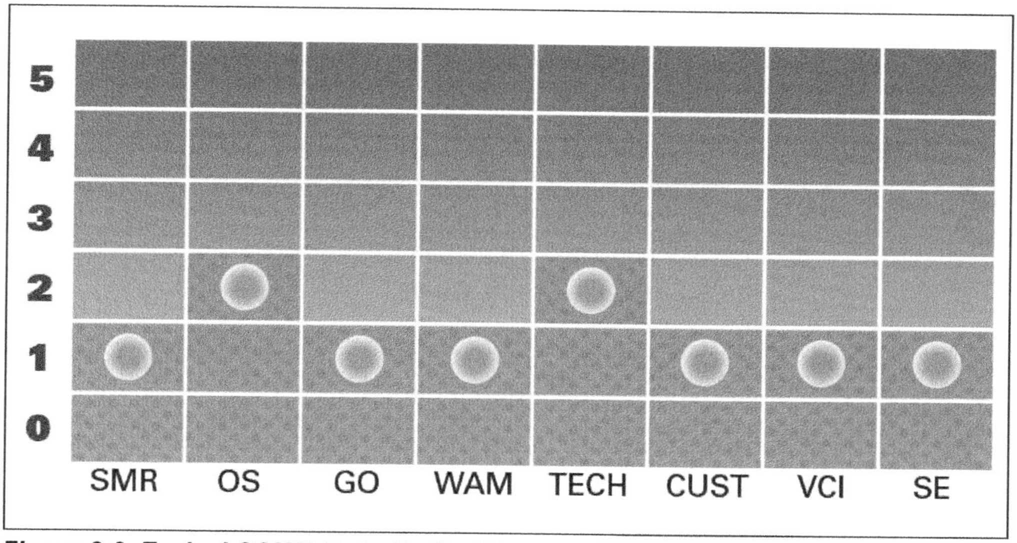


**Figure 8-9. Typical SGMM Maturity Profile**
*Source: The Smart Grid Maturity model (2011)*

The SGMM Matrix document provides a tabular instantiation of the SGMM model for easy reference. It is presented in Table8-2.

**Table 8-2. SGMM Model Matrix**

*Source: The Smart Grid Maturity Model, Model Matrix Document (2011)*

| | **Strategy, Management, and Regulatory (SMR)**<br>***vision, planning, governance, stakeholder collaboration*** | **Organization and Structure (OS)**<br>***culture, structure, training, communications, knowledge management*** |
|---|---|---|
| **PIONEERING 5** | 1 Smart grid strategy capitalizes on smart grid as a foundation for the introduction of new services and product offerings.<br>2 Smart grid business activities provide sufficient financial resources to enable continued investment in smart grid sustainment and expansion.<br>3 New business model opportunities emerge as a result of smart grid capabilities and are implemented. | 1 The organizational structure enables collaboration with other grid stakeholders to optimize overall grid operation and health.<br>2 The organization is able to readily adapt to support new ventures, products, and services that emerge as a result of smart grid.<br>3 Channels are in place to harvest ideas, develop them, and reward those who help shape future advances in process, workforce competencies, and technology. |
| **OPTIMIZING 4** | 1 Smart grid vision and strategy drive the organization's strategy and direction.<br>2 Smart grid is a core competency throughout the organization.<br>3 Smart grid strategy is shared and revised collaboratively with external stakeholders. | 1 Management systems and organizational structure are capable of taking advantage of the increased visibility and control provided by smart grid.<br>2 There is end-to-end grid observability that can be leveraged by internal and external stakeholders.<br>3 Decision making occurs at the closest point of need as a result of an efficient organizational structure and the increased availability of information due to smart grid. |
| **INTEGRATING 3** | 1 The smart grid vision, strategy, and business case are incorporated into the vision and strategy.<br>2 A smart grid governance model is established.<br>3 Smart grid leaders with explicit authority across functions and lines of business are designated to ensure effective implementation of the smart grid strategy.<br>4 Required authorizations for smart grid investments have been secured. | 1 The smart grid vision and strategy are driving organizational change.<br>2 Smart grid measures are incorporated into the measurement system.<br>3 Performance and/or compensation are linked to smart grid success.<br>4 Leadership is consistent in communication and actions regarding smart grid.<br>5 A matrix or overlay structure to support smart grid activities is in place.<br>6 Education and training are aligned to exploit smart grid capabilities. |
| **ENABLING 2** | 1 An initial smart grid strategy and a business plan are approved by management.<br>2 A common smart grid vision is accepted across the organization.<br>3 Operational investment is explicitly aligned to the smart grid strategy.<br>4 Budgets are established specifically for funding the implementation of the smart grid vision.<br>5 There is collaboration with regulators and other stakeholders regarding implementation of the smart grid vision and strategy.<br>6 There is support and funding for conducting proof-of-concept projects to evaluate feasibility and alignment. | 1 A new vision for a smart grid begins to drive change and affect related priorities.<br>2 Most operations have been aligned around end-to-end processes.<br>3 Smart grid implementation and deployment teams include participants from all impacted functions and LOBs.<br>4 Education and training to develop smart grid competencies have been identified and are available.<br>5 The linking of performance and/or compensation plans to achieve smart grid milestones is in progress. |
| **INITIATING 1** | 1 Smart grid vision is developed with a goal of operational improvement.<br>2 Experimental implementations of smart grid concepts are supported.<br>3 Discussions have been held with regulators about the organization's smart grid vision. | 1 The organization has articulated its need to build smart grid competencies in its workforce.<br>2 Leadership has demonstrated a commitment to change the organization in support of achieving smart grid.<br>3 Smart grid awareness efforts to inform the workforce of smart grid activities have been initiated. |
| **DEFAULT 0** | | |

**Table 8-2. SGMM Model Matrix**

*Source: The Smart Grid Maturity Model, Model Matrix Document (2011)*

| | **Grid Operations (GO)** ***reliability, efficiency, security, safety, observability, control*** | **Work and Asset Management (WAM)** ***asset monitoring, tracking and maintenance, mobile workforce*** |
|---|---|---|
| **PIONEERING 5** | 1 Self-healing capabilities are present.<br>2 System-wide, analytics-based, and automated grid decision-making is in place. | 1 The use of assets between and across supply chain participants is optimized with processes defined and executed across the supply chain.<br>2 Assets are leveraged to maximize utilization, including just-in-time asset retirement, based on smart grid data and systems. |
| **OPTIMIZING 4** | 1 Operational data from smart grid deployments is being used to optimize processes across the organization.<br>2 Grid operational management is based on near real-time data.<br>3 Operational forecasts are based on data gathered through smart grid.<br>4 Grid operations information has been made available across functions and LOBs.<br>5 There is automated decision-making within protection schemes that is based on wide-area monitoring. | 1 A complete view of assets based on status, connectivity, and proximity is available to the organization.<br>2 Asset models are based on real performance and monitoring data.<br>3 Performance and usage of assets is optimized across the asset fleet and across asset classes.<br>4 Service life for key grid components is managed through condition-based and predictive maintenance, and is based on real and current asset data. |
| **INTEGRATING 3** | 1 Smart grid information is available across systems and organizational functions.<br>2 Control analytics have been implemented and are used to improve cross-LOB decision-making.<br>3 Grid operations planning is now fact-based using grid data made available by smart grid capabilities.<br>4 Smart meters are important grid management sensors.<br>5 Grid data is used by an organization's security functions.<br>6 There is automated decision-making within protection schemes. | 1 Performance, trend analysis, and event audit data are available for components of the organization's systems.<br>2 CBM programs for key components are in place.<br>3 Remote asset monitoring capabilities are integrated with asset management.<br>4 Integration of remote asset monitoring with mobile workforce systems, in order to automate work order creation, is underway.<br>5 An integrated view of GIS and asset monitoring is in place.<br>6 Asset inventory is being tracked using automation.<br>7 Modeling of asset investments for key components is underway. |
| **ENABLING 2** | 1 Distribution substations are automated and linked to some form of remote distribution automation.<br>2 Advanced outage restoration schemes are being implemented, which resolve or reduce the magnitude of unplanned outages.<br>3 Aside from SCADA, piloting of remote asset monitoring of key grid assets to support manual decision making is underway.<br>4 Investment in and expansion of data communications networks in support of grid operations is underway. | 1 An approach to track, inventory, and maintain event histories of assets is in development.<br>2 An integrated view of GIS for asset monitoring based on location, status, and interconnectivity (nodal) has been developed.<br>3 An organization-wide mobile workforce strategy is in development. |
| **INITIATING 1** | 1 Business cases for new equipment and systems related to smart grid are approved.<br>2 New sensors, switches, and communications technologies are evaluated for grid monitoring and control.<br>3 Proof-of-concept projects and component testing for grid monitoring and control are underway.<br>4 Outage and distribution management systems linked to substation automation are being explored and evaluated.<br>5 Safety and security (physical and cyber) requirements are considered. | 1 Enhancements to work and asset management have been built into approved business cases.<br>2 Potential uses of remote asset monitoring are being evaluated.<br>3 Asset and workforce management equipment and systems are being evaluated for their potential alignment to the smart grid vision. |
| **DEFAULT 0** | | |

**Table 8-2. SGMM Model Matrix**

*Source: The Smart Grid Maturity Model, Model Matrix Document (2011)*

| | **Technology (TECH)** ***IT architecture, standards, infrastructure, integration, tools*** | **Customer (CUST)** ***pricing, customer participation and experience, advanced services*** |
|---|---|---|
| **PIONEERING 5** | 1 Autonomic computing and machine learning are implemented.<br>2 The enterprise Information infrastructure can automatically identify, mitigate, and recover from cyber incidents. | 1 Customers can manage their end-to-end energy supply and usage levels.<br>2 There is automatic outage detection at the premise or device level.<br>3 Plug-and-play, customer-based generation is supported.<br>4 Security and privacy for all customer data is assured.<br>5 The organization plays a leadership role in industry-wide information sharing and standards development efforts for smart grid. |
| **OPTIMIZING 4** | 1 Data flows end-to-end from customer to generation.<br>2 Business processes are optimized by leveraging the enterprise IT architecture.<br>3 Systems have sufficient wide-area situational awareness to enable real-time monitoring and control for complex events.<br>4 Predictive modeling and near real-time simulation are used to optimize support processes.<br>5 Performance is improved through sophisticated systems that are informed by smart grid data.<br>6 Security strategy and tactics continually evolve based on changes in the operational environment and lessons learned. | 1 Support is provided to customers to help analyze and compare usage against all available pricing programs.<br>2 There is outage detection and proactive notification at the circuit level.<br>3 Customers have access to near real-time data on their own usage.<br>4 Residential customers participate in demand response and/or utility-managed remote load control programs.<br>5 Automatic response to pricing signals for devices within the customer's premise is supported.<br>6 In-home net billing programs are enabled.<br>7 A common customer experience has been integrated. |
| **INTEGRATING 3** | 1 Smart grid-impacted business processes are aligned with the enterprise IT architecture across LOBs.<br>2 Systems adhere to an enterprise IT architectural framework for smart grid.<br>3 Smart grid-specific technology has been implemented to improve cross-LOB performance.<br>4 The use of advanced distributed intelligence and analytical capabilities are enabled through smart grid technology.<br>5 The organization has an advanced sensor plan.<br>6 A detailed data communication strategy and corresponding tactics that cross functions and LOBs are in place. | 1 The organization tailors programs to customer segments.<br>2 Two-way meter communication has been deployed.<br>3 A remote connect/disconnect capability is deployed.<br>4 Demand response and/or remote load control is available to residential customers.<br>5 There is automatic outage detection at the substation level.<br>6 Residential customers have on-demand access to daily usage data.<br>7 A common experience has been implemented across two or more customer interface channels.<br>8 Customer education on how to use smart grid services to curtail peak usage is provided.<br>9 All customer products and services have built-in standards based on security and privacy controls. |
| **ENABLING 2** | 1 Tactical IT investments are aligned to an enterprise IT architecture within an LOB.<br>2 Changes to the enterprise IT architecture that enable smart grid are being deployed.<br>3 Standards are selected to support the smart grid strategy within the enterprise IT architecture.<br>4 A common technology evaluation and selection process is applied for all smart grid activities.<br>5 There is a data communications strategy for the grid.<br>6 Pilots based on connectivity to distributed IEDs are underway.<br>7 Security is built into all smart grid initiatives from the outset. | 1 Pilots of remote AMI/AMR are being conducted or have been deployed.<br>2 The organization has frequent (more than monthly) knowledge of residential customer usage.<br>3 The organization is modeling the reliability of grid equipment.<br>4 Remote connect/disconnect is being piloted for residential customers.<br>5 The impact on the customer of new services and delivery processes is being assessed.<br>6 Security and privacy requirements for customer protection are specified for smart grid-related pilot projects and RFPs. |
| **INITIATING 1** | 1 An enterprise IT architecture exists or is under development.<br>2 Existing or proposed IT architectures have been evaluated for quality attributes that support smart grid applications.<br>3 A change control process is used for applications and IT infrastructure.<br>4 Opportunities are identified to use technology to improve departmental performance.<br>5 There is a process to evaluate and select technologies in alignment with smart grid vision and strategies. | 1 Research is being conducted on how to use smart grid technologies to enhance the customer's experience, benefits, and participation.<br>2 Security and privacy implications of smart grid are being investigated.<br>3 A vision of the future grid is being communicated to customers.<br>4 The utility consults with public utility commissions and/or other government organizations concerning the impact on customers. |
| **DEFAULT 0** | | |

**Table 8-2. SGMM Model Matrix**

*Source: The Smart Grid Maturity Model, Model Matrix Document (2011)*

| | **Value Chain Integration (VCI)** ***demand and supply management leveraging market opportunities*** | **Societal and Environmental (SE)** ***responsibility, sustainability, critical infrastructure, efficiency*** |
|---|---|---|
| **PIONEERING 5** | 1 The optimization of energy assets is automated across the full value chain.<br>2 Resources are adequately dispatchable and controllable so that the organization can take advantage of granular market options.<br>3 Automated control and resource optimization schemes consider and support regional and/or national grid optimization. | 1 Triple bottom line goals align with local, regional, and national objectives.<br>2 Customers control their energy-based environmental footprints through automatic optimization of their end-to-end energy supply and usage level (energy source and mix).<br>3 The organization is a leader in developing and promoting industry-wide resilience best practices and/or technologies for protection of the national critical infrastructure. |
| **OPTIMIZING 4** | 1 Energy resources (including Volt/VAR, DG, and DR) are dispatchable and tradable.<br>2 Portfolio optimization models that encompass available resources and real-time markets are implemented.<br>3 Secure two-way communications with Home Area Networks (HANs) are available.<br>4 Visibility and potential control of customer's large-demand appliances to balance demand and supply is available. | 1 The organization collaborates with external stakeholders to address environmental and societal issues.<br>2 A public environmental and societal scorecard is maintained.<br>3 Programs are in place to shave peak demand.<br>4 End-user energy usage and devices are actively managed through the utility's network.<br>5 The organization fulfills its critical infrastructure assurance goals for resilience, and contributes to those of the region and the nation. |
| **INTEGRATING 3** | 1 An integrated resource plan is in place and includes new targeted resources and technologies.<br>2 Customer premise energy management solutions with market and usage information are enabled.<br>3 Additional resources are available and deployed to provide substitutes for market products to support reliability or other objectives.<br>4 Security management and monitoring processes are deployed to protect the interactions with an expanded portfolio of value chain partners. | 1 Performance of societal and environmental programs are measured and effectiveness is demonstrated.<br>2 Segmented and tailored information that includes environmental and societal benefits and costs is available to customers.<br>3 Programs to encourage off-peak usage by customers are in place.<br>4 The organization regularly reports on the societal and environmental impacts of its smart grid programs and technologies. |
| **ENABLING 2** | 1 Support is provided for energy management systems for residential customers.<br>2 The value chain has been redefined based on its smart grid capabilities.<br>3 Pilots to support a diverse resource portfolio have been conducted.<br>4 Secure interactions have been piloted with an expanded portfolio of value chain partners. | 1 Smart-grid strategies and work plans address societal and environmental issues.<br>2 Energy efficiency programs for customers have been established.<br>3 The organization considers a "triple bottom line" view when making decisions.<br>4 Environmental proof-of-concept projects are underway that demonstrate smart grid benefits.<br>5 Increasingly granular and more frequent consumption information is available to customers. |
| **INITIATING 1** | 1 Assets and programs necessary to facilitate load management are identified.<br>2 Distributed generation sources and the capabilities needed to support them are identified.<br>3 Energy storage options and the capabilities needed to support them are identified.<br>4 There is a strategy for creating and managing a diverse resource portfolio.<br>5 Security requirements to enable interaction with an expanded portfolio of value chain partners have been identified. | 1 The smart grid strategy addresses the organization's role in societal and environmental issues.<br>2 The environmental benefits of the smart grid vision and strategy are publicly promoted.<br>3 Environmental compliance performance records are available for public inspection.<br>4 The smart grid vision or strategy specifies the organization's role in protecting the nation's critical infrastructure. |
| **DEFAULT 0** | | |

## Automation Maturity Model

In a publication[6] by the California-Nevada Section of the American Water Works Association (AWWA), Philip Gaberdiel of Westin Engineering proposed an Automation Maturity Model™ (AMM) to measure the capability of an organization's industrial automation and control system. The capability also implies an increasing level of automation system security. The model defines the following six maturity levels of an automation system:

- **Level 0** – Local hardwired control
- **Level 1** – Local digital control
- **Level 2** – Remote manual control
- **Level 3** – Remote auto control
- **Level 4** – Enhanced operations
- **Level 5** – Optimized operations

The Automation Maturity Model is illustrated in Figure 8-10.

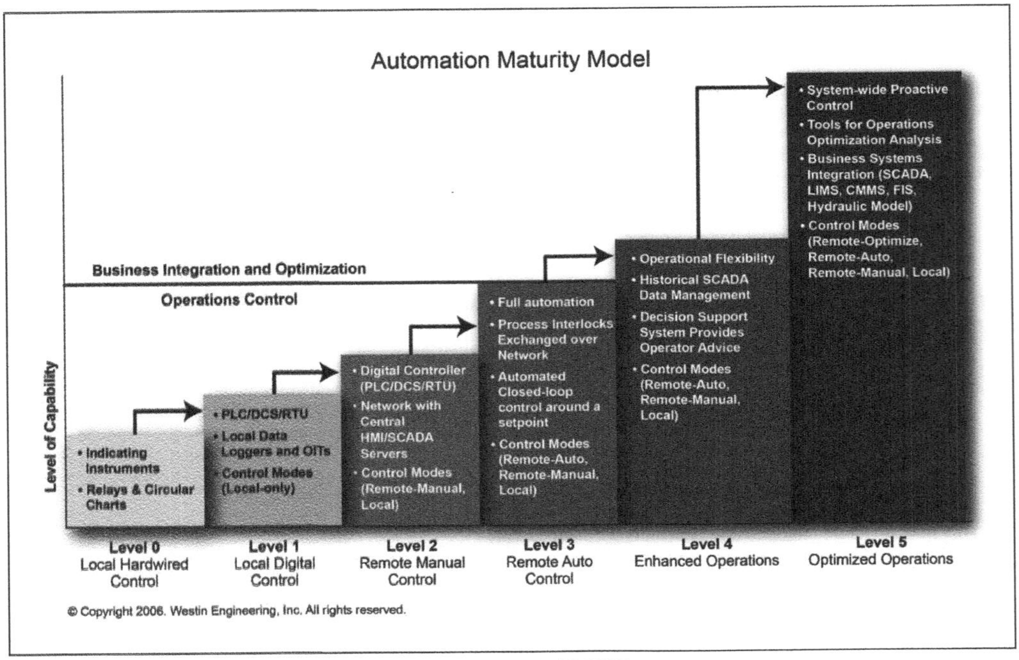


**Figure 8-10. The Automation Maturity Model (AMM)**
*Source: Phil Gaberdiel, Westin Engineering*

In the model, an automation system at Level 0 is based on relays and instruments read locally, comprising hardware control. Level 1 describes an automation system with local digital control, using PLCs, RTUs, and data loggers, while Level 2 comprises digital controllers and networked HMI/SCADA servers providing remote manual control. A Level 3 remote auto control system includes full automation with automated closed-loop controls. Level 4, enhanced operations, has characteristics of operational flexibility, an application of decision support systems, and historical SCADA data management. Level 5 achieves optimized operation through the use of operations optimization analysis tools, business systems integration, and systems-wide, proactive control.

## Future Smart Grid Issues and Automation Security Issues

Because of the increased number of computational components and sensors that are associated with manufacturing systems, processing plants, and the Smart Grid, the electromagnetic environment will be of increased concern in the future. Even though the following material provides examples relating to the Smart Grid, the issues and countermeasures apply equally to automation systems.

In addition, government/industry initiatives such as NIST 7628 are emerging to address future specific Smart Grid cybersecurity and privacy issues. These two areas are explored in the followingsections.

### Smart Grid Electromagnetic Radiation Issues

As discussed in Chapter 5, electromagnetic fields have the potential of causing serious interruptions and malfunctions not only in substations, but in "green" energy sources and production facilities. The increased number of transmitting stations will create interference that can affect sensitive instrumentation and home-based devices.

In a paper titled *High Power Electromagnetic (HPEM) Threats to the Smart Grid,*[3] William Radasky describes ongoing efforts to understand and mitigate the effects of electromagnetic radiation on future instantiations of the Smart Grid. Toward this end, an Electromagnetic Interoperability Issues Working Group (EMIIWG) was formed in 2010. The EMIIWG is a subgroup of the Smart Grid Interoperability Panel (SGIP), which was established to investigate the interoperability among interconnected components of the Smart Grid. The paper describes the scope of the EMIIWG and addresses the following issues:

- Enhancing the immunity of Smart Grid devices and systems to the detrimental effects of natural and man-made electromagnetic interference, both radiated and conducted
- Developing recommendations for the application of standards and testing criteria to ensure electromagnetic compatibility (EMC) for the Smart Grid, with a particular focus on issues directly related to interoperability of Smart Grid devices and systems
- Determining impacts, avoidance, generation and mitigation of and immunity to electromagnetic interference

Specific tasks of the EMIIWG associated with these areas of concern include:

1. Review potential electromagnetic issues and the existing state of EMC of the power grid and associated systems, including current and proposed Smart Grid enhancements.
2. Segment the Smart Grid devices, systems, and electromagnetic environments into a minimal set of categories for which electromagnetic issues and EMC requirements can be identified. These categories should be compatible with the environment classifications of IEC 61000-2-5,[4] where possible.
3. Prioritize or rank the categories in task 2 according to the potential impact on Smart Grid reliability. The priorities should consider the extent and severity of possible failures and the availability requirements for the relevant interface as defined in NISTIR 7628[5] (see discussion of NIST 7628 in the next section).
4. Identify and/or propose EMC terminology and definitions applicable to the Smart Grid and compatible with international standards.
5. Identify and compile the source-victim matrix for each category identified in task 2.
6. Identify or develop EMC performance metrics for systems in each category identified in task 2.
7. Identify appropriate EMC standards and requirements to meet performance metrics.

8. Identify areas where EMC standards are not available and identify appropriate standards developing organizations (SDOs) where such standards should be developed.

9. Identify and propose priority action plans to address standards or guidelines in high-priority categories if needed.

10. Propose strategic recommendations for EMC of Smart Grid systems, beginning with the highest priority categories. These recommendations should reflect a long-term strategy to maintain EMC as the Smart Grid evolves.

11. Consider the need and, if appropriate, the nature of a conformity assessment program for EMC for coordination with the SGIP Smart Grid Testing and Certification Committee.

Task 2 in the list above is critical because it is basic to the determination of the appropriate EM environment for Smart Grid equipment, which is supported by IEC 61000-2-5. The paper *High Power Electromagnetic (HPEM) Threats to the Smart Grid* describes the process steps of IEC 61000-2-5 as the following:

1. Define the location classes with regard to the types of exposures (both conducted and radiated) and the distances expected from particular types of emitters.

2. Compile a comprehensive list of radiated and conducted phenomena and disturbance levels along with formulas to compute field levels where appropriate (include formulas for near-field exposure).

3. Combine the results to develop recommendations for disturbance levels for a given "location" for all applicable phenomena.

Related and important guidance documents that will affect the future characteristics of the Smart Grid as well as industrial control systems are NIST Interagency or Internal Reports (NISTIRs), particularly those of NISTIR 7628, cited earlier in this chapter.

## NIST 7628

NIST 7628, *Introduction to NISTIR 7628 Guidelines for Smart Grid Cybersecurity*, is a three-volume document that provides a methodical approach to implementing customized cybersecurity and privacy protection strategies for Smart Grid implementations. It was developed by the Cybersecurity Working Group of the Smart Grid Interoperability Panel and is designed to apply to a variety of Smart Grid–related domains, such as electric vehicle charging stations, utilities, manufacturers, and providers of energy management services. Its approach is also directly applicable to process plants, manufacturing operations, and automation systems in general. It brings together techniques from information system security, information technology, control systems, telecommunications, and manufacturing.

Volume 1 of NIST 7628 addresses the following topics:

- Risk assessment processes to identify high-level security requirements
- High-level architectures
- Logical interface architectures used to identify and define categories of interfaces within and across the Smart Grid domains
- High-level security requirements for the logical interface categories
- Technical cryptographic and key management issues across the scope of Smart Grid systems and devices

Volume II deals with privacy issues in homes and electric vehicles, including:

- Evolving Smart Grid technologies and their relationships with individuals
- Behaviors of groups of individuals and individuals in their dwellings
- Behaviors of groups of individuals and individuals in electric vehicles
- Privacy issues affecting individuals in their premises and electric vehicles
- Recommended privacy practices for participants within the Smart Grid
- Recommendations to consumers involving Smart Grid privacy issues and associated mitigation approaches

Volume III of NIST 7628 is an examination of Smart Grid high-level security requirements, such as:

- Vulnerability categories

- Research and development topics focused on achieving increased security and resilience characteristics for the Smart Grid
- Description of the Cybersecurity Working Group (CSWG) effort to support the evaluation of proposed standards relative to the security needs developed in NIST 7628

## Summary

The future environment of industrial automation and control systems will be affected by advances in a number of related and seemingly unrelated fields. Breakthroughs in physics, semiconductors, software engineering, artificial intelligence, communications, information technology, and so on will translate into opportunities and challenges in automation and control systems. No longer can technical personnel rely on narrow control disciplines and ignore the wider technological community if they want to take advantage of progress in those arenas.

In this book, a number of proven, effective techniques have been presented and explored to provide management and technical personnel associated with industrial automation and control systems with the background and tools needed to significantly affect the security of our critical automation systems. These methodologies must be embraced and coupled with the latest technological knowledge in order to maximize the security and efficiency of tomorrow's automation and control systems.

## Review Questions for Chapter 8

8.1 The technology that eliminates the binding of software to specific hardware platforms is known as which of the following?

A. Artificial intelligence

B. Virtualization

C. Cloud computing

D. Software as a service

8.2 A hypervisor is which of the following?

A. A virtual machine monitor

B. An advanced HMI

C. A manager

D. An alarm mechanism

8.3 NIST SP 800-137, *Information Security Continuous Monitoring (ISCM) for Federal Information Systems and Organizations*, defines the process for developing an ISCM strategy and implementing an ISCM program. Which one of the following best describes the ISCM strategy?

A. Providing security mechanisms to reduce risks to automation systems

B. Developing security policies to specify management intent for information systems and automation systems

C. Maintaining ongoing awareness of information security, vulnerabilities, and threats to support organizational risk management decisions

D. Maintaining ongoing awareness of risk mitigation alternatives

8.4 Which of the following is NOT one of the ISCM program steps?

A. Define

B. Establish

C. Respond

D. Mitigate

8.5 In the ISCM process, the status of automation security in a facility can be determined by establishing appropriate metrics. Which of the following activities is NOT used to define the metrics?

A. Evaluating applicable mitigation approaches

B. Assessing all security controls

C. Providing actionable communication of security status across all tiers of the organization

D. Collecting, correlating, and analyzing security-related information

8.6 NIST SP 800-137 defines the process for managing information security and information system–related risk as which of the following?

A. Risk Management Framework (RMF)

B. Risk-Impact Criteria (RIC)

C. Vulnerability Analysis (VA)

D. Risk-Vulnerability Matrix (RVM)

8.7 Security control volatility, system categorizations/impact levels, organizational risk tolerance, and security controls with identified weaknesses are associated with what characteristic discussed in NIST SP 800-137?

A. The necessity of adding compensating controls

B. The risk impact levels

C. The level of risk

D. The frequency of monitoring security controls for effectiveness

8.8 Appendix D of NIST SP 800-137 defines a term as "an information security area that includes a grouping of tools, technologies, and data." What is this term?

A. Security automation domain

B. Compensating domain

C. Risk management domain

D. Risk mitigation domain

8.9 The SEI-maintained Smart Grid Maturity Model (SGMM) is described best by which of the following statements?

A. Provides a framework for understanding the current state of Smart Grid deployment and capability within an electric utility and provides a context for establishing future strategies and work plans as they pertain to Smart Grid implementations

B. Provides a management perspective of Smart Grid deployment and capability within an electric utility and develops guidelines for Smart Grid implementations

C. Provides a targeted hierarchical methodology applicable to Smart Grid evaluation and maintenance within an electric utility and provides a basis for implementing effective Smart Grid controls

D. Provides a Smart Grid assurance model and algorithms for future deployment and implementation of Smart Grid systems

8.10 Which of the following is NOT one of the eight domains of the SGMM?

A. Strategy, Management, and Regulatory (SMR)

B. Organization and Structure (OS)

C. Metering and Monitoring (MAM)

D. Work and Asset Management (WAM)

8.11 Which of the following is NOT one of the six maturity levels defined in the SGMM?

A. Default

B. Evaluating

C. Initiating

D. Integrating

8.12 The SGMM maturity rating for each domain in an organization is determined by which of the following means?

A. The SGMM Domain Survey

B. The SGMM Risk Survey

C. The SGMM Level Survey

D. The SGMM Compass Survey

8.13 NIST 7628, *Introduction to NISTIR 7628 Guidelines for Smart Grid Cybersecurity,* is a three-volume document that provides a methodical approach to implementing customized cybersecurity and privacy protection strategies for Smart Grid implementations. Which of the following areas is NOT addressed in the volumes comprising NISTIR 7628?

A. Risk assessment processes

B. Privacy issues

C. Cost evaluation matrices

D. Research and development topics

8.14 Which of the following describes the optimizing level (Level 4) of the SGMM?

A. Focuses on implementing features that will enable an organization to achieve and sustain grid modernization

B. Smart Grid implementations within a given domain are being fine-tuned and used to further increase organizational performance

C. Smart Grid deployments are being integrated across the organization

D. Organizations are achieving new results with a given domain and advancing the state of the practice

8.15 What entity in the SGMM is defined as "the stages of an organization's progress toward achieving its Smart Grid vision in terms of automation, efficiency, reliability, energy and cost savings, integration of alternative energy sources, improved customer interaction, and access to new business opportunities and markets"?

A. Maturity

B. Capability

C. Characteristics

D. Value

8.16 In the ISCM process, the status of automation security in a facility can be determined by establishing appropriate metrics. The metrics are defined through the following activities, except one, which is not applicable. Which of the following is NOT applicable?

A. Maintaining an understanding of threats and threat activities

B. Conducting security awareness training

C. Collecting, correlating, and analyzing security-related information

D. Active management of risk by organizational officials

8.17 What technology reduces the effort required to maintain hardware and software systems by eliminating the binding of software to specific hardware platforms through the logical partitioning of physical computing resources into multiple execution environments, including servers, applications, and operating systems?

A. Virtualization

B. Cloud computing

C. Distributed computing

D. Multi-core chips

8.18 Which of the following statements is FALSE regarding penetration testing of industrial automation and control systems?

A. If not properly conducted, penetration testing can cause damage to automation systems and disrupt operations.

B. Penetration testing tools do not create additional traffic on automation networks and do not affect systems with limited memory capacity.

C. If executed properly, penetration testing can provide important information regarding the security of automation systems.

D. It is useful to use nonintrusive methods when conducting penetration testing.

8.19 Penetration testing conducted with no prior knowledge of the automation system is known as which of the following?

A. Black box

B. White box

C. Grey box

D. Auto box

8.20 Which of the following is NOT a component of the reconnaissance phase of penetration testing?

A. Footprinting

B. Scanning

C. Controlling

D. Enumerating

8.21 Which of the following is NOT one of the maturity levels of the Automation Maturity Model?

A. Local digital control

B. Remote auto control

C. Optimized operations

D. Enabled control

8.22 What level of the Automation Maturity Model is characterized by full automation and automated closed-loop control around a set point?

A. Local digital control

B. Enhanced operations

C. Remote auto control

D. Optimized operations

## References

1. NIST SP 800-137. *Information Security Continuous Monitoring (ISCM) for Federal Information Systems and Organizations*. Washington, DC: NIST (National Institute of Standards and Technology), 2011.

2. Smart Grid Maturity Model Team. *Smart Grid Maturity Model, Model Definition: A Framework for Smart Grid Transformation*. Technical Report CMU/SEI-2011-TR-025. Version 1.2. Carnegie Mellon. September 2011.

3. Radasky, W. "EMC and the Smart Grid." *Interference Technology, 2011 EMC Test and Design Guide*, 2011.

4. IEC/TS 61000-2-5. *Electromagnetic Compatibility (EMC) - Part 2: Environment - Section 5: Classification of Electromagnetic Environments*. Geneva 20 – Switzerland: IEC (International Electrotechnical Commission), 2011.

5. The Smart Grid Interoperability Panel Cybersecurity Working Group. *Introduction to NISTIR 7628, Guidelines for Smart Grid Cybersecurity*. Technical paper. September 2010.

6. Gaberdiel, P. "Taking Your SCADA System to the Next Level." *SOURCE* 25, no. 3 (2011): CA-NV Section, AWWA.

# 9

# Emerging Approaches to Industrial Automation and Control System Security

*Chapter 5 imparted important risk management methodologies and their application to industrial automation and control systems and Chapter 6 expanded these results to develop the foundation of security controls that are pertinent to these systems. Chapter 6 detailed the most effective controls, some specifically designed to secure industrial automation and control systems and some that could be adapted to such systems. A result of this approach is the consolidation of best practices to provide a single reference source as a baseline for securing industrial automation and control systems.*

*Chapter 9 builds on the knowledge presented in the previous chapters to introduce new and emerging technologies and approaches related to securing industrial automation and control systems that will provide practitioners in the field with the latest applicable methodologies. With this advanced information, critical industries can be aware of the evolving procedures and plan for the evaluation and eventual implementation of the most promising practices.*

## Internet of Things

The Internet of Things (IoT) refers to objects such as sensors, humans, animals, machines, or other entities that each have a unique IP address and can transmit and receive data over a network without human intervention. One of the popular instantiations of IoT has been with machine-to-machine (M2M) communication involving "smart" devices in industrial and process settings. In an article in the automation magazine *InTech*, author Bill Lydon states, "The IoT vision is of a massively instrumented world of intelligent sensors (analog and digital) and actuators (analog and digital) communicating using Internet Protocol (IP) to improve performance and efficiency...The goal is frictionless communications and interaction from manufacturing field input/output (I/O), including sensors, actuators, analyzers, drives, vision, video, and robotics, for increased manufacturing performance and flexibility. This revolution will drive intelligence to the edge of the system with the ultimate goal of all industrial devices supporting IP, including field I/O."[1]

The data from the IoT objects will, in most cases, be handled by third-party applications running in a data center and will require significant data processing and storage upgrades. Objects on the IoT will create terabytes of data that have to be processed by an enterprise or cloud broker and distilled into a form that can be used as the basis for decision-making. The adoption of M2M communication will increase requirements for event-based processing and real-time platforms.

The proliferation of IoT devices has been made possible by the large number of IP addresses provided by IPv6. Specifically, IPv6 has an address space of $2^{128}$ or 340,282,366,920,938,463,463,374,607,431,768,211,456 addresses. This number of addresses supports assigning a unique address to all objects on Earth.

Threat vectors in the M2M arena are varied and are present in a variety of domains, including health care, transportation, personal mobile devices, and industrial automation.[2] In particular, typical attack targets include:

- Smart meters
- Home automation
- Plant automation
- SCADA systems
- Wearable smart devices such as smart watches
- Communication systems
- Telemedicine

Information sent from "things" and processed with specialized software can develop predictive models relative to automation systems. This trend will be accelerated by the availability of low-cost sensors that can be integrated into a variety of products and transmit information on all aspects of processes, production systems, critical infrastructure components, as well as individuals' behaviors, health, and personal habits.

## Open Platform Communications Unified Architecture

An important entity that supports the IoT paradigm is the Open Platform Communications Unified Architecture (OPC UA) (IEC 62541). The IoT coupled with OPC UA provides the capability to manage large amounts of sensor data in applications to improve operations and efficiency.

OPC UA was developed and is maintained by the OPC Foundation (https://opcfoundation.org/). The OPC Foundation characterizes OPC UA as "the interoperability standard for secure and reliable information exchange in industrial automation and the enterprise" and as a specification that defines "an interface between clients and servers, and servers-to-servers, so system components like PLCs, HMIs, and any OPC-aware device can share data without having to develop custom software device interface applications."

OPC UA was preceded by the OPC Classic specifications, which were based on Microsoft Windows object linking and embedding (OLE) technology for process control using the Component Object Model/Distributed Component Object Model (COM/DCOM) for the exchange of data between software components. These specifications define the interface between clients and servers, including access to real-time data, monitoring of alarms and events, and access to historical data and other applications. Responding to the new requirements of service-oriented architectures (SOA) in automation systems, the OPC Foundation developed the OPC UA specifications that were scalable to address new architectures. OPC UA is a service-oriented architecture that has the following salient characteristics:

- **Comprehensive Information Modeling** – Using complete object-oriented capabilities, modeling can be used to evaluate complex multilevel structures and information stores. In addition to compatibility with OPC Classic models, OPC UA incorporates data models from a number of other standards groups, including the following:
  - Field Device Integration (FDI)
  - The Standards Leadership Council (SLC)

- The Drilling System Automation Technical Section (DSATS)
- Production Markup Language (PRODML)
- MCS-DCS Interface Standardization (MDIS)
- ISA-95 Common Object Model ISA95 (ANSI/ISA-95/IEC 62264)

- **Extensible** – Systems have the ability to add new features without affecting existing applications.
- **Functional Equivalence** – COM OPC Classic specifications DA (Data Access), Alarms and Events (A&E), OPC Historical Data Access (HDA), commands, complex data, and object types are mapped to OPC UA specifications.
- **Platform Independence** – OPC UA is designed to be independent of platforms ranging from local controllers to the cloud and including portable ANSI C, Java, and .NET implementations.
- **Secure** – SOA authentication, encryption, and auditing are supported in OPC UA.
- **Reliability** – OPC UA provides for communication failure recovery, redundancy among applications, configurable timeouts, and error detection.
- **Protocols** – OPC UA supports two protocols: the binary protocol, opc.tcp://Server, and the web service protocol, http://Server.

## Industry 4.0

In manufacturing, the IoT is generating a fourth Industrial Revolution embodied in the concept known as *Industry 4.0*.[3] The first stage of the Industrial Revolution began in the 1800s with mechanical production powered by water and steam power. The second stage incorporated mass production methods using electrical energy, as applied by Henry Ford in the beginning of the 1900s. With the advent of electronics and IT in the 1970s, Industry 3.0 was defined. Industry 4.0 is characterized by the interconnection of highly intelligent cyber-physical systems (CPS) to create the IoT. In this fourth stage, organizations will employ global networks that incorporate production facilities, smart machines, and storage systems and that can autonomously communicate and initiate control functions in an independent manner.

This capability can also be referred to as *informed manufacturing*,[4] which is defined as:

> An envisioned state of operations in which all relevant and synthesized information is made available when, where, and in the form in which it is needed across the manufacturing supply chain to all stakeholders (people, process, products and infrastructure).

In this paradigm, informed manufacturing sees the following four essential elements of manufacturing converging to create an advanced production environment:[5]

- **Products** – Advanced sensors, controls, and software applications working together to obtain and share real time as finished goods are produced. "Informed products" will enable machines to take autonomous actions.
- **People** – By connecting people across all business functions and geographies, and providing them with relevant information in real time, "informed people" will provide intelligent design, operations, and maintenance, as well as higher quality service and safety.
- **Processes** – By emphasizing bidirectional information-sharing across the global manufacturing value chain—from supplier to customer—"informed processes" will lead to a flexible and adaptable supply chain.
- **Infrastructure** – Using smart infrastructure components that interface with mobile devices, products, and people, "informed infrastructure" will better manage complexities and enable more efficient manufacturing.

## Security and Privacy

The communication among things brings up the obvious issues of data security and privacy, especially when dealing with objects that historically were not connected to a network and did not have intrinsic intelligence. Because equipment, personnel, machines, and a wide variety of devices will now incorporate numerous sensors that transmit critical information, there is a potential for data interception and modification. In many instances, existing installations will require extensive alterations to be part of the IoT. This situation is particularly significant where devices make determinations and execute corresponding actions automatically without the involvement of humans. This environment brings to mind personnel safety, the use and misuse of analytics in deriving secondary data, and liability issues.

Another not commonly discussed issue is the effect of countless devices transmitting information on the RF spectrum. As described in an *IEEE Spectrum* article,[6] "The coming Internet of Things is going to make things worse, much worse. It will do so by adding complex RF-control chips to countless common devices (like door locks, light switches, appliances of every type, our cars, and maybe even our bodies), which will enable them to connect to the Internet. Each of these chips is a potential source of noise." Regarding the power grid, the article expresses further concerns, in that "Other sources, such as the power grid, are expanding as wind farms and solar households connect to it. Such devices need to switch large amounts of DC power at a 60-hertz or an even faster rate whenever they feed excess generated power back into the grid. If not done properly, this could also feed large amounts of noise into the power grid. This risk is magnified when the solar and wind systems operate without expert maintenance inside millions of ordinary homes."

## OWASP IoT Security Categories

The IoT has a number of attack surfaces that must be considered when security is involved. For example, the Open Web Application Security Project (OWASP) has developed its "Internet of Things Top Ten"[7] security categories. The areas identified in their report are:

1. **Insecure Web Interface**
   Internal and external users who can access IoT web interfaces that have weak credential requirements—situations where credentials can be exposed in network traffic or do not have account lockout can be sources of attacks that can result in data corruption, account takeover, denial of service, and so on.

2. **Insufficient Authentication/Authorization**
   Successful attacks can be mounted by trying weak passwords, exploiting insecure password recovery mechanisms, accessing weakly-protected credentials, or taking advantage of the lack of granular access control. Attacks based on insufficient authentication/authorization can produce data loss or corruption, lack of accountability, or denial of access.

3. **Insecure Network Services**
   Insecure network services are susceptible to buffer overflow attacks or attacks that create a denial-of-service condition. They also can be the basis of denial-of-service attacks against other users. Vulnerabilities in open ports and having unnecessary open ports are examples of insecure network services.

4. **Lack of Transport Encryption**
   Lack of transport encryption allows data to be viewed as it travels over local networks or the Internet. It is prevalent on local networks, particularly on wireless networks where misconfiguration of the network can make traffic visible to anyone within range of that wireless network. Many issues with transport encryption are easy to discover by monitoring network traffic and searching for readable data. Automated tools can also look for proper implementation of common transport encryption, such as SSL and TLS.

5. **Privacy Concerns**
   Privacy concerns result from the collection of personal data and the lack of proper protection of that data. Privacy issues can be determined by reviewing the data being collected as the user sets up and activates a device. Automated tools can also look at specific patterns of data that may indicate collection of personal data or other sensitive data. Attackers might use multiple vectors such as insufficient authentication, lack of transport encryption, or insecure network services to view personal and sensitive data.

6. **Insecure Cloud Interface**
   An insecure cloud interface is present when it is possible to guess credentials that are used or where enumeration of an account is possible. These interfaces can be discovered by reviewing the connection to the interface and identifying if SSL is in use or by using the password reset mechanism to identify valid accounts that can be the targets of enumeration.

7. **Insecure Mobile Interface**
   An insecure mobile interface exists when it is easy to determine credentials that are used or where account enumeration is possible. Insecure mobile interfaces can be detected by reviewing the connection to the wireless networks and identifying if SSL is in use or

by using the password reset mechanism to identify valid accounts that can be the targets of enumeration.

8. **Insufficient Security Configurability**
Insufficient security configurability is present when users of a device have limited or no ability to alter security controls. Insufficient security configurability is apparent when the web interface of the device does not have an option for creating granular user permissions (e.g., forcing the use of strong passwords). Manual review of the web interface and its available options will reveal these deficiencies.

9. **Insecure Software/Firmware**
The lack of the ability of a device to be updated presents a security weakness on its own. Devices should have the ability to be updated when vulnerabilities are discovered. Software/firmware updates can be insecure when the updated files themselves and the network connection they are delivered on are not protected. Software/firmware updates can also be insecure if they contain hardcoded sensitive data such as credentials. Security issues with software/firmware can be discovered by inspecting the network traffic during the update to check for encryption or using a hex editor to inspect the update file.

10. **Poor Physical Security**
Physical security weaknesses are present when an attacker can disassemble a device to easily access the storage medium and any data stored on that medium. Weaknesses are also present when USB or other external ports can be used to access the device using features intended for configuration or maintenance.

In order to address these and other security issues, it is generally agreed that organizations developing IoT products should implement strong security in these products. The Federal Trade Commission (FTC) created a report[8] that outlines its recommendations for best practices that companies should adopt in developing and applying IoT devices. These best practices are summarized as follows:

- Companies should build security into the devices at the outset rather than as an afterthought.
- As part of the security by design process, companies should:
  - Conduct a privacy or security risk assessment

  - Minimize the data they collect and retain
  - Test their security measures before launching their products
- Companies should train all employees in effective security practices.
- Companies should ensure that security issues are addressed at the appropriate level of responsibility within the organization.
- Companies should retain service providers that are capable of maintaining reasonable security and provide reasonable oversight for these service providers.
- Companies should implement a defense-in depth approach, in which they consider implementing security measures at several levels.
- Companies should consider implementing reasonable access control measures to limit the ability of an unauthorized person to access a consumer's device, data, or even the consumer's network.
- Companies should continue to monitor products throughout the life cycle and, to the extent feasible, patch known vulnerabilities.

Another critical area that the FTC report[9] identifies relative to IoT is data minimization. Data minimization refers to the concept that companies should limit the data they collect and retain, and dispose of it once they no longer need it.

## Big Data Analytics and the Industrial Internet of Things

Big Data refers to a collection of data sets so large or complex that it exceeds the capability of traditional data processing technologies. A primary advantage of Big Data is the ability to apply analytics to its contents. Big Data analytics is the process of analyzing and mining Big Data, which can produce valuable information that can be used for a variety of purposes.

A detailed perspective on Big Data is provided in the *NIST Big Data Interoperability Framework* (draft). The Framework consists of seven volumes, each of which addresses a specific key topic, resulting from the work of the NIST Big Data Public Working Group (NBD-PWG). The Framework states that:[10]

> Big Data refers to the inability of traditional data architectures to efficiently handle the new datasets. Characteristics of Big Data that force

new architectures are volume (i.e., the size of the dataset) and variety (i.e., data from multiple repositories, domains, or types), and the data in motion characteristics of velocity (i.e., rate of flow) and variability (i.e., the change in other characteristics). These characteristics—volume, variety, velocity, and variability—are known colloquially as the *Vs* of Big Data.

The *NIST Big Data Interoperability Framework: Volume 4, Security and Privacy*,[11] identifies a number of ways that security and privacy in Big Data projects can be different from traditional implementations. The following seven principles are considered representative of a larger set of differences:

1. Big Data projects often encompass heterogeneous components in which a single security scheme has not been designed from the outset.
2. Most security and privacy methods have been designed for batch or online transaction processing systems. Big Data projects increasingly involve one or more streamed data sources that are used in conjunction with data at rest, creating unique security and privacy scenarios.
3. The use of multiple Big Data sources not originally intended to be used together can compromise privacy, security, or both. Approaches to de-identify personally identifiable information (PII) that were satisfactory prior to Big Data may no longer be adequate.
4. An increased reliance on sensor streams, such as those anticipated with the IoT (e.g., process industries, manufacturing plants, power plants, smart cities, and smart homes), can create vulnerabilities that were more easily managed before they were amassed to Big Data scale.
5. Certain types of data thought to be too big for analysis, such as geospatial and video imaging, will become commodity Big Data sources. These uses were not anticipated and/or may not have implemented security and privacy measures.
6. Issues of veracity, provenance, and jurisdiction are greatly magnified in Big Data. Multiple organizations, stakeholders, legal entities, governments, and an increasing number of citizens will find data about themselves included in Big Data analytics.

7. Volatility is significant because Big Data scenarios envision that data is permanent by default. Security is a fast-moving field with multiple attack vectors and countermeasures. Data may be preserved beyond the lifetime of the security measures designed to protect it.

The NBD-PWG identifies veracity, validity, and volatility as additional key characteristics of Big Data that affect security and privacy. These elements are summarized below with respect to their impact on Big Data security and privacy.

- **Veracity** – Sometimes referred to as *provenance*, veracity encompasses information assurance for the methods through which information is collected. It addresses the problem of understanding the data's original source, such as through metadata, and establishing the pedigree and chain of custody, including calibration, errors, missing data (e.g., time stamp, location, equipment serial number, transaction number, and authority).
- **Validity** – Validity refers to the accuracy and correctness of data. Traditionally, this is referred to as *data quality*.
- **Volatility** – Volatility refers to how data management changes over time. For example, systems may produce indefinitely persisting data—data that outlives the instruments on which it was collected. Security and privacy requirements can shift accordingly.

The effects of these characteristics result in a number of differences between what is new for Big Data and the requirements of previous big system security and privacy. These differences, discussed in volume 4 of the *NIST Big Data Interoperability Framework*,[12] are summarized as follows:

- Big Data may be gathered from diverse end points and from more sources than traditional providers and consumers have used, including mobile users, industrial plants, field operations, and social network users.
- Data aggregation and dissemination must be secured inside the context of a formal, understandable framework. Big Data systems may be operational outside formal, readily understood frameworks, where such frameworks have been unsystematically composed.

- Data search and selection can lead to privacy or security policy concerns. There is a lack of systematic understanding of the capabilities that should be provided by a data provider in this respect. A combination of well-educated users, well-educated architects, and system protections, such as excluding databases or limiting queries that may be foreseen as enabling re-identification, may be needed.

- Privacy-preserving mechanisms are needed for Big Data, such as for personally identifiable information (PII). Because there may be disparate, potentially unanticipated processing steps between the data owner, provider, and data consumer, the privacy and integrity of data coming from end points should be protected at every stage.

- Big Data is pushing beyond traditional definitions for information trust, openness, and responsibility. Governance, previously consigned to static roles and typically employed in larger organizations, is becoming an increasingly important intrinsic design consideration for Big Data systems.

- Information assurance and disaster recovery for Big Data systems may require unique and emergent practices. Because of its extreme scalability, Big Data presents challenges for information assurance (IA) and disaster recovery (DR) practices that were not previously addressed in a systematic way. Traditional backup methods may be impractical for Big Data systems. In addition, test, verification, and provenance assurance for Big Data replicas may not be completed in time to meet temporal requirements that were readily accommodated in smaller systems.

- Big Data creates potential targets of increased value. The effort required to consummate system attacks will be scaled to meet the opportunity value. Big Data systems will present concentrated, high-value targets to adversaries. As Big Data becomes ubiquitous, such targets become more numerous.

- Risks have increased for de-anonymization and transfer of PII without consent traceability. Security and privacy can be compromised through unintentional lapses or malicious attacks on data integrity. Managing data integrity for Big Data presents additional challenges related to all the Big Data characteristics but especially for PII. While there are technologies available to develop methods for de-identification, some experts caution that equally powerful methods can leverage Big Data to re-identify personal information.

- If not properly managed, data identification, metadata tagging, aggregation, and segmentation—widely anticipated for data science and open datasets—may have degraded veracity because they are derived and are not primary information sources. Retractions of peer-reviewed research due to inappropriate data interpretations may become more commonplace as researchers leverage third-party Big Data.

## Industrial Internet of Things

An important emerging application of Big Data analytics in the industrial sector is the Industrial Internet of Things or IIoT. The IIoT, through the use of sensors, advanced analytics, and intelligent decision-making, forms the basis of significant improvements in the methods that field assets use to communicate with enterprise activities. IIoT strategies will focus on pushing data into a centralized cloud platform that can bridge the gap between industrial and IT domains. The capabilities that are necessary in supporting the IIoT are:

- Cloud computing and storage
- Big Data analytics tools
- Necessary application software
- Supporting data link hardware and software

The IIoT must interface to the different network protocols and topologies, and automatically discover and identify data associated with industrial devices as well as accommodate high-frequency updates. The data sent to the cloud from the industrial networks originates from devices that have their own particular characteristics and operating parameters. In this scenario, the following emerging standards are relevant:

- **Asynchronous Messaging Queuing Protocol (AMQP)** – AMQP is a means for high-quality and safe delivery of messages between applications and processes by creating interoperability between clients and brokers (i.e., messaging middleware). It enables a wide range of different applications and systems to communicate and cooperate.

- **Message Queueing Telemetry Transport (MQTT)** – MQTT is a M2M/IoT protocol supporting connections with devices such as sensors in remote locations where network bandwidth is at a premium. It is useful for mobile applications because of its small size, low power usage, minimized data packets, and efficient distribution of information to one or many receivers.
- **Constrained Application Protocol (CoAP)** – CoAP is defined in the Internet Engineering Task Force (IETF) Request for Comments (RFC) as:[13]

  > The Constrained Application Protocol (CoAP) is a specialized web transfer protocol for use with constrained nodes and constrained (e.g., low-power, lossy) networks. The nodes often have 8-bit microcontrollers with small amounts of ROM and RAM, while constrained networks such as IPv6 over Low-Power Wireless Personal Area Networks (6LoWPANs) often have high packet error rates and a typical throughput of tens of Kbit/s. The protocol is designed for machine-to-machine (M2M) applications such as smart energy and building automation.

- **Data Distribution Services (DDS)** - The Data Distribution Service (DDS™) is a middleware protocol and API standard for data-centric connectivity from the Object Management Group® (OMG®).[14] It provides network interoperability for linked machines, enterprise systems, and mobile devices, providing low-latency data connectivity, extreme reliability, and a scalable architecture that support business and mission-critical IoT applications. It can be used in Big Data applications, smart grid management, and other IIoT functions as well as supporting Quality of Service (QoS).

Big Data technologies can be divided into two groups: batch processing, which is analytics on data at rest, and stream processing, which is analytics on data in motion. Data-driven information security dates back to bank fraud detection and anomaly-based intrusion detection systems.

Big Data technologies will transform security analytics by:[15]

- Collecting data at a massive scale from many internal enterprise sources and external sources, such as vulnerability databases
- Performing deeper analytics on the data

- Providing a consolidated view of security-related information
- Achieving real-time analysis of streaming data

It is important to note that Big Data tools still require system architects and analysts to have a deep knowledge of their system in order to properly configure the Big Data analysis tools.

In order to take advantage of Big Data in cybersecurity, algorithms and systems must be designed and implemented in order to identify actionable security information from large enterprise data sets and reduce false positive rates. Accordingly, there are a number of issues that have to be addressed in applying Big Data analytics, including legal, privacy, analysis, collection, and presentation.

The Cloud Security Alliance has categorized a number of promising areas of application of Big Data analytics to cybersecurity[16] and has established projects to explore these areas. Two of the major projects are summarized as follows:

- **NetFlow Monitoring to Identify Botnets** – This leverages the MapReduce paradigm for analyzing enormous quantities of NetFlow data to identify infected hosts participating in a botnet. (MapReduce is a primary component of the Apache™ Hadoop software framework that supports distributed processing and storage of Big Data on large clusters of commodity servers. It provides for massive data storage and high-speed processing with a high degree of fault tolerance.) The rationale for using MapReduce for this project stemmed from the large amount of NetFlow data collected for data analysis; 720 million NetFlow records (77GB) were collected in only 23 hours. Processing this data with traditional tools is challenging. However, Big Data solutions like MapReduce greatly enhance analytics by enabling an easy-to-deploy distributed computing paradigm.
- **Detecting Advanced Persistent Threats (APTs)** – The focus of this project is to identify targeted APT attacks against a high-value asset or a physical system that can be extended over a long period of time without the knowledge of the infected organization. APTs can gain access to customer information, steal intellectual property, poison data, hold an organization for blackmail, and disrupt the business of an organization. Because detecting APTs requires searching through very

large amounts of data, Big Data analytics can be a valuable tool in uncovering these threats.

The Cloud Security Alliance has identified a number of challenges[17] that must be overcome to realize the true potential of Big Data analytics. These items are:

- **Data Provenance** – Data provenance refers to the authenticity and integrity of data used for analytics. As Big Data expands the sources of data it can use, the trustworthiness of each data source needs to be verified, and the inclusion of ideas such as adversarial machine learning must be explored in order to identify maliciously inserted data.
- **Privacy** – Regulatory incentives and technical mechanisms are needed to minimize the amount of inferences that Big Data users can make that might result in privacy compromises.
- **Securing Big Data Stores** – The security of the Big Data repository is protected by identifying applicable best practices.
- **Human–Computer Interaction** – Focusing on the human-computer interaction with Big Data, such as providing visualization tools, helps analysts understand the data of their systems.

## The NIST Cyber-Physical Systems (CPS) Framework

In mid-2014, in order to better manage the development and implementation of CPS across multiple application domains (including smart manufacturing, transportation, energy, and healthcare) NIST established the CPS Public Working Group (CPS PWG). In September 2015, the CPS PWG published a draft *Framework for Cyber-Physical Systems.*[18] The focus of the Framework is to develop a CPS analysis methodology and a vocabulary that describes it. The Framework defines CPS as systems that "integrate computation, communication, sensing, and actuation with physical systems to fulfill time-sensitive functions with varying degrees of interaction with the environment, including human interaction." Thus, a CPS provides a service that consists of one or more functions, and the CPS Model properties are properties of that service.

Additional key concepts addressed by the Framework include the following:[19]

- CPS are frequently system of systems (SoS), and the architectural constructs should be able to be applied recursively or iteratively to support this nested nature of CPS.
- CPS should be characterized by well-defined components using standardized semantics and syntax.
- CPS architectures should support application and domain flexibility. To do this, the definition of the components should be flexible and open ended.
- CPS frequently perform critical applications, so the CPS architecture must support the level of reliability needed to meet requirements.
- Security is a necessary feature of the CPS architecture to ensure that CPS capabilities are not compromised by malicious agents and that the information processed, stored, and transferred has its integrity preserved and is kept confidential where needed.
- Data exchange is a prominent dimension of CPS operation. The nature of data and its reliability, type, identity, and discovery are all key attributes that allow for a common understanding of data conveyed through communications in and among CPS.
- Components that contain sensors and/or actuators should have an appropriate level of awareness of physical location and time.
- CPS architectures should support legacy component integration and migration. Legacy components should be integrated in a way that ensures that security and other essential performance and functional requirements are met.

The key elements comprising the Framework are described as follows:

- **Domains** – Domains represent the different application areas of CPS.
- **Concerns** – A concern is an “interest in a system relevant to one or more of its stakeholders.” Concerns that are conceptually equivalent or related are grouped into aspects, which are addressed by activities within the facets.
- **Aspects** – Aspects consist of groupings of conceptually equivalent or related concerns.

- **Facets** – Facets are views on CPS encompassing identified responsibilities in the systems engineering process. Each facet contains a set of well-defined activities and artifacts (outputs) for addressing concerns. There are three identified facets: conceptualization, realization, and assurance. A typical facet model is given in Figure 9-1.

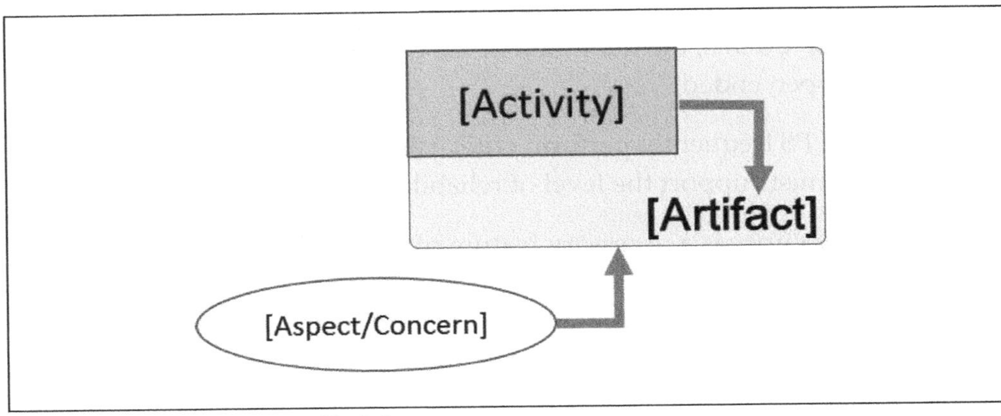


**Figure 9-1. Typical Facet Model**
*Source: Cyber Physical Systems Public Working Group (2015)*

Using the facet model, the three Framework facets of conceptualization, realization, and assurance are developed as shown in Figure 9-2.

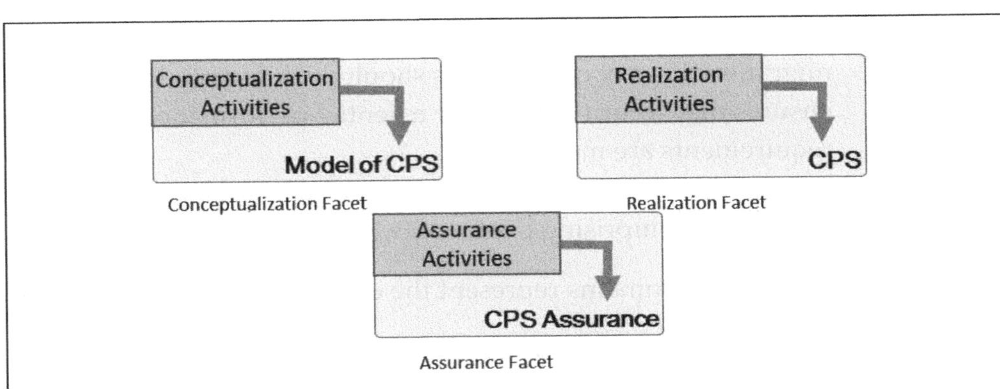


**Figure 9-2. Three Primary Framework Facets**
*Source: Cyber Physical Systems Public Working Group (2015)*

The conceptualization facet is responsible for defining the stakeholders and user-oriented view of the CPS-desired capabilities and includes the set of activities that address the following areas:

- The problems or the opportunity with respect to the organization's business, mission, vision, strategic goals, and objectives
- The needs and requirements of the major stakeholders (customers, users, administrators, owners, and regulators) who have an interest in the CPS throughout its life cycle
- The analysis and decomposition into abstract functional elements and their logical rather than their engineering interactions, which are addressed in the realization facet

The realization facet is responsible for the following activities:

- Transforming the outputs of the conceptualization facet into an effective instance of CPS
- Enabling consistent reproduction of the CPS
- Operating and maintaining the CPS while providing the required services
- Disposing of the CPS when those services are no longer needed

The realization facet incorporates key conceptual layers associated with concepts, components, and notional architectures that can be instantiated into layer- and domain-specific CPS architectures. These layers comprise the following:

- Business Case Analysis
- Life Cycle Management
- Design
- Manufacturing/Implementation
- Operations
- Disposal
- Cyber-Physical Abstraction Layer Formation

- Physical Layer Realization

The assurance facet then provides the assurance that the properties gathered in conceptualization are satisfied with a defined level of confidence and the conceptualization was indeed realized. The assurance activities are summarized as follows:

- Identifying the assurance objectives
- Defining the strategy to achieve those objectives
- Managing/controlling the assurance evidence
- Analyzing that evidence
- Forming the assurance argument
- Providing the estimate of confidence that assurance objectives have been met
- Assessing the product configuration for completeness and integrity
- Assessing the requirements and test results
- Performing certification and compliance testing

Figure 9-3 summarizes these concepts.

The Framework itself is developed through an analysis process comprised of the following steps:

1. Identify domains of CPS; these are the areas of deployment of CPS, such as manufacturing, transportation, and energy, in which stakeholders may have domain-specific and cross-domain concerns.
2. Identify cross-cutting concerns, like societal, business, technical, etc. Stakeholders can have concerns that overlap or are instances of broader conceptual concerns.
3. Analyze cross-cutting concerns to produce aspects, or to group conceptually equivalent or related concerns.
4. Address concerns (aspects) through activities and artifacts organized within three fundamental facets: conceptualization, realization, and assurance.

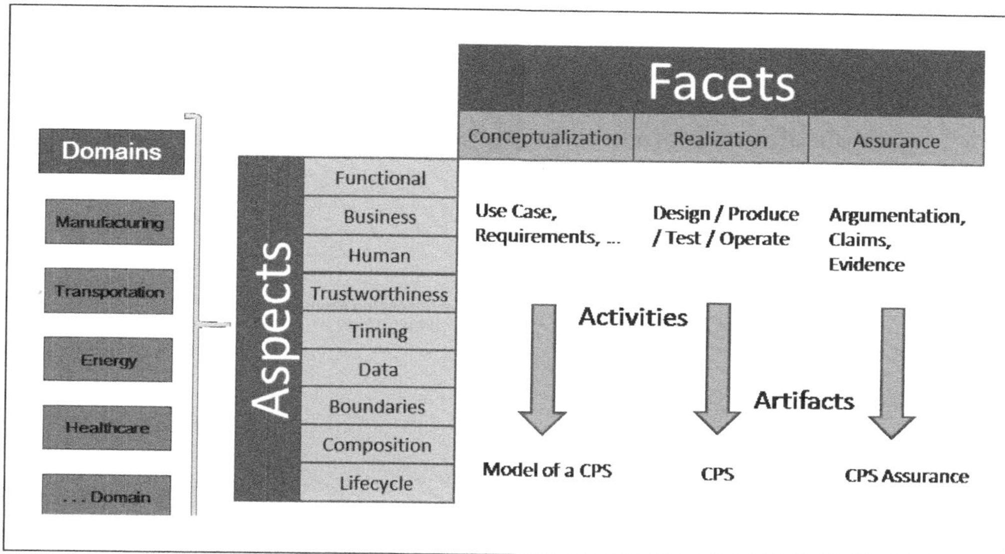


**Figure 9-3. CPS Framework – Domains, Facets, Aspects**
*Source: Cyber Physical Systems Public Working Group (2015)*

Facet activities can be viewed as a template and need to be tailored to a CPS domain by defining which of the CPS aspects apply to that domain and by updating the facet activities and artifacts for each applicable aspect.

Depending on the application, CPS can be categorized into the following six different complexities:

- Critical tightly-coupled CPS for vital infrastructures such as the energy grid would involve a thorough and detailed process based on hard requirements and assurances using the Framework
- Loosely-coupled CPS for applications that build on the capabilities of existing CPS and integrate them in new approaches with more emphasis on functional goals and less stress on hard requirements
- Shallow analysis focusing on concepts or reviewing different approaches to CPS issues where small subsets of the Framework might be used
- Single CPS devices such as a video camera or robot where the analysis would focus on the robustness of the design
- System or subsystem of individual cyber, physical, and cyber-physical devices such as an HVAC system

- System of systems (SoS) comprising a system of interconnected systems

The CPS domain space, the application areas where CPS devices exist, has multiple dimensions as shown in the Beecham Research diagram in Figure 9-4.

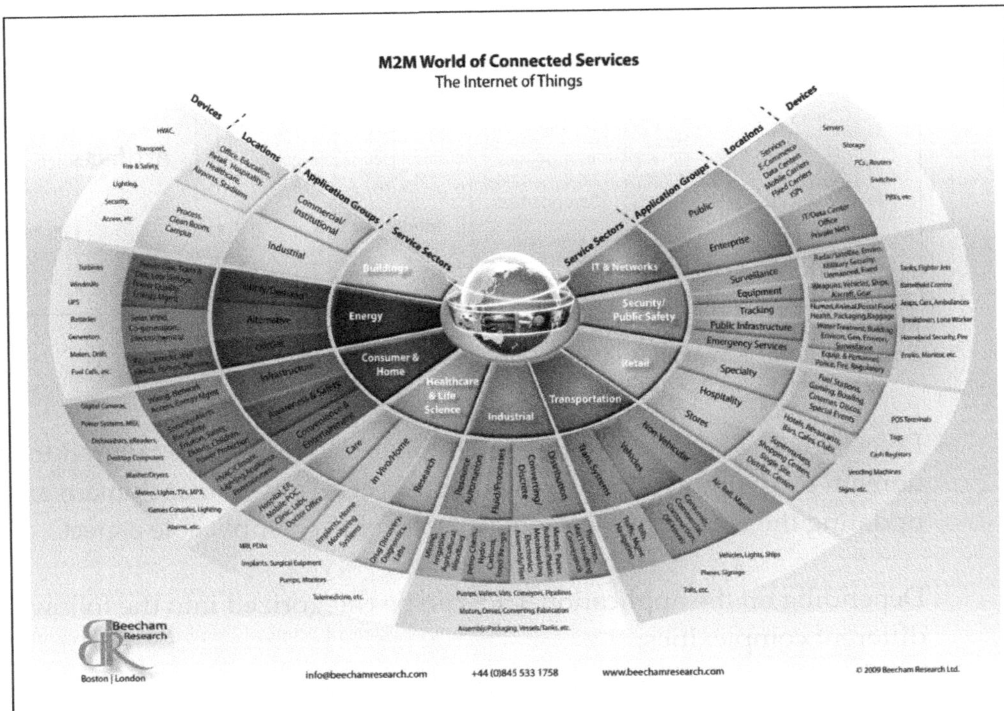


**Figure 9-4. Dimensionality of the CPS Domain Space Instrumented Functions versus Other Process Functions**
*Source: Beecham Research Ltd., www.beechamresearch.com (2009)*

Some specific example domains of the CPS include:

- Aerospace
- Agriculture
- Buildings
- Communities
- Defense
- Disaster resilience

- Emergency response
- Energy
- Entertainment/sports
- Environmental monitoring
- Financial services
- Healthcare
- Infrastructure (communications, power, water)
- Manufacturing
- Science
- Transportation
- Weather

## CPS and Cybersecurity

Because of the many real-time, always-up, safety issues associated with CPS, cybersecurity poses significant challenges to those systems. As stated in the CPS PWG *Draft Framework for Cyber-Physical Systems*:[20]

> CPS cybersecurity must protect operational goals from the impacts of malicious cyberattack, enabling continuing safe operations even in compromised conditions. Cybersecurity for CPS must address how a system can continue to function correctly when under attack, provide mechanisms that support fault-tolerance and/or graceful degradation in accordance with mission- or business-driven priorities, and enable the system to fail-safe in those circumstances in which resilience cannot be provided in the face of threat.

The following items are the aspects that shape resilience in understanding the multidisciplinary nature of CPS issues:[21]

- **Unexpected Condition Adaptation**
  - The ability to have large-scale, integrated supervisory control methodologies implements graceful degradation.
  - Widely distributed, dynamic, control system elements organize to prevent destabilization of the controlled system.

- **Human Interaction Challenges**
  - Humans possess great capability based on knowledge and skill, but are not always operating at the same performance level.
  - The ability to recognize and mitigate cyberattacks is necessary to ensure the integrity of the control system.
- **Goal Conflicts**
  - Potentially conflicting goals and flawed understanding of the factors affects system behavior.
  - One conflict is a lack of state awareness where raw data must be translated to information on the condition of the process and the control system components.

Additional issues that have to be considered regarding the CPS interaction with the physical world include:

- Impact of attacks on a CPS can be physically catastrophic and can adversely impact product quality, operational safety, and product performance.
- The time scale for addressing an attack cannot be in minutes.
- CPS are sometimes deployed in ways that preclude physically securing all of their components.
- Continuous, autonomous, real-time monitoring, detection, and response is needed.
- CPS often exist on resource-constrained platforms. As a result, security mechanisms must be lightweight in terms of storage space, memory use, processor use, network connectivity, and electrical power consumption.
- Operational situations that require immediate access to control systems are not conducive to complex passwords.
- "Always on" makes rebooting and patching nonviable strategies for many systems.
- The software that executes processes in a number of these systems is often old and does not meet new safety requirements

- Traditional key management is difficult when the set of valid users is constantly changing.
- Infrastructure CPS typically have very long lifetimes and are difficult and expensive to change.
- Many legacy CPS have devices and software for which there is no support and that might have security vulnerabilities.

These issues are summarized in Figure 9-5.

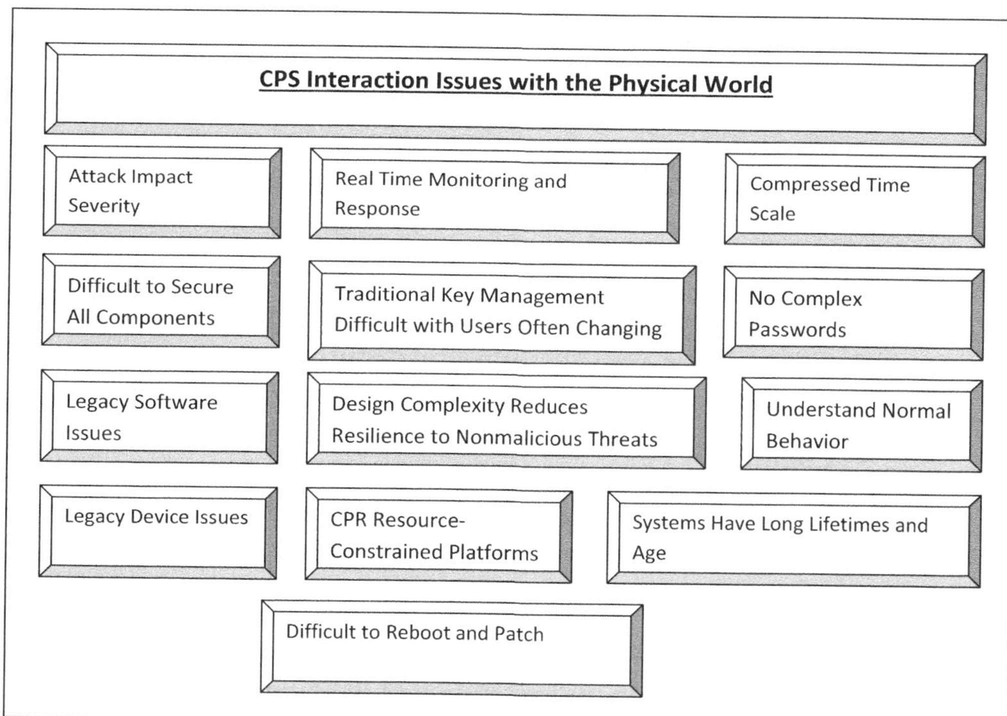


**Figure 9-5. CPS Physical World Interaction Issues**

Keeping these issues in mind, as a consequence, design activities should be based on the following key areas:

- **Threats to Resilience**
  - System design should incorporate diverse methods for maintaining awareness and adapting the system's configuration to transform system behaviors.

  - Constantly evolving methodologies are needed to predict the nature of potential threats, their specific objectives, and their specific effects on each system.

- **Human-in-the-Loop Systems** may not be capable of sufficiently timely response to some threats, but humans should still be kept aware of the system's state.
  - Adversaries may have many opportunities to compromise CPS digital technology through the global supply chain.
  - Increased design obfuscation and complexity provides resilience to malicious man-made threats but reduces resilience to nonmalicious, man-made, and natural threats.
- **Cyber-Physical**
  - Minimize shared interdependencies between networked systems to reduce the potential for unrecognized interactions that lead to cascading failures.
  - Localize the process and security dynamics of system operation to stabilize any cyber-physical disturbance and prevent cascading failures.
- **Cognitive**
  - Tailor CPS information to the benign human-based role and expected action to ensure a reproducible response irrespective of background or experience.
  - Gain cognitive understanding of expected operational behavior based on an understanding of the difference between normal behavior and abnormal behavior.
  - Obfuscate and abstract CPS information available, accessible by the potentially malicious human, in response to a perceived threat.
- **Cyber-Physical-Cognitive**
  - CPS cyber-physical degradation must be quantified within the distributed architecture to ensure an effective response.

– Concepts of trust are earned, based on conformance metrics associated with the design.

These design activities are summarized in Figure 9-6.

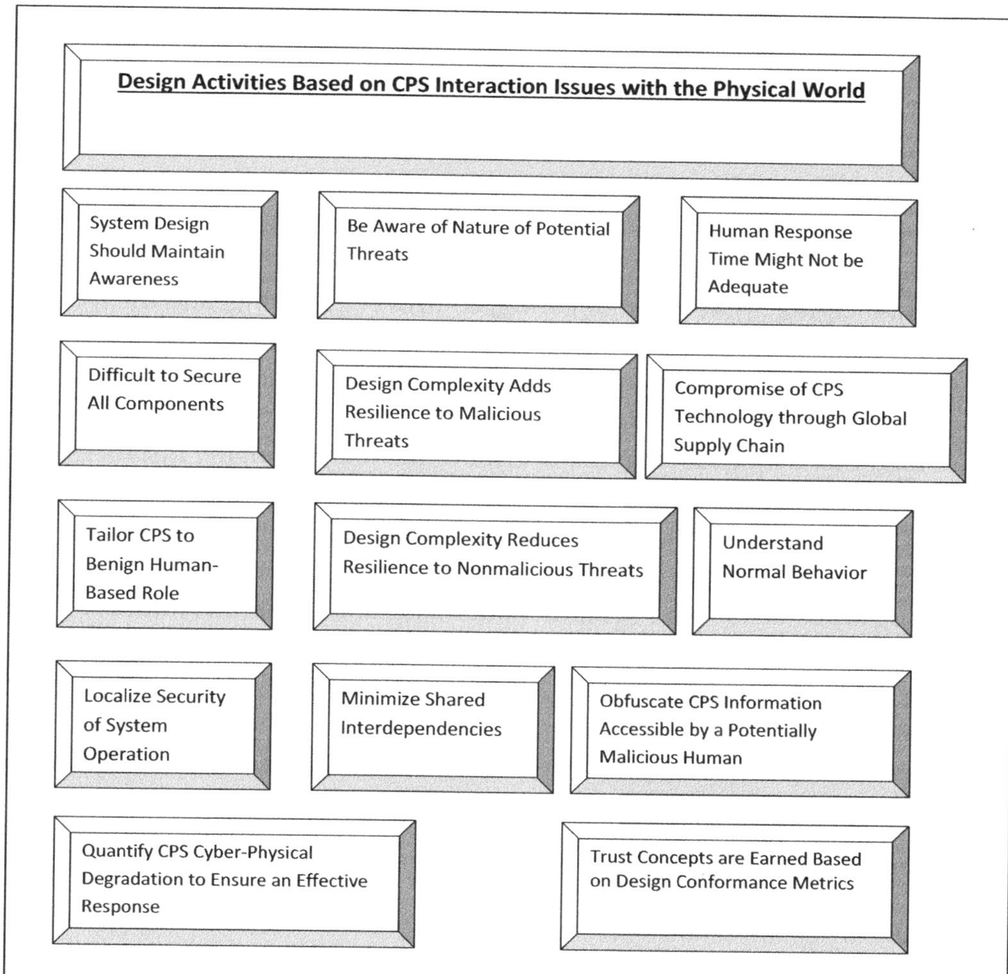


**Figure 9-6. Design Activities Based on CPS Physical World Interaction Issues**

In summary, application of the concepts of facets and aspects in the CPS analysis methodology provides the means to address concerns in the CPS design, development, and implementation cycles. The methodology supports the understanding and development of new and existing CPS, including those designed to interact with other CPS and function in multiple interconnected infrastructure environments.

## Critical Infrastructure Security

Protecting the nation's critical infrastructure is becoming increasingly important because of the many hostile threats and rogue nations intent on executing cyberattacks against these facilities. Thus, it has become imperative to develop standards and methodologies to counter such threats. In doing so, however, it is necessary to understand the culture of the entities comprising the critical infrastructure and their differences and similarities relative to the information technology arena.

Typically, a large number of the personnel populating the industrial automation and control system field come from engineering backgrounds, with training in such areas as electrical engineering, chemical engineering, mechanical engineering, systems engineering, control engineering, and the like. On the other hand, most IT personnel have origins in computer science, software engineering, systems analysis, database administration, network management, and so on. For example, software quality and process improvement methods widely used in the IT environment are, in many instances, foreign to control engineers and, in fact, are viewed as cumbersome in implementing SCADA and control algorithms. In addition, the performance of a process in a plant is critical, and inadequate performance in production areas can result in huge financial losses, equipment damage, and personnel injuries. These severe consequences of operational errors are not usually a common occurrence in IT facilities. Similarly, safety is a very critical concern in a production environment and computer malfunctions can result in fires or explosions in some instances. Thus, in a production environment, safety and performance usually take precedence over information security, whereas this is not the case in an information system paradigm.

In order to address the protection of the critical infrastructure, a number or organizations have proposed and developed guidelines and approaches that can be used to mitigate associated risks. One approach was initiated by Executive Order (EO) 13636, "Improving Critical Infrastructure Cybersecurity," on February 12, 2013, which established that "it is the policy of the United States to enhance the security and resilience of the Nation's critical infrastructure and to maintain a cyber environment that encourages efficiency, innovation, and economic prosperity while promoting safety, security, business confidentiality, privacy, and civil liberties." In enacting this policy, the Executive Order called for the development of a voluntary risk-based Cybersecurity Framework – a set of industry standards and best practices to help organizations manage cybersecurity risks. The resulting Framework, created through collaboration between government and the private sector, uses a common language to address and manage cybersecurity risk in a cost-effective way based on business needs without placing additional regulatory requirements on businesses. The Framework development was a year-long collaborative process in which NIST served as a convener for industry, academia, and government stakeholders. That effort resulted in the *NIST Framework for Improving Critical Infrastructure Cybersecurity*, Version 1.0, on 12 February 2014. The Framework defines critical infrastructure as "systems and assets, whether physical or virtual, so vital to the United States that the incapacity or destruction of such systems and assets would have a debilitating impact on security, national economic security, national public health or safety, or any combination of those matters." The Framework complements NIST SP 800-82, *Guide to Industrial Control Systems*, and can be leveraged to address the CSF functions in the context of an organization's risk management program.

### Framework Fundamentals

The goal of the Framework is to provide a common taxonomy and mechanism for organizations to:

1. Describe their current cybersecurity posture
2. Describe their target state for cybersecurity
3. Identify and prioritize opportunities for improvement within the context of a continuous and repeatable process
4. Assess progress toward the target state
5. Communicate among internal and external stakeholders about cybersecurity risk

The Framework comprises three parts: the Framework Core, the Framework Implementation Tiers, and a Framework Profile, summarized as follows:

- **Framework Core** – This is a set of cybersecurity activities, desired outcomes, and applicable references that are common across critical infrastructure sectors. The Core presents industry standards, guidelines, and practices in a manner that allows for communication of cybersecurity activities and outcomes across the organization from the executive level to the implementation/operations level. The Core itself is comprised of four elements: Functions, Categories, Subcategories, and Informative References, listed as follows:

- **Five Functions** – Identify, Protect, Detect, Respond, and Recover. When considered together, these Functions provide a high-level, strategic view of the life cycle of an organization's management of cybersecurity risk. These Core Functions provide a set of activities to achieve specific cybersecurity outcomes and reference examples of guidance to achieve those outcomes as summarized below:
    - **Identify** – Develop the organizational understanding to manage cybersecurity risk to systems, assets, data, and capabilities. The activities in the Identify Function are foundational for effective use of the Framework. Understanding the business context, the resources that support critical functions, and the related cybersecurity risks enables an organization to focus and prioritize its efforts, consistent with its risk management strategy and business needs.
    - **Protect** – Develop and implement the appropriate safeguards to ensure delivery of critical infrastructure services. The Protect Function supports the ability to limit or contain the impact of a potential cybersecurity event.
    - **Detect** – Develop and implement the appropriate activities to identify the occurrence of a cybersecurity event. The Detect Function enables timely discovery of cybersecurity events.
    - **Respond** – Develop and implement the appropriate activities to take action regarding a detected cybersecurity event. The Respond Function supports the ability to contain the impact of a potential cybersecurity event.
    - **Recover** – Develop and implement the appropriate activities to maintain plans for resilience and to restore any capabilities or services that were impaired due to a cybersecurity event. The Recover Function supports timely recovery to normal operations to reduce the impact from a cybersecurity event.

- **Categories** – Subdivisions of a Function into groups of cybersecurity outcomes, such as Access Control, that are closely tied to programmatic needs and particular activities.
- **Subcategories** – Further divide a Category into specific outcomes of technical and/or management activities.
- **Informative References** – Specific sections of standards, guidelines, and practices common among critical infrastructure sectors that illustrate a method to achieve the outcomes associated with each Subcategory.

Figure 9-7 illustrates the relationship among these concepts.

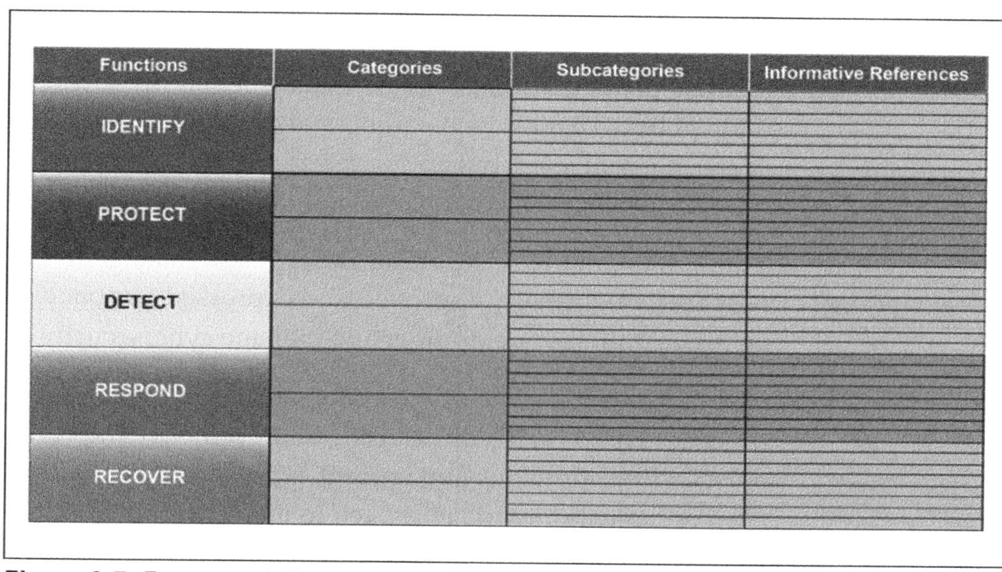


**Figure 9-7. Framework Core**
*Source: NIST Framework (2014)*

- **Framework Implementation Tiers** – These provide context on how an organization views cybersecurity risk and the processes in place to manage that risk. Tiers describe the degree to which an organization's cybersecurity risk management practices exhibit the characteristics defined in the Framework (e.g., risk and threat aware, repeatable, and adaptive). The tiers characterize an organization's practices over a range, from Partial (Tier 1) to Adaptive (Tier 4). These tiers reflect a progression from informal, reactive responses to approaches that are agile and risk-informed, summarized as follows:
  - **Tier 1: Partial** – Organizational cybersecurity risk management practices are not formalized, and risk is managed in an ad hoc and sometimes reactive manner. There is limited awareness of cybersecurity risk at the organizational level.
  - **Tier 2: Risk Informed** – Risk management practices are approved by management but may not be established as organizational-wide policy. There is an awareness of cybersecurity risk at the organizational level, but an organization-wide approach to managing cybersecurity risk has not been established.
  - **Tier 3: Repeatable** – The organization's risk management practices are formally approved and expressed as policy. There is an organization-wide approach to manage cybersecurity risk.
  - **Tier 4: Adaptive** – The organization adapts its cybersecurity practices based on lessons learned and predictive indicators derived from previous and current cybersecurity activities. There is an organization-wide approach to managing cybersecurity risk that uses risk-informed policies, processes, and procedures to address potential cybersecurity events.
- **Framework Profile** – This represents the outcomes based on business needs that an organization has selected from the Framework Categories and Subcategories. The Profile can be characterized as the alignment of standards, guidelines, and practices to the Framework Core in a particular implementation scenario. Profiles can be used to identify opportunities for improving cybersecurity posture by comparing a Current Profile (the "as is" state) with a Target Profile (the "to be" state). To develop a Profile, an organization can review all of the Categories and Subcategories and, based on business drivers and a risk assessment, determine which are most important.

The Framework is designed to complement existing business and cybersecurity operations and provides a means of expressing cybersecurity requirements to business partners and customers. It also assists in identifying gaps in an organization's cybersecurity practices and provides a general set of considerations and processes for considering privacy and civil liberties implications in the context of a cybersecurity program.

The Framework describes how it can be applied by organizations in the following ways:

- **Basic Review of Cybersecurity Practices** – The Framework can be used to compare an organization's current cybersecurity activities with those outlined in the Framework Core. Through the creation of a Current Profile, organizations can examine the extent to which they are achieving the outcomes described in the Core Categories and Subcategories, aligned with the five high-level Functions: Identify, Protect, Detect, Respond, and Recover.
- **Establishing or Improving a Cybersecurity Program** – The following steps illustrate how an organization could use the Framework to create a new cybersecurity program or improve an existing program. These steps should be repeated as necessary to continuously improve cybersecurity.

    **Step 1:** **Prioritize and Scope** – The organization identifies its business/mission objectives and high-level organizational priorities. With this information, the organization makes strategic decisions regarding cybersecurity implementations and determines the scope of systems and assets that support the selected business line or process.

    **Step 2:** **Orient** – Once the scope of the cybersecurity program has been determined for the business line or process, the organization identifies related systems and assets, regulatory requirements, and the overall risk approach. The organization then identifies threats to, and vulnerabilities of, those systems and assets.

    **Step 3:** **Create a Current Profile** – The organization develops a Current Profile by indicating which Category and Subcategory outcomes from the Framework Core are currently being achieved.

**Step 4:** **Conduct a Risk Assessment** – This assessment could be guided by the organization's overall risk management process or previous risk assessment activities. The organization analyzes the operational environment in order to discern the likelihood of a cybersecurity event and the impact that the event could have on the organization.

**Step 5:** **Describe the Organization's Desired Cybersecurity Outcomes** – The organization creates a Target Profile that focuses on the assessment of the Framework Categories and Subcategories describing the organization's desired cybersecurity outcomes.

**Step 6:** **Determine, Analyze, and Prioritize Gaps** – The organization compares the Current Profile and the Target Profile to determine gaps. Next it creates a prioritized action plan to address those gaps that draws upon mission drivers, a cost/benefit analysis, and understanding of risk to achieve the outcomes in the Target Profile.

**Step 7:** **Implement Action Plan** – The organization determines which actions to take in regards to the gaps, if any, identified in the previous step. It then monitors its current cybersecurity practices against the Target Profile.

An organization may repeat the steps as needed to continuously assess and improve its cybersecurity.

- **Communicating Cybersecurity Requirements with Stakeholders** – The Framework provides a common language to communicate requirements among interdependent stakeholders responsible for the delivery of essential critical infrastructure services.
- **Identifying Opportunities for New or Revised Informative References** – The Framework can be used to identify opportunities for new or revised standards, guidelines, or practices where additional Informative References would help organizations address emerging needs.

- **Methodology to Protect Privacy and Civil Liberties** – This topic describes a methodology (as required by the Executive Order to address individual privacy and civil liberties implications) that may result from cybersecurities operations.

## Framework Feedback

On 11 December 2015, NIST issued its third request for information (RFI) to receive feedback on the Framework. Respondents were asked to evaluate the main elements of the Framework, and NIST analysts sought out themes regarding answers to RFI questions Ten response themes were identified; a summary of each is given below including a specific example related to each theme.

- **Framework Update Timeline** – There were diverse comments on whether an update is necessary or desirable.

  ***Example Response:*** In the short and intermediate term, perhaps it is best to not change the Framework, but instead focus on making it easier to understand and incrementally implement. The Framework is valuable as is and, while it may not be perfect, it can help get industry on the road to higher security.

- **Update to Framework Content** – Many respondents had specific suggestions of ways to update and expand the Framework.

  ***Example Response:*** [The organization] recommends adding a section to the Framework document that specifically addresses how small, less mature organizations can implement the Framework consistent with the recommendations outlined in NISTIR 7621, *Small Business Information Security: The Fundamentals.*

- **Update Process** – The Framework should be updated through a collaborative process with minimal disruption to current industry use.

  ***Example Response:*** Suggested changes should go through a change management process, approved with an RFI published to get comments. Once finalized, there should be a "transition" window (1–2 years for example) to allow organizations to evaluate the changes and impact, and then justify the inclusion or exclusion of the additional/ updated controls.

- **Framework Governance** – Respondents are comfortable with NIST's continued leadership in the Framework process, though transition should be considered at a later date.

  ***Example Response:*** [The organization's] members urge NIST to continue to play a key role in facilitating further Framework development activities. Given the competence demonstrated throughout the effort and the proven ability to bring a broad array of stakeholders to the table, it would be premature at this time to transfer responsibilities to the private sector.

- **Optimal Industry Leadership** – Any possible future steward of the Framework should be a respected, internationally recognized, neutral, third-party organization.

  ***Example Response:*** [The organization] would prefer that NIST retain responsibility for the Framework, however if the responsibility is transitioned, it should be transitioned to an organization that is funded by the government or is in academia.

- **Industry Resources** – Industry resources are useful but additional guidance is needed, especially for small and medium-sized businesses.

  ***Example Response:*** Collecting a variety of case studies and sector-implementation approaches for implementing the Framework and making them available would be useful to organizations using the Framework.

- **Challenges in Sharing Best Practices** – There is a need for additional sharing of best practices surrounding use of the Framework.

  ***Example Response:*** Sharing best practices requires a time commitment and not every organization or every individual is able to afford to break away from their daily work. Limited human bandwidth and availability of expertise is one of the main reasons why best practices are not shared.

- **Regulation** – Many users of the Framework say that regulation is a necessary consideration in the development of their cybersecurity programs and they caution about the potential negative impact of additional regulatory requirements.

  ***Example Response:*** Broader adoption of the Framework as a regulatory tool, not as a regulation, will have a significantly positive impact [on] the level of effort an organization deploys against the regulatory burden. As intended, the Framework enables an organization to assess once and then to report to multiple regulators.

- **International Alignment** – The Framework is gaining traction internationally, but still needs continued outreach.

  ***Example Response:*** Add a section in the NIST document that compares, contrasts, and ultimately integrates the ISO/IEC 27001:2013 information security management system requirements into the Framework.

- **Awareness** – Much progress has been made in spreading Framework awareness, but more is still needed.

  ***Example Response:*** [The organization] urges NIST to explore ways to leverage the proposed budget increase and the widespread recognition of the need for cyber awareness to promote the widespread use of the Cybersecurity Framework.

As a result of additional, subsequent feedback from Framework stakeholders, NIST is considering clarifying the Framework's Implementation Tiers and adding guidance for applying the Framework for supply chain risk management. It is also developing a tool to help an organization assess its cybersecurity risk management process. The Cybersecurity Excellence Builder will be based on the Cybersecurity Framework and key concepts from the internationally recognized Baldrige Performance Excellence Program developed by NIST.

Further recommendations for NIST include the following:

- Publish a governance process that outlines the process of Framework maintenance and evolution, and defines the role of stakeholders and how they will continue to work together in the future.
- Remain as convener of Framework stakeholders.
- Continue Framework outreach and focus on international, small, and medium-sized businesses and regulators.

In summary, the Framework, created through collaboration between the government and the private sector, uses a common language to address and manage cybersecurity risk in a cost-effective way based on business needs without placing additional regulatory requirements on businesses.

## Software-Defined Elements

Software definition and characterization of computing, storage, and networking elements provides for flexibility in the delivery and upgrading of IT functionality. One prime example of this paradigm is software-defined networking (SDN). SDN can be used to program the interconnections of switches, routers, firewalls, and network security appliances. Physical connections can be software defined, thus enabling the dynamic deployment and provision of network resources and real-time integration of a variety of services, mobile devices, cloud applications, and storage.

Another application of software definition is software-defined storage in which storage can be dynamically provisioned, optimized, and shared. Software-defined storage combined with SDN can result in a programmable IT environment that can be tailored to the needs at hand.

Relative to cybersecurity, "software defined" refers to the decoupling and abstraction of infrastructure elements that were previously tightly coupled in data centers, such as servers, storage, networking, security, and so on. As in computing, storage, and networking, software definition will change the present views of cybersecurity. Software-defined security (SDS) still involves security hardware but more of the intelligence is transferred to software. SDS is a type of security model in which the information security in a computing environment is software-managed, policy-driven, and governed security where most of the security controls, such as intrusion detection, network segmentation, and access controls, are automated and monitored through software.

One SDS example is the concept of a software-defined perimeter (SDP). The SDP paradigm incorporates security standards from NIST and the Department of Defense (DoD), including NIST SP 800-53. The Cloud Security Alliance (CSA) has established a project to produce a public SDP standard.[22] Instead of the traditional approach, where an internal network is separated from the external world by a perimeter comprising firewall functions that block external users, an SDP provides the ability to deploy perimeters at locations of their choosing, such as on a corporate network, in a cloud, and on the Internet. SDP employs standard tools such as SAML, IPsec, PKI, and so on. The SDP aims to give application owners the ability to deploy perimeter functionality where needed. SDPs replace physical appliances with logical components that operate under the control of the application owner.

According to the Cloud Security Alliance,[23] "Connectivity in an SDP is based on a need-to-know model, in which device posture and identity are verified before access to application infrastructure is granted. Application infrastructure is effectively 'black' without visible DNS information or IP addresses. SDP mitigates the most common network-based attacks, including: server scanning, denial of service, SQL injection, OS and application vulnerability exploits, password cracking, man-in-the-middle, cross-site scripting (XSS), cross-site request forgery (CSRF), pass-the-hash, pass-the-ticket, and many others."

An example of a specific implementation[24] of SDP is a security infrastructure that is broken into the following three interconnected layers:

- An Enforcement Layer that is based on physical and virtual security enforcement points and that segments the network, as well as executes the protection logic in high demand environments
- A Control Layer that analyzes different sources of threat information and generates protections and policies to be executed by the Enforcement Layer
- A Management Layer that orchestrates the infrastructure and brings the highest degree of agility to the entire architecture

To summarize, some of the advantages of SDS are:

- **Automated Implementation** – Software-defined security can be computer controlled across virtual servers and zones, operating under the enterprise security policy.
- **Ease of Use** – In lieu of specialized hardware, software-defined security implements logical policies through software and can address security issues at different nodes as required.
- **Expense Management** – Implementing security in a virtual environment reduces the need for expensive hardware, supports utilization of only the resources that are needed at the time, and allows for a "pay as you use" model.

- **Flexibility** – Because of implementation through virtualization, security can be adjusted to meet the requirements of the information system both in upscaling resources and in addressing changes in architecture and physical locations.

This approach provides operational resilience as well as proactive incident prevention for the continually changing threat landscape. Designed to be forward-looking, the SDP architecture supports traditional network security and access control policy requirements, as well as the threat prevention needed by modern enterprises that embrace new technologies such as mobile computing and SDNs.

## Summary

Chapter 9 develops the latest thinking in the protection of industrial automation and control systems. Moreover, it provides management and application personnel with a fresh perspective of emergent methods and techniques that, in the near future, will be incorporated into the planning and implementation of more effective and reliable security methodologies.

## Review Questions for Chapter 9

9.1 What term refers to objects such as sensors, humans, animals, machines, or other entities that each have a unique IP address and can transmit and receive data over a network without human intervention?

A. Artificial intelligence

B. Virtualization

C. Intelligent resolution

D. Internet of Things

9.2 The interoperability standard for secure and reliable information exchange in industrial automation and the enterprise is which of the following?

A. Open Platform Communications Unified Architecture (OPC UA)

B. Advanced HMI

C. Production Markup Language (PRODML)

D. M2M communication

9.3 In manufacturing, the Internet of Things is generating a next generation Industrial Revolution embodied in the concept known as which of the following?

A. Additive manufacturing

B. Industry 4.0

C. Machine-to-machine manufacturing

D. Industry 5.0

9.4 An envisioned state of operations in which all relevant and synthesized information is made available when, where, and in the form in which it is needed across the manufacturing supply chain to all stakeholders (people, process, products, and infrastructure) is known as which one of the following?

A. Additive manufacturing

B. Evolving manufacturing

C. Responsive manufacturing

D. Informed manufacturing

9.5 The four essential elements of manufacturing that converge to create an advanced production environment are which of the following?

A. People, processes, infrastructure, mitigation

B. People, processes, infrastructure, analysis

C. Products, people, processes, infrastructure

D. Collection, correlation, processes, infrastructure

9.6 Which of the following are in the Open Web Application Security Project (OWASP) "Internet of Things Top Ten" security categories?

A. Lack of Transport Encryption

B. Insecure Web Interface

C. Insecure Cloud Interface

D. All of the above

9.7 The concept that companies should limit the data they collect and retain, and dispose of it once they no longer need it, is known as?

A. Information protection

B. Information preservation

C. Data minimization

D. Data conservation

9.8 Big Data refers to which one of the following?

A. A collection of data sets so large or complex that it exceeds the capability of traditional data processing technologies

B. A collection of large data sets that exceeds the traditional risk management paradigms

C. A collection of data sets so complex that it exceeds the capability of traditional security analytics

D. A collection of data sets so large or complex that it exceeds the capability of traditional risk analysis

9.9 The process of analyzing and mining Big Data, which can produce valuable information that can be used for a variety of purposes, is which of the following?

A. Big Data minimization

B. Big Data analytics

C. Big Data execution

D. Big Data solutions

9.10 Which of the following is NOT one of the "Vs" of Big Data?

A. Volume

B. Variety

C. Visibility

D. Velocity

9.11 Volume 4, *Security and Privacy*, of the *NIST Big Data Interoperability Framework* identifies a number of ways that security and privacy in Big Data projects can be different from traditional implementations. Which one of the following is NOT one of these differences?

A. The use of multiple Big Data sources not originally intended to be used together can compromise privacy, security, or both.

B. Big Data projects increasingly involve one or more streamed data sources that are used in conjunction with data at rest, creating unique security and privacy scenarios.

C. An increased reliance on sensor streams can create vulnerabilities that were more easily managed before they were amassed to Big Data scale.

D. Big Data projects often encompass homogeneous components in which a single security scheme has not been designed from the outset.

9.12 The additional characteristic of Big Data that is sometimes referred to as provenance and that addresses the problem of understanding the data's original source is known as which one of the following?

A. Veracity

B. Validity

C. Volatility

D. Volume

9.13 Which of the following describe differences between what is new for Big Data and the requirements of previous big system security and privacy, as discussed in volume 4 of the *NIST Big Data Interoperability* Framework?

A. Big Data may be gathered from diverse end points and from more sources than traditional providers and consumers.

B. Big Data systems may be operational outside formal, readily understood frameworks.

C. Data search and selection can lead to privacy or security policy concerns.

D. All of the above

9.14 The use of sensors, advanced analytics, and intelligent decision-making that forms the basis of significant improvements in the methods that field assets communicate with enterprise activities is known as which of the following?

A. Informed Internet of Things (IIoT)

B. Industrial Internet of Things (IIoT)

C. Information Internet of Things (IIoT)

D. Inclusive Internet of Things (IIoT)

9.15 Strategies that focus on pushing data into a centralized cloud platform that can bridge the gap between industrial and IT domains is known as which of the following?

A. Informed Internet of Things (IIoT)

B. Industrial Internet of Things (IIoT)

C. Information Internet of Things (IIoT)

D. Inclusive Internet of Things (IIoT)

9.16 The data sent to the cloud from industrial networks originates from devices that have their own particular characteristics and operating parameters. In this scenario, which of the following emerging standards are relevant?

A. Message Queueing Telemetry Transport (MQTT)

B. Asynchronous Messaging Queuing Protocol (AMQP)

C. Constrained Application Protocol (COAP)

D. All of the above

9.17 Big Data technologies can be divided into which of the following two groups?

A. Batch processing and stream processing

B. Distributed processing and stream processing

C. Multi-core processing and stream processing

D. Batch processing and distributed processing

9.18 Which of the following statements is FALSE in regards to how Big Data technologies will transform security analytics?

A. By collecting data at a massive scale from many internal enterprise sources and external sources such as vulnerability databases

B. By using nonintrusive methods when conducting penetration testing

C. By providing a consolidated view of security-related information

D. By achieving real-time analysis of streaming data

9.19 Targeted attacks against a high-value asset or a physical system that can be extended over a long period of time without the knowledge of the infected organization is known as which of the following?

A. Black box threats

B. Advanced persistent threats

C. Zero day threats

D. Automatic threats

9.20 Authenticity and integrity of data used for analytics is known as which of the following?

A. Data provenance

B. Data enumeration

C. Data optimization

D. Data resolution

9.21 Systems that integrate computation, communication, sensing, and actuation with physical systems to fulfill time-sensitive functions with varying degrees of interaction with the environment, including human interaction, are defined as which of the following?

A. Analog control systems

B. Remote control systems

C. Cyber-physical systems

D. Enabled control systems

9.22 The NIST *Framework for Cyber-Physical Systems* comprises which of the following key elements?

A. Domains

B. Aspects

C. Facets

D. All of the above

9.23 A facet in the NIST *Framework for Cyber-Physical Systems* comprises Views encompassing identified responsibilities in the systems engineering process. Each facet contains a set of well-defined activities and artifacts (outputs) for addressing concerns. The three identified facets in the View model are which of the following?

A. Conceptualization, realization, and assurance

B. Conceptualization, realization, and confirmation

C. Conceptualization, identification, and assurance

D. Evaluation, realization, and assurance

9.24 Depending on the application, a CPS can be categorized into which of the following complexities?

A. Critical tightly-coupled

B. Loosely-coupled

C. System of Systems (SoS)

D. All of the above

9.25 On 12 February 2013, Executive Order (EO) 13636, "Improving Critical Infrastructure Cybersecurity," established that "it is the policy of the United States to enhance the security and resilience of the Nation's critical infrastructure" and called for the development of a voluntary risk-based Cybersecurity Framework. The goal of the Framework, developed by NIST, is to provide a common taxonomy and mechanism for organizations to do which of the following?

A. Describe their current cybersecurity posture

B. Describe their target state for cybersecurity

C. Assess progress toward the target state

D. All of the above

9.26 The Cybersecurity Framework comprises which of the following three parts?

A. The Framework Core, Framework Elements, and a Framework Profile

B. The Framework Architecture, Framework Implementation Tiers, and a Framework Profile

C. The Framework Core, Framework Implementation Tiers, and a Framework Profile

D. The Framework Core, Framework Implementation Tiers, and the Framework Security Posture

9.27 The NIST Cybersecurity Framework describes how it can be applied by organizations in which of the following ways?

A. Basic review of cybersecurity practices

B. Establishing or improving a cybersecurity program

C. Communicating cybersecurity requirements with stakeholders

D. All of the above

9.28 The means that can be used to program the interconnections of switches, routers, firewalls, and network security appliances is known as which of the following?

A. Software-implemented networking

B. Software-defined networking

C. Elemental networking

D. Software-defined specification

## References


1. Lydon, Bill. "Internet of Things." *InTech Magazine*, March/April, 2014. https://www.isa.org/standards-and-publications/isa-publications/intech-magazine/2014/mar-apr/cover-story-internet-of-things/.

2. CTIA, The Wireless Association. *Mobile Cybersecurity and the Internet of Things Empowering M2M Communication*. White paper, CTIA, 2014.

3. The German Federal Ministry of Education and Research. *Recommendations for Implementing the Strategic Initiative INDUSTRIE 4.0*. Final report of the Industrie 4.0 Working Group, 2013.

4. Setlurleads, Badrinath, Ganesh Iyer, and Shanmuga Varadan. "Informed Manufacturing: The Next Industrial Revolution." *Cognizant*, April 2014.

5. Ibid.

6. McHenry, Mark A., Dennis Roberson, and Robert J. Matheson. "Electronic Noise Is Drowning Out the Internet of Things." *IEEE SPECTRUM*, September 17, 2015. http://spectrum.ieee.org/telecom/wireless/electronic-noise-is-drowning-out-the-internet-of-things.

7. OWASP (Open Web Application Security Project). "OWASP Internet of Things Project." Last modified February 5, 2016. https://www.owasp.org/index.php/OWASP_Internet_of_Things_Top_Ten_Project.

8. FTC (Federal Trade Commission). "Internet of Things." FTC Staff Report, January 2015.

9. Ibid.

10. NIST Big Data Public Working Group (NBD-PWG), Security and Privacy Subgroup. *DRAFT NIST Big Data Interoperability Framework: Volume 1, Definitions*. Draft Version 1. Washington, DC: NIST (National Institute of Standards and Technology), 2015.

11. NIST Big Data Public Working Group (NBD-PWG). Security and Privacy Subgroup. *DRAFT NIST Big Data Interoperability Framework: Volume 4, Security and Privacy*. Draft Version 1, Washington, DC: NIST (National Institute of Standards and Technology), 2015.

12. NIST Big Data Public Working Group (NBD-PWG). Security and Privacy Subgroup. *DRAFT NIST Big Data Interoperability Framework: Volume 4, Security and Privacy*. Draft Version 1, Washington, DC: NIST (National Institute of Standards and Technology), 2015.

13. Internet Engineering Task Force (IETF). *Request for Comments: 7252*. ISSN: 2070-1721. Universitaet Bremen, June 2014.

14. OMG (Object Management Group). "What is DDS?" Object Management Group, 2015. http://portals.omg.org/dds/what-is-dds-3/.

15. Big Data Working Group. "Big Data Analytics for Security Intelligence." White paper. CSA (Cloud Security Alliance), September 2013.

16. Ibid.

17. Ibid.

18. Cyber Physical Systems Public Working Group. *DRAFT Framework for Cyber-Physical Systems*. Release 0.8. Washington, DC: CPS PWG, September 2015.

19. Ibid.

20. Ibid.

21. IEEE-IES. "ReSia Challenge." http://ieee-ies.org/resources/media/members/committees/resia/challenges/MultidisciplinaryProjectforReSia.pdf.

22. CSA (Cloud Security Alliance) Software Defined Perimeter Working Group. "Software Defined Perimeter," CSA, December 2013. https://downloads.cloudsecurityalliance.org/initiatives/sdp/Software_Defined_Perimeter.pdf

23. Ibid.

24. Check Point Software Technologies Ltd. *Software-Defined Protection, Enterprise Security Blueprint*. White paper. https://www.checkpoint.com/downloads/product-related/Ebook/Software-defined%20Protection(2).pdf

# Appendix A
# Review Questions and Answers

## Review Questions for Chapter 1

1.1 According to NIST SP 800-82, the term *industrial control system* encompasses which of the following?

A. SCADA systems

B. Distributed control systems (DCSs)

C. Programmable logic controllers (PLCs)

D. All of the above

**Answer: D**

1.2 According to ANSI/ISA-62443-1-1 (99.01.01)-2007, "a collection of personnel, hardware, and software that can affect or influence the safe, secure, and reliable operation of an industrial process" is a(n):

A. Industrial automation and control system

B. Distributed control system

C. Discrete control system

D. Programmable logic controller

**Answer: A**

1.3 A "type of loosely coupled distributed monitoring and control system commonly associated with electric power transmission and distribution systems, oil and gas pipelines, and water and sewage systems" is defined by ANSI/ISA-62443-1-1 (99.01.01)-2007 as which of the following?

A. Distributed control system

B. SCADA system

C. Safety instrumented system (SIS)

D. Human-machine interface (HMI)

**Answer: B**

1.4 Which one of the following items is NOT a component of a classical SCADA system model?

A. Human-machine interface (HMI)

B. Remote terminal unit (RTU)

C. Enterprise resource planning (ERP)

D. Programmable logic controller (PLC)

**Answer: C. ERP is a resource used by management systems.**

1.5 Which of the following is NOT a characteristic of a SCADA master terminal unit (MTU)?

A. Has two-way communication with field devices

B. Usually located at the master control center

C. Has high-bandwidth communication requirements

D. Communicates through telephone, VHF/UHF radio, spread spectrum radio, satellite, and/or microwave

**Answer: C. A MTU has low-bandwidth communication requirements.**

1.6 In a typical SCADA application, what component serves as a data concentrator and is an interface between the MTU and field devices?

A. HMI

B. RTU

C. Data historian

D. Relational database

**Answer: B**

1.7 IEC standard 61131-3 was designed to make it easier to implement control logic functions. IEC 61131-3 is which of the following?

A. Vendor-independent international standard for PLC programming languages

B. Standard guide to industrial control systems

C. Safety instrumented systems (SIS) standard for the process industry

D. Integrated enterprise-wide risk management standard

**Answer: A**

1.8 Which of the following is NOT a characteristic of a data historian?

A. Provides for prompt recovery of data

B. Performs data compression

C. Supports interactive storage and retrieval of detailed production information

D. Provides system histories over given time periods

**Answer: C. Interactive storage and retrieval are functions of real-time relational databases.**

1.9 A yield accounting system used in conjunction with a SCADA system provides which of the following functions?

A. Plant material movement

B. Inventory discrepancies

C. Material balances

D. All of the above

**Answer: D**

1.10 ANSI/ISA-62443-1-1 (99.01.01)-2007 defines which of the following as "a type of control system in which the system elements are dispersed but operated in a coupled manner"?

A. Distributed control system

B. Discrete control system

C. Dispersed control system

D. Coupled control system

**Answer: A. Answer B, a discrete control system, is one in which parameters at one or more points may change only at discrete values of time. Answers C and D are made-up distracters.**

1.11 ANSI/ISA-84.00.01-2004 Part 1(IEC 61511-1 Mod) defines which of the following as an "instrumented system used to implement one or more safety instrumented functions (SIF)"?

A. Safety surety system

B. Failure proof system

C. Safety instrumented system

D. Safety function system

**Answer: C. The other answers are made-up distracters.**

1.12 Necessary activities involved in the implementation of safety instrumented functions occurring during a period of time that starts at the concept phase of a project and finishes when all the safety instrumented functions are no longer available for use is known as which of the following?

A. Safety duration

B. Safety life cycle

C. Safety boundary

D. Safety period

**Answer: B**

1.13 The safety integrity level (SIL) is a discrete level for specifying the safety integrity requirements of the safety instrumented functions to be allocated to the safety instrumented systems. Which of the following statements is TRUE?

A. Safety integrity level 4 has the lowest level of safety integrity; safety integrity level 1 has the highest.

B. Safety integrity level 4 has the highest level of safety integrity; safety integrity level 1 has the lowest.

C. Safety integrity level 3 has the highest level of safety integrity; safety integrity level 1 has the lowest.

D. Safety integrity level 3 has the lowest level of safety integrity; safety integrity level 1 has the highest.

**Answer: B**

1.14 The Open Systems Interconnection (OSI) reference model has how many layers?

A. Six

B. Five

C. Seven

D. Four

**Answer: C. The layers are application, presentation, session, transport, network, data link, and physical.**

1.15 Which layer of the OSI model converts packets into electrical or optical signals for sending on the transmission media?

A. Application

B. Presentation

C. Physical

D. Session

**Answer: C**

1.16 What are the four layers of the TCP/IP model?

A. Application, Host-to-Host, Internet, and Network Access

B. Application, Presentation, Internet, and Network Access

C. Application, Network, Internet, and Network Access

D. Application, Host-to-Host, Internet, and Physical

**Answer: A**

1.17 NIST SP 800-82 defines which of the following as "a set of open standards developed to promote interoperability between disparate field devices, automation/control, and business systems"?

A. Modbus/TCP

B. OPC

C. DNP3

D. PROFIBUS

**Answer: B**

1.18 The protocol that supports the communication of safety-related data over a variety of industrial networks and ensures data transfer integrity is known as which of the following?

A. Safety instrumented system (SIS)

B. Safety instrumented function (SIF)

C. openSAFETY

D. ClosedSafety

**Answer: C**

1.19 Which of the following statements relative to the IEC-layered architecture standard IEC 61850 for substation automation is FALSE?

A. It is a set of protocols for electric utilities.

B. IEC-enabled devices can obtain power grid condition data via an Ethernet process bus.

C. Merge units provide interfaces to field devices.

D. It is not compatible with legacy protocols.

**Answer: D**

1.20 Which of the following are versions of PROFIBUS?

A. PROFIBUS PA, PROFIBUS CP, PROFIBUS CIP

B. PROFIBUS PA, PROFIBUS DP, PROFIBUS FMS

C. PROFIBUS PA, PROFIBUS CP, PROFIBUS FMS

D. PROFIBUS PA, PROFIBUS CIP, PROFIBUS FMS

**Answer: B**

## Review Questions for Chapter 2

2.1 Information system security is defined as comprising which of the following three basic elements?

A. Confidentiality, integrity, and authorization

B. Confidentiality, integrity, and availability

C. Audit, integrity, and availability

D. Security, integrity, and availability

**Answer: B**

2.2 Protecting documents and messages from unauthorized disclosure refers to which of the following?

A. Confidentiality

B. Availability

C. Authorization

D. Integrity

**Answer: A**

2.3 An attack that overloads the resources of a computing system is an attack against which of the following?

A. Integrity

B. Availability

C. Confidentiality

D. Authentication

**Answer: B**

2.4 Which of the following items refers to the act of verifying a user's identity and confirming that the user is who he or she professes to be?

A. Authentication

B. Authorization

C. Registration

D. Accountability

**Answer: A**

2.5 Which of the following is a detective activity used to determine if violations of information system security have occurred?

A. Confirming

B. Authenticating

C. Auditing

D. Authorizing

**Answer: C**

2.6 Ensuring that the sender of a message or contract cannot later deny sending the message or contract is known as which of the following?

A. Nonrepudiation

B. Denial

C. Validation

D. Confirmation

**Answer: A**

2.7 A security control that minimizes the effect of an attack and the degree of resulting damage is known as which type of control?

A. Corrective

B. Preventive

C. Deterrent

D. Detective

**Answer: A**

2.8 The Information Assurance Technical Framework Forum (IATF) Document (Table 2-1) defines which of the following attacks as an "attempt to circumvent or break protection features, introduce malicious code, or steal or modify information"?

A. Distribution

B. Close-in

C. Active

D. Passive

**Answer: C**

2.9 The act of establishing numerous layers of protection wherein a subsequent layer will provide protection if a previous layer is breached is known as which of the following?

A. Defense in depth

B. Complete mediation

C. Least privilege

D. Open design

**Answer: A**

2.10 Which of the following is noncompulsory?

A. Standards

B. Guidelines

C. Policies

D. Procedures

**Answer: B**

2.11 A self-replicating or self-reproducing program that spreads by inserting copies of itself into other executable code or documents is known as which of the following?

A. Worm

B. Mobile code

C. Back door

D. Virus

**Answer: D**

2.12 An attack that focuses on the authentication protocol between a claiming party and a verifying party communicating with each other is known as which of the following?

A. Meet-in-the-middle

B. Man-in-the-middle

C. Social engineering

D. Brute force

**Answer: B**

2.13 NIST SP 800-41 defines which of the following as "devices or programs that control the flow of network traffic between networks or hosts that employ differing security postures"?

A. Firewalls

B. Demilitarized zones

C. Trap doors

D. Browsers

**Answer: A**

2.14 What type of firewall filters packets based on firewall access control lists (ACLs), which specify the packets that can be sent to particular destination addresses or on specific application port destinations?

A. Stateful inspection

B. Packet filtering

C. Application proxy

D. Screened-host

**Answer: B**

2.15 What type of firewall serves as an intermediary between networks and masks the source of a message?

A. Application

B. Packet filtering

C. Stateful inspection

D. Application proxy

**Answer: D**

2.16 What type of firewall defines a demilitarized zone (DMZ) or perimeter network, which is a third network that lies between the internal trusted network and the external untrusted network as an added layer of protection?

A. Dual-homed host

B. Screened-subnet

C. Screened-host

D. Application

**Answer: B**

2.17 What type of cryptography uses public and private keypairs?

A. Symmetric

B. Asymmetric

C. Secret key

D. Transposition

**Answer: B**

2.18 The Advanced Encryption Standard (AES) is an example of which of the following types of cryptography?

A. Symmetric key

B. Asymmetric key

C. Public key

D. Dual key

**Answer: A**

2.19 The result of a variable length message being fed into a hash function is a shorter, fixed-length output digital stream, known as which of the following?

A. Ciphertext

B. Plaintext

C. Message digest

D. Cryptovariable

**Answer: C**

2.20 A digital signature uses asymmetric key encryption to achieve which of the following?

A. Nonrepudiation

B. Nonrepudiation and authentication

C. Authorization

D. Confidentiality

**Answer: B**

2.21 A cryptographic attack in which the attacker uses multiple samples of encrypted messages that have been encrypted with the same algorithm to attempt to find the key is known as which of the following?

A. Chosen ciphertext

B. Ciphertext only

C. Adaptive chosen ciphertext

D. Known plaintext

**Answer: B**

2.22 A private network that operates as an overlay on a public infrastructure is known as which of the following?

A. Virtual private network

B. Dual-band network

C. 4G network

D. Demilitarized zone

**Answer: A**

2.23 IPsec provides which of the following two modes of operation?

A. Symmetric key mode and transport mode

B. Authentication mode and tunnel mode

C. Transport mode and tunnel mode

D. Transport mode and transparent mode

**Answer: C**

2.24 A VPN that provides secure communications over a public network between two trusted networks is known as which of the following?

A. Host-to-host

B. Gateway-to-gateway

C. Host-to-gateway

D. Host-to-demilitarized zone

**Answer: B**

## Review Questions for Chapter 3

3.1 Which of the following statements is generally TRUE regarding an industrial automation and control system?

A. Installation of software patches can be performed routinely and frequently.

B. Encryption of data can sometimes lead to problematic delays.

C. Penetration testing can be conducted routinely and frequently.

D. Confidentiality is a key concern in automation systems as opposed to integrity and availability.

**Answer: B**

3.2 In both IT and automation and control systems, which of the following is the primary concern in the event of an emergency or malicious event?

A. Equipment safety

B. Preservation of documentation

C. Personnel safety

D. Facility protection

**Answer: C**

3.3 Which of the following statements is FALSE?

A. Flash drives and other portable memory devices can be sources of malware injections into control systems.

B. Maintenance hooks and trap doors installed in automation and control systems for remote maintenance can be easy entry points to modify critical software and firmware with negative consequences.

C. In many control system environments, control engineers, in general, do not have multiple responsibilities, such that the security principle of separation of duties is not normally violated.

D. Many facilities house legacy systems with outdated technology, minimal memory and computing power, and little thought to security.

**Answer: C**

3.4 Which of the following actions is the most likely to result in blockages and lack of system availability in automation and control systems?

A. Remote access

B. Life cycle design

C. Accountability

D. Port scanning

**Answer: D**

3.5 Which threat source is motivated by revenge, ego, and dissatisfaction?

A. Insider

B. Espionage

C. Criminal

D. Hacker

**Answer: A**

3.6 What is a source of a possible disruption of control system functions that is not normally considered?

A. Changes from digital to analog systems

B. Upgrades from analog to digital systems

C. Malware

D. Attacks

**Answer: B**

3.7 In general, what distinguishes analog control equipment from digital control equipment?

A. Analog controls generate more high-frequency peak voltages than digital controls.

B. Digital controls generate more high-frequency peak voltages than analog controls.

C. Analog controls generate essentially the same number of high-frequency peak voltages as digital controls.

D. Digital controls generate essentially the same number of high-frequency peak voltages as analog controls.

**Answer: B**

3.8 Which of the following is more likely to be performed in an IT environment than in an automation and control system environment?

A. Security architecture analysis

B. Information security testing

C. Quantitative risk analysis

D. Change management

**Answer: D**

3.9 Which of the follow statements is FALSE?

A. In many IACS environments, control engineers have multiple responsibilities that, in many instances, violate the security principle of separation of duties.

B. Installing patches and upgrades in process control systems can lead to serious and sometimes dangerous situations in production facilities.

C. Life-cycle design disciplines common in the IT field are not widely used in industrial automation and control systems.

D. Many IACS vendors combine safety mechanisms with security mechanisms, which results in increased system reliability.

**Answer: D**

3.10 Which of the following threat sources is motivated by economic exploitation and competitive advantage, and uses social engineering?

A. Insider

B. Terrorist

C. Industrial espionage

D. Computer criminal

**Answer: C**

3.11 What is a detective control that is more frequently applied in IT systems than in control and automation systems?

A. Firewall

B. Separation of duties

C. Biometrics

D. Auditing

**Answer: D**

3.12 Which of the following is NOT a usual reason for an organization's reluctance to disclose successful attacks against it?

A. Hope the attack will harm competitors

B. Embarrassment

C. Effect on reputation

D. Possible loss of customers

**Answer: A**

3.13 Which of the following can lead to a single point of failure in an industrial automation and control system?

A. Separation of duties

B. Disk redundancy

C. Combination of safety and security mechanisms

D. Use of authentication with identification

**Answer: C**

3.14 What is a typical characteristic of industrial automation and control systems?

A. Have excess computing cycles

B. Have limited extra computing cycles

C. Have excess memory

D. Computational speed is not an issue

**Answer: B**

3.15 What is a typical characteristic of an automation and control system supplier?

A. Usually ensures maintenance hooks are never left enabled without the customer's approval

B. Usually ensures default passwords are never duplicated from one customer to another

C. Usually provides unmodified off-the-shelf hardware and software

D. Usually provides modified hardware and software

**Answer: D**

## Review Questions for Chapter 4

4.1 Which of the following is an important element in connecting to smart meters?

A. Home area network

B. Energy storage devices

C. Cloud computing

D. Printers

**Answer: A**

4.2 Energy storage devices are especially useful in which of the following cases?

A. Fossil fuel energy sources

B. Nonrenewable energy sources

C. Intermittent renewable energy sources

D. Constant energy sources

**Answer: C**

4.3 The application of data mining and analysis to large amounts of data for the purposes of discovering knowledge and making intelligent predictions and decisions is known as which of the following?

A. Artificial intelligence

B. Analytics

C. Data aggregation

D. Predictive calculus

**Answer: B**

4.4 Which of the following techniques is used for condition monitoring of high-voltage electrical equipment and identifying areas of potential failure?

A. Analytics

B. Artificial intelligence

C. Robotics

D. Scanning

**Answer: A**

4.5 Data from a variety of sources that is analyzed in real time or near real time to perform predictive modeling and support physical, logical, and administrative security is known as which of the following?

A. Data aggregation

B. Cyber analytics

C. Cybersecurity

D. Sensitivity analysis

**Answer: B**

4.6 What of the following is NOT a characteristic of cloud computing?

A. Measured service

B. Resource pooling

C. On-demand self-service

D. Minimal elasticity

**Answer: D**

4.7 Which of the following is NOT one of the three cloud service models?

A. Cloud Application as a Service

B. Cloud Software as a Service

C. Cloud Platform as a Service

D. Cloud Infrastructure as a Service

**Answer: A**

4.8 Which of the following is NOT one of the five basic privacy principles?

A. Access

B. Choice

C. Preemption

D. Notice

**Answer: C**

4.9 Which of the following items is NOT one of the European Union (EU) privacy principles?

A. Data should be collected in accordance with the law.

B. Data should be used only for the purposes for which it was collected, and it should be used only for a reasonable period of time.

C. Transmission of personal information to locations where equivalent personal data protection cannot be assured is permitted.

D. Individuals have the right to correct errors contained in their personal data.

**Answer: C**

4.10 The ability of an entity to exchange information with another entity is known as which of the following?

A. Compatibility

B. Commonality

C. Interchangeability

D. Interoperability

**Answer: D**

4.11 Which of the following is NOT a distinguishing characteristic of the Smart Grid?

A. Development and incorporation of demand response, demand-side resources, and energy efficiency resources

B. Deployment and integration of distributed resources and generation, excluding renewable resources

C. Provision to consumers of timely information and control options

D. Increased use of digital information and controls technology to improve reliability, security, and efficiency of the electric grid

**Answer: B**

4.12 Which of the following is NOT an expected result of Smart Grid implementation?

A. Higher grid resiliency

B. Improved energy delivery efficiency

C. Reduced operating costs

D. Possible loss of customers

**Answer: D**

4.13 Which of the following is NOT a domain in the NIST Smart Grid framework?

A. Control

B. Customer

C. Operations

D. Bulk generation

**Answer: A**

4.14 Which domain of the Smart Grid framework provides energy generated from sources such as coal, gas, water from dams, nuclear reactions, geothermal, the sun, and wind?

A. Distribution

B. Transmission

C. Bulk generation

D. Service provider

**Answer: C**

4.15 Which domain of the Smart Grid delivers electrical power to the distribution domain and balances the required electrical load?

A. Transmission

B. Service provider

C. Operations

D. Bulk generation

**Answer: A**

4.16 Which domain of the Smart Grid handles customer problems and manages business accounts?

A. Markets

B. Service provider

C. Customer

D. Operations

**Answer: B**

4.17 What subsystem of the Smart Grid comprises hardware and meter data management (MDM) software applications that provide the normal meter reading functions as well as supporting two-way communications that can exchange energy information and commands with customer's devices through a home area network?

A. Advanced Metering Infrastructure (AMI)

B. Energy Management System (EMS)

C. Energy Services Interface (ESI)

D. Smart Grid Interface (SGI)

**Answer: A**

4.18 In Smart Grid terminology, DER stands for which of the following terms?

A. Delayed energy reduction

B. Determined energy requirements

C. Distributed energy requirements

D. Distributed energy resources

**Answer: D**

4.19 Which domain of the Smart Grid framework has a home area network and also might have a requirement for energy storage for intermittent sources of electricity?

A. Customer

B. Distribution

C. Operations

D. Markets

**Answer: A**

4.20 Which domain of the Smart Grid framework handles large amounts of customer personally identifiable information (PII) and must ensure that adequate privacy protections are in place?

A. Customer

B. Markets

C. Service provider

D. Operations

**Answer: C**

4.21 Which of the following evolving technologies can increase communications, collaboration, and innovation, but also be a source of compromise of PII and other sensitive information?

A. Social networks

B. Interoperability

C. Cloud computing

D. Analytics

**Answer: A**

4.22 Which domain of the Smart Grid framework does not have an energy storage requirement for intermittent sources of electricity?

A. Customer

B. Transmission

C. Bulk generation

D. Markets

**Answer: D**

4.23 Which domain of the Smart Grid framework has the most potential for benefiting from the use of cloud computing?

A. Customer

B. Transmission

C. Distribution

D. Operations

**Answer: D**

## Review Questions for Chapter 5

5.1 In ANSI/ISA-62443-1-1 (99.01.01)–2007, a "potential for violation of security, which exists when there is a circumstance, capability, action, or event that could breach security and cause harm," is which of the following?

A. Threat

B. Vulnerability

C. Weakness

D. Risk

**Answer: A**

5.2 The "expectation of loss expressed as the probability that a particular threat will exploit a particular vulnerability with a particular consequence" is which of the following?

A. Consequence

B. Threat source

C. Weakness

D. Risk

**Answer: D**

5.3 The program and supporting processes to manage information security risk to organizational operations (i.e., mission, functions, image, reputation), organizational assets, individuals, other organizations, and the nation are defined as which of the following?

A. Risk assessment

B. Risk management

C. Risk mitigation

D. Risk association

**Answer: B**

5.4 Which of the following is considered an asset in risk management?

A. Computer hardware

B. A control system

C. A professional with unique knowledge

D. All of the above

**Answer: D**

5.5 The ANSI/ISA-62443-2-1 (99.02.01)-2009 Cyber Security Management System (CSMS) comprises which of the following three main categories?

A. Risk analysis, addressing the risk, and monitoring and improving the CSMS

B. Risk mitigation, addressing the risk, and monitoring and improving the CSMS

C. Risk analysis, addressing the risk, and monitoring and improving the automation system

D. Risk analysis, eliminating the risk, and monitoring and improving the CSMS

**Answer: A**

5.6 In the CSMS, a business rationale is established. Which of the following is NOT one of the elements of the business rationale?

A. Establishing a basis for securing the industrial automation and control system

B. Obtaining management support for securing automation systems

C. Identifying and understanding the consequences of a successful attack on automation systems

D. Confirmation of the value of the automation system

**Answer: D**

5.7 The risk identification, classification, and assessment activities of the CSMS do NOT include which of the following?

A. Conduct a high-level risk assessment

B. Develop simple network diagrams

C. Mitigate the risk

D. Perform a detailed vulnerability assessment

**Answer: C**

5.8 Which of the following is NOT one of the areas included in "Addressing risk with the CSMS?"

A. Security policy, organization, and awareness

B. Correction

C. Selected security countermeasures

D. Implementation

**Answer: B**

5.9 Subdividing a network into zones that have similar security requirements and characteristics and are separated by protective devices to screen traffic from one network segment to another is known as which of the following?

A. Network segmentation

B. Network isolation

C. Network masking

D. Network fragmentation

**Answer: A**

5.10 In the CSMS, access control is subdivided into which of the following three areas?

A. Account verification, authentication, and authorization

B. Account administration, authentication, and authorization

C. Account administration, audit, and authorization

D. Account administration, authentication, and activation

**Answer: B**

5.11 Risk management, system development and maintenance, and incident planning are part of which of the following areas of addressing risk in the CSMS?

A. Security policy, organization, and awareness

B. Selected security countermeasures

C. Implementation

D. Scope

**Answer: C**

5.12 In the monitoring and improving category of the CSMS, which element verifies that an automation system is compliant with appropriate policies, procedures, and regulations?

A. Review

B. Conformance

C. Improve

D. Maintain

**Answer: B**

5.13 NIST SP 800-39 views which of the following as a comprehensive process that requires organizations to: (1) frame risk (i.e., establish the context for risk-based decisions); (2) assess risk; (3) respond to risk once determined; and (4) monitor risk on an ongoing basis?

A. Risk management

B. Risk assessment

C. Risk monitoring

D. Risk evaluation

**Answer: A**

5.14 NIST SP 800-39 defines three tiers. Which of the following includes those three?

A. Policy, mission/business processes, information systems

B. Organization, policy, information systems

C. Organization, mission/business processes, objectives

D. Organization, mission/business processes, information systems

**Answer: D**

5.15 Categorizing organizational information systems and properly allocating security controls to those systems to support the organization's mission/business processes is performed in which tier of the multi-tiered risk management architecture?

A. Tier 1

B. Tier 2

C. Tier 3

D. Tier 4

**Answer: C**

5.16 According to NIST SP 800-39, risk framing, risk assessment, risk response, and risk monitoring are components of which of the following?

A. Risk valuation

B. Risk determination

C. Risk management

D. Risk acceptance

**Answer: C**

5.17 Making assumptions about threats, vulnerabilities, consequences/impact, and likelihood of occurrence are part of which of the following, according to NIST SP 800-39?

A. Risk assessment

B. Risk framing

C. Risk response

D. Risk monitoring

**Answer: B**

5.18 Which of the following is NOT an element of responding to risk, according to NIST SP 800-39?

A. Evaluating the effectiveness of implemented risk response measures

B. Developing alternative courses of action

C. Selecting the best courses of action

D. Implementing selected responses

**Answer: A**

5.19 According to NIST SP 8000-37, which of the following "provides a disciplined and structured process that integrates information security and risk management activities into the system development life cycle?"

A. Risk Monitoring Framework (RMF)

B. Risk Assessment Framework (RAF)

C. Risk Management Framework (RMF)

D. Risk Evaluation Framework (REF)

**Answer: C**

5.20 Which of the following groups of elements is presented in NIST SP 800-37 to manage risk?

A. Categorize, select, implement, authorize

B. Categorize, evaluate, implement, monitor

C. Determine, select, implement, authorize

D. Determine, evaluate, implement, monitor

**Answer: A**

5.21 Which of the following is NOT characteristic of an insider threat?

A. Many insider attacks are conducted by disgruntled insiders.

B. Most insider attacks do not result in serious losses or harm.

C. Many insider attacks are conducted remotely.

D. Many inside attackers have privileged access to computer systems.

**Answer: B**

5.22 Which of the following best describes Stuxnet?

A. A worm that was designed to change control outputs on specific PLCs and conceal its existence from control room observers

B. A worm that was designed to modify email messages and conceal its existence from control room observers

C. A virus that was designed to modify disk storage and send random messages to users' computers

D. A virus that was designed to crash computers when email attachments were opened

**Answer: A**

5.23 Which of the following is NOT true regarding Stuxnet?

A. Attacks both networked and non-networked PCs

B. Installs device drivers using valid, digital certificates

C. Infects Apple Macintosh computers

D. Infection is accomplished through USB flash drives

**Answer: C**

5.24 The process that only uses software that is "clean" and not compromised and is verified before running is known as which of the following?

A. Confirming

B. Whitelisting

C. Cleanlisting

D. Prechecking

**Answer: B**

5.25 Modular computer malware that burrowed into Iran's oil ministry, severing Internet links to rigs and the hub for nearly all the country's crude exports, causing widespread data loss is known as which of the following?

A. Night Dragon

B. Stuxnet

C. Flame

D. Havex

**Answer: C**

5.26 An electromagnetic pulse generated by a high-altitude nuclear detonation in space at heights above 30 km is known as which of the following?

A. Intentional electromagnetic interference (IEMI)

B. Space generated EMP (SGEMP)

C. Intensive electromagnetic interference (IEMI)

D. High-altitude EMP (HEMP)

**Answer: D**

5.27 Which one of the following is NOT a component of a high-power electromagnetic (HPEM) burst?

A. Final-time (E4) – Several hundred seconds after the burst

B. Early-time (E1) – Within 10 nanoseconds of the burst

C. Intermediate-time (E2) – Between 1 microsecond and 1 second after the burst

D. Late-time (E3) – Between 1 second and several hundred seconds after the burst

**Answer: A**

5.28 Solar-generated EMP is the result of which of the following actions?

A. Changes in the sun's orbit

B. Solar eclipses

C. Geomagnetic storms on the sun

D. Changes in Earth's orbit

**Answer: C**

5.29 Which of the following statements is NOT true regarding EMP?

A. High-frequency pulses can induce currents in floating wires.

B. High-frequency pulses can penetrate through windows and gaps in metal shields.

C. High-frequency grounding with filters and surge arrestors can be used to protect sensors, smart meters, and communications systems.

D. Fiber-optic cabling cannot minimize coupling.

**Answer: D**

## Review Questions for Chapter 6

6.1 NIST SP 800-82, *Guide to Industrial Control Systems (ICS) Security,* partitions security controls into which one of the following groups?

A. Management controls, operational controls, and technical controls

B. Management controls, administrative controls, and technical controls

C. Management controls, operational controls, and physical controls

D. Logical controls, operational controls, and technical controls

**Answer: A**

6.2 NIST SP 800-82 defines five major security objectives for an industrial control system (ICS) implementation. Which of the following is NOT one of those objectives?

A. Restricting logical access to the ICS network and network activity

B. Restricting physical access to the ICS network and devices

C. Protecting individual ICS components from exploitation

D. Shutting down the ICS after an incident

**Answer: D**

6.3 In NIST SP 800-82, policies and procedures for controlling modifications to hardware, firmware, software, and documentation to ensure the information system is protected against improper modifications prior to, during, and after system implementation is known as which of the following?

A. Configuration management

B. Risk management

C. System management

D. System maintenance

**Answer: A**

6.4 In the NIST SP 800-82 controls, the process of granting or denying specific requests for obtaining and using information and related information processing services for physical access to areas within the information system environment is known as which of the following?

A. Authentication

B. Audit and accountability

C. Access control

D. Identification

**Answer: C**

6.5 The controls of ANSI/ISA-62443-1-1 (99.01.01)-2007 are organized into six categories. Which of the following is NOT one of those categories?

A. Encryption technologies and data validation

B. Risk mitigation technologies

C. Authentication and authorization technologies

D. Filtering/blocking/access control technologies

**Answer: B**

6.6 ANSI/ISA-62443-1-1 (99.01.01)-2007 describes which of the following as "the initial step in protecting an industrial automation and control system (IACS) and its critical assets from unwanted breaches. It is the process of determining who and what should be allowed into or out of a system"?

A. Authorization

B. Authentication

C. Identification

D. Confirmation

**Answer: A**

6.7 Which of the following are the major components of authentication and authorization technologies spelled out in ANSI/ISA-62443-1-1 (99.01.01)-2007?

A. Role-based, password, and challenge response

B. Rule-based, user ID, and challenge response

C. Role-based, password, and call-back

D. Rule-based, password, and call-back

**Answer: A**

6.8 Which of the following does ANSI/ISA-62443-1-1 (99.01.01)-2007 identify as the three main types of software that have to be considered in industrial automation and control system software?

A. Mobile operating systems, real-time and embedded operating systems, and web servers and Internet technologies

B. Server and workstation operating systems, real-time and embedded operating systems, and wireless technologies

C. Server and workstation operating systems, real-time and embedded operating systems, and web servers and Internet technologies

D. Server and workstation operating systems, real-time and embedded operating systems, and mobile technologies

**Answer: C**

6.9 In ANSI/ISA-62443-1-1 (99.01.01)-2007, what are the three main categories of physical security elements?

A. Active, identification and monitoring devices, and passive

B. Active, real time, and passive

C. Active, identification and monitoring devices, and real time

D. Reactive, identification and monitoring devices, and passive

**Answer: A**

6.10 The North American Electric Reliability Corporation (NERC) Critical Infrastructure Protection (CIP) Cyber Security Standards are divided into standards CIP-002-4 through CIP-009-4 and address "the identification and protection of critical cyber assets to support reliable operation of the Bulk Electric System (BES)." Which of the following is NOT one of the eight standards?

A. Risk assessment

B. Critical cyber asset identification

C. Security management controls

D. Electronic security perimeters

**Answer: A**

6.11 Which NERC CIP standard "ensures the identification, classification, response, and reporting of cybersecurity incidents related to critical cyber assets?"

A. Physical security of critical cyber assets

B. Systems security management

C. Recovery plans for critical cyber assets

D. Incident reporting and response planning

**Answer: D**

6.12 In NIST SP 800-53, the security controls are divided into 18 families, each with a unique, two-character identifier. The control families are themselves partitioned into three classes. Which of the three classes does the control family titled CP (contingency planning) fall under?

A. Technical

B. Management

C. Operational

D. Administrative

**Answer: C**

6.13 The NIST SP 800-53 controls are considered baseline controls, which can be tailored to more closely meet the needs of a particular organization and automation system. Which of the following is NOT one of the elements of the NIST SP 800-53 tailoring process?

A. Selecting (or specifying) compensating security controls, if needed, to adjust the preliminary set of controls to obtain an equivalent set deemed to be more feasible toimplement

B. Applying scoping guidance to the initial baseline security controls to obtain a preliminary set of applicable controls for the tailored baseline

C. Specifying organization-defined parameters in the security controls via explicit assignment and selection statements to complete the definition of the tailored baseline

D. Developing custom parameters to meet the needs ofthe control system

**Answer: D**

6.14 The NIST SP 800-53 tailoring process refers to guidance "designed to ensure that only the controls appropriate to the particular mission and requirements of the organization are used." This guidance is referred to as which of the following?

A. Correlating

B. Assessing

C. Evaluating

D. Scoping

**Answer: D**

6.15 NIST SP 800-53 provides for the use of security controls, which the publication defines as "a management, operational, or technical control (i.e., safeguard or countermeasure) employed by an organization in lieu of a recommended security control … that provides an equivalent or comparable level of protection for an information system and the information processed, stored, or transmitted by that system." This type of additional control is known as which of the following?

A. Supporting

B. Supplemental

C. Compensating

D. Custom

**Answer: C**

6.16 Recognizing some of the special requirements of industrial automation and control systems, NIST SP 800-53 defines supplements to the baseline control systems to address automation and control issues. Which one of the following is NOT one of these supplements?

A. Risk evaluation

B. Predictable failure prevention

C. Emergency power

D. Access enforcement

**Answer: A**

6.17 The Department of Homeland Security's *Catalog of Control Systems Security* is based on which of the following security controls documents?

A. NIST SP 800-82

B. NIST SP 800-53

C. NIST SP 800-39

D. The NERC CIP Cyber Security Standards

**Answer: B**

6.18 The Advanced Metering Infrastructure (AMI) Security Requirements are characterized by letters. Which of the following is NOT one of the AMI security requirements categories?

A. "R" are risk requirements

B. "F" are functional requirements

C. "S" are supporting services to functional requirements

D. "A" are assurance requirements

**Answer: A**

6.19 The AMI security requirement FIN refers to which of the following security principles?

A. Inventory

B. Final evaluation

C. Integrity

D. Internal

**Answer: C**

6.20 The recording of activity by actors/elements throughout the system and providing the means to perform a successful audit of events that occur on the system refers to which of the following AMI security controls?

A. Accounting

B. Anomaly detection

C. Authentication

D. Authorization

**Answer: A**

6.21 A major change in Version 5 of the NERC CIP standard was a focus on which of the following?

A. The risk to a Bulk Electric System (BES)

B. The risk to a registered entity's operations

C. The risk to physical and environmental factors

D. The risk to personnel

**Answer: A**

6.22 The North American Electric Reliability Corporation (NERC) Critical Infrastructure Protection (CIP) Cyber Security Standards, Version 3, are divided into standards CIP-002-4 through CIP-009-4 and address which of the following?

A. Availability of technologies to support reliable operation of the Bulk Electric System (BES)

B. The identification and protection of critical cyber assets to support reliable operation of the Bulk Electric System (BES)

C. Integrity of software to support reliable operation of the Bulk Electric System (BES)

D. Accountability of methodologies for the protection of critical cyber assets to support reliable operation of the Bulk Electric System (BES)

**Answer: B**

6.23 In NIST SP 800-82, *Guide to Industrial Control Systems (ICS) Security*, risk assessment, planning, systems and services acquisition, certification, accreditation, and security assessments fall under what category of controls?

A. Technical

B. Logical

C. Management

D. Operational

**Answer: C**

6.24 In NIST SP 800-82, *Guide to Industrial Control Systems (ICS) Security*, controls such as personnel security and contingency planning fall under which of the following control categories?

A. Management

B. Operational

C. Technical

D. Physical

**Answer: B**

6.25 NIST SP 800-53 provides guidance to organizations in selecting security controls through the concept of baseline controls, which should be chosen based on the security category and associated impact level of information systems determined in accordance with which of the following publications?

A. NIST 800-37 and FIPS 200

B. NIST 800-37 and NIST 800-39

C. FIPS 109 and FIPS 200

D. FIPS 199 and FIPS 200

**Answer: D**

6.26 ANSI/ISA-62443-1-1 (99.01.01)-2007, *Security Technologies for Industrial Automation and Control Systems*, defines which of the following principles as "the process of positively identifying potential network users, hosts, applications, services, and resources, using a combination of identification factors or credentials. The result of this process then becomes the basis for permitting or denying further actions"?

A. Authentication

B. Authorization

C. Accountability

D. Accessibility

**Answer: A**

6.27 ANSI/ISA-62443-1-1 (99.01.01)-2007, *Security Technologies for Industrial Automation and Control Systems*, divides authentication into which of the following two areas?

A. User authentication and group authentication

B. Group authentication and network service authentication

C. User authentication and network service authentication

D. Platform authentication and user authentication

**Answer: C**

6.28 ANSI/ISA-62443-1-1 (99.01.01)-2007 defines which of the following as something that "can be used to distinguish communication by devices that are not part of the desired network. This feature is generally attractive for SCADA and other process control systems that wish to allow little or no access to the control network, but is difficult to deploy in systems requiring unrestricted Internet access"?

A. Asymmetric key encryption

B. Digital signature

C. Biometrics

D. Symmetric key encryption

**Answer: D**

6.29 ANSI/ISA-62443-1-1 (99.01.01)-2007 states that encryption can also be employed in a virtual private network (VPN), which provides for authentication, integrity, and protection from unauthorized disclosure. Which of the following is NOT one of the most common VPN implementations?

A. Hypertext Transfer Protocol (HTTP)

B. Internet Protocol Security (IPSec)

C. Secure Shell (SSH)

D. Secure Sockets Layer (SSL)

**Answer: A**

6.30 What standards address "the identification and protection of critical cyber assets to support reliable operation of the Bulk Electric System (BES)"?

A. ANSI/ISA-62443-1-1 (99.01.01)-2007

B. NIST SP 800-53

C. NIST SP 800-82

D. NERC CIP Cyber Security Standards

**Answer: D**

6.31 Which of the following is NOT one of the NERC CIP Cyber Security Standards?

A. Risk Management

B. Critical Cyber Asset Identification

C. Security Management Controls

D. Personnel and Training

**Answer: A**

6.32 The following are characterized as what type of guidance in NIST SP 800- 53, which includes determining the primary security objectives, which system components the controls are applicable to, and what technologies are employed in the system?

A. Custom

B. Complementary

C. Scoping

D. Supplemental

**Answer: C**

6.33 NIST SP 800-53 security controls are organized into how many families?

A. 15

B. 17

C. 18

D. 20

**Answer: C**

6.34 The Organizational Security controls in the Department of Homeland Security *Catalog of Control Systems Security* relate to which of the following control categories in NIST SP 800-53?

A. Personnel Security (PS)

B. System and Services Acquisition (SA)

C. Access Control (AC) and Program Management (PM)

D. Contingency Planning (CP) and Media Protection (MP)

**Answer: C**

6.35 The Monitoring and Reviewing Control System Security Policy controlsin the *Catalog of Control Systems Security* relate to which of the following control categories in NIST SP 800-53?

A. Audit and Accountability (AU)

B. Program Management (PM)

C. Security Assessment and Authorization (CA)

D. System and Information Integrity (SI)

**Answer: C**

6.36 The AMI System Security Requirement that addresses "the proof of identity of an actor" is which of the following?

A. Authentication

B. Identification

C. Authorization

D. Nonrepudiation

**Answer: A**

6.37 The AMI System Security Requirement that is "based on security auditing and recognizing, recording, storing, and analyzing information related to security relevant activities" is which of the following?

A. Authorization

B. Authentication

C. Accountability

D. Integrity

**Answer: C**

## Review Questions for Chapter 7

7.1 What organizations host the National SCADA Test Bed (NSTB) program?

A. NIST and the NSA

B. The Idaho National Laboratory and the Sandia National Laboratories

C. ISA and SANS

D. US-CERT and NIST

**Answer: B**

7.2 Which of the following is NOT a function of the National SCADA Test Bed (NSTB)?

A. Assess selected control systems and control system components to identify cyber vulnerabilities

B. Develop new standards and guidelines for governmental agencies

C. Provide control system security training through workshops that describe common cyber vulnerabilities found in control systems

D. Share information that can be used to support the development of improved industry standards applicable to control system security with the appropriate standard's organizations

**Answer: B**

7.3 What organization developed and teaches the application of 62443-2-1 (99.02.01)-2009, *Security for Industrial Automation and Control Systems: Establishing an Industrial Automation and Control Systems Security Program*?

A. SANS

B. NIST

C. ISA

D. NSA

**Answer: C**

7.4 In a training program hands-on exercise, what is the function of the Red Team?

A. Mitigating

B. Scanning

C. Defending

D. Attacking

**Answer: D**

7.5 In a training program hands-on exercise, what is the function of the Blue Team?

A. Defending

B. Attacking

C. Scanning

D. Mitigating

**Answer: A**

7.6 What is fuzzing?

A. Modifying memory locations

B. An attack using random or malformed bad data

C. Scanning to determine vulnerabilities

D. Responding to an attack by counterattacking

**Answer: B**

7.7 The National Initiative for Cybersecurity Education (NICE) is a partnership of government, industrial, educational, and professional organizations. Which of the following is NOT one of the goals of NICE?

A. Raise awareness among the American public about the risks of online activities

B. Broaden the pool of skilled workers capable of supporting a cyber-secure nation

C. Develop and maintain an unrivaled, globally competitive cybersecurity workforce

D. Develop secure digital control systems

**Answer: D**

7.8 Which of the following organizations is NOT a participant in the NICE initiative?

A. NIST

B. NASA

C. Department of Homeland Security (DHS)

D. NSA

**Answer: B**

7.9 The *NICE Cybersecurity Workforce Framework* develops a vocabulary that can be used by any type of organization in categorizing cybersecurity work. Which of the following is NOT a category in this framework?

A. Penetration testing

B. Design

C. Operation

D. Maintenance

**Answer: A**

7.10 The U.S. National Centers of Academic Excellence initiative is sponsored by which of the following organizations?

A. NIST and the Department of Energy

B. SANS and the Department of Education

C. NSA and the Department of Homeland Security

D. The Department of Defense and the US-CERT

**Answer: C**

7.11 The designation of an institution as a National Center of Academic Excellence has a duration of how many years before the institution has to reapply?

A. Two years

B. Ten years

C. Five years

D. Seven years

**Answer: C**

7.12 Which of the following is NOT one of the critical steps in the life cycle of a security awareness and training program as described in NIST SP 800-50, *Building an Information Technology Security Awareness and Training Program*?

A. Awareness and training program design

B. Program implementation

C. Post-implementation

D. Pre-implementation needs review

**Answer: D**

7.13 Which of the following is NOT one of the three common models used in managing a security training function as described in NIST SP 800-50?

A. Centralized

B. Partially Decentralized

C. Partitioned

D. Fully Decentralized

**Answer: C**

7.14 According to NIST SP 800-50, learning is a continuum comprising which of the following elements?

A. Awareness, skill, knowledge

B. Awareness, training, education

C. Training, education, skill

D. Education, awareness, skill

**Answer: B**

7.15 NIST SP 800-16, *Information Security Training Requirements: A Role- and Performance-Based Model (Draft)*, defines the following as what type of training?

Training that "focuses on providing the knowledge, skills, and abilities specific to an individual's roles and responsibilities relative to information systems. At this level, training recognizes the differences between beginning, intermediate, and advanced skill requirements."

A. Role-based training

B. Task-oriented training

C. Responsibility training

D. Knowledge and skills training

**Answer: A**

7.16 Which one of the following is NOT one of the five phases of the industrial design model described in NIST SP 800-16?

A. Analysis

B. Training

C. Design

D. Evaluation

**Answer: B**

7.17 What are the two types of evaluations in the evaluation phase of the industrial design model described in NIST SP 800-16?

A. Formative and summative

B. Formative and intuitive

C. Summative and descriptive

D. Descriptive and formative

**Answer: A**

## Review Questions for Chapter 8

8.1 The technology that eliminates the binding of software to specific hardware platforms is known as which of the following?

A. Artificial intelligence

B. Virtualization

C. Cloud computing

D. Software as a service

**Answer: B**

8.2 A hypervisor is which of the following?

A. A virtual machine monitor

B. An advanced HMI

C. A manager

D. An alarm mechanism

**Answer: A**

8.3 NIST SP 800-137, *Information Security Continuous Monitoring (ISCM) for Federal Information Systems and Organizations*, defines the process for developing an ISCM strategy and implementing an ISCM program. Which one of the following best describes the ISCM strategy?

A. Providing security mechanisms to reduce risks to automation systems

B. Developing security policies to specify management intent for information systems and automation systems

C. Maintaining ongoing awareness of information security, vulnerabilities, and threats to support organizational risk management decisions

D. Maintaining ongoing awareness of risk mitigation alternatives

**Answer: C**

8.4 Which of the following is NOT one of the ISCM program steps?

A. Define

B. Establish

C. Respond

D. Mitigate

**Answer: D**

8.5 In the ISCM process, the status of automation security in a facility can be determined by establishing appropriate metrics. Which of the following activities is NOT used to define the metrics?

A. Evaluating applicable mitigation approaches

B. Assessing all security controls

C. Providing actionable communication of security status across all tiers of the organization

D. Collecting, correlating, and analyzing security-related information

**Answer: A**

8.6 NIST SP 800-137 defines the process for managing information security and information system–related risk as which of the following?

A. Risk Management Framework (RMF)

B. Risk-Impact Criteria (RIC)

C. Vulnerability Analysis (VA)

D. Risk-Vulnerability Matrix (RVM)

**Answer: A**

8.7 Security control volatility, system categorizations/impact levels, organizational risk tolerance, and security controls with identified weaknesses are associated with what characteristic discussed in NIST 800-137?

A. The necessity of adding compensating controls

B. The risk impact levels

C. The level of risk

D. The frequency of monitoring security controls for effectiveness

**Answer: D**

8.8 Appendix D of NIST SP 800-137 defines a term as "an information security area that includes a grouping of tools, technologies, and data." What is this term?

A. Security automation domain

B. Compensating domain

C. Risk management domain

D. Risk mitigation domain

**Answer: A**

8.9 The SEI-maintained Smart Grid Maturity Model (SGMM) is described best by which of the following statements?

A. Provides a framework for understanding the current state of Smart Grid deployment and capability within an electric utility and provides a context for establishing future strategies and work plans as they pertain to Smart Grid implementations

B. Provides a management perspective of Smart Grid deployment and capability within an electric utility and develops guidelines for Smart Grid implementations

C. Provides a targeted hierarchical methodology applicable to Smart Grid evaluation and maintenance within an electric utility and provides a basis for implementing effective Smart Grid controls

D. Provides a Smart Grid assurance model and algorithms for future deployment and implementation of Smart Grid systems

**Answer: A**

8.10 Which of the following is NOT one of the eight domains of the SGMM?

A. Strategy, Management, and Regulatory (SMR)

B. Organization and Structure (OS)

C. Metering and Monitoring (MAM)

D. Work and Asset Management (WAM)

**Answer: C**

8.11 Which of the following is NOT one of the six maturity levels defined in the SGMM?

A. Default

B. Evaluating

C. Initiating

D. Integrating

**Answer: B**

8.12 The SGMM maturity rating for each domain in an organization is determined by which of the following means?

A. The SGMM Domain survey

B. The SGMM Risk survey

C. The SGMM Level survey

D. The SGMM Compass survey

**Answer: D**

8.13 NIST 7628, *Introduction to NISTIR 7628 Guidelines for Smart Grid Cybersecurity,* is a three-volume document that provides a methodical approach to implementing customized cybersecurity and privacy protection strategies for Smart Grid implementations. Which of the following areas is NOT addressed in the volumes comprising NISTIR 7628?

A. Risk assessment processes

B. Privacy issues

C. Cost evaluation matrices

D. Research and development topics

**Answer: C**

8.14 Which of the following describes the optimizing level (Level 4) of the SGMM?

A. Focuses on implementing features that will enable an organization to achieve and sustain grid modernization

B. Smart Grid implementations within a given domain are being fine-tuned and used to further increase organizational performance

C. Smart Grid deployments are being integrated across the organization

D. Organizations are achieving new results with a given domain and advancing the state of the practice

**Answer: B**

8.15 What entity in the SGMM is defined as "the stages of an organization's progress toward achieving its Smart Grid vision in terms of automation, efficiency, reliability, energy and cost savings, integration of alternative energy sources, improved customer interaction, and access to new business opportunities and markets?"

A. Maturity

B. Capability

C. Characteristics

D. Value

**Answer: A**

8.16 In the ISCM process, the status of automation security in a facility can be determined by establishing appropriate metrics. The metrics are defined through the following activities, except one, which is not applicable. Which of the following is NOT applicable?

A. Maintaining an understanding of threats and threat activities

B. Conducting security awareness training

C. Collecting, correlating, and analyzing security-related information

D. Active management of risk by organizational officials

**Answer: B**

8.17 What technology reduces the effort required to maintain hardware and software systems by eliminating the binding of software to specific hardware platforms through the logical partitioning of physical computing resources into multiple execution environments, including servers, applications, and operating systems?

A. Virtualization

B. Cloud computing

C. Distributed computing

D. Multi-core chips

**Answer: A**

8.18 Which of the following statements is FALSE regarding penetration testing of industrial automation and control systems?

A. If not properly conducted, penetration testing can cause damage to automation systems and disrupt operations.

B. Penetration testing tools do not create additional traffic on automation networks and do not affect systems with limited memory capacity.

C. If executed properly, penetration testing can provide important information regarding the security of automation systems.

D. It is useful to use nonintrusive methods when conducting penetration testing.

**Answer: B**

8.19 Penetration testing conducted with no prior knowledge of the automation system is known as which of the following?

A. Black box

B. White box

C. Grey box

D. Auto box

**Answer: A**

8.20 Which of the following is NOT a component of the reconnaissance phase of penetration testing?

A. Footprinting

B. Scanning

C. Controlling

D. Enumerating

**Answer: C**

8.21 Which of the following is NOT one of the maturity levels of the Automation Maturity Model?

A. Local digital control

B. Remote auto control

C. Optimized operations

D. Enabled control

**Answer: D**

8.22 What level of the Automation Maturity Model is characterized by full automation and automated closed-loop control around a set point?

A. Local digital control

B. Enhanced operations

C. Remote auto control

D. Optimized operations

**Answer: C**

## Review Questions for Chapter 9

9.1 What term refers to objects such as sensors, humans, animals, machines, or other entities that each have a unique IP address and can transmit and receive data over a network without human intervention?

A. Artificial intelligence

B. Virtualization

C. Intelligent resolution

D. Internet of Things

**Answer: D**

9.2 The interoperability standard for secure and reliable information exchange in industrial automation and the enterprise is which of the following?

A. Open Platform Communications Unified Architecture (OPC UA)

B. Advanced HMI

C. Production Markup Language (PRODML)

D. M2M communication

**Answer: A**

9.3 In manufacturing, the Internet of Things is generating a next generation Industrial Revolution embodied in the concept known as which of the following?

A. Additive manufacturing

B. Industry 4.0

C. Machine-to-machine manufacturing

D. Industry 5.0

**Answer: B**

9.4 An envisioned state of operations in which all relevant and synthesized information is made available when, where, and in the form in which it is needed across the manufacturing supply chain to all stakeholders (people, process, products, and infrastructure) is known as which one of the following?

A. Additive manufacturing

B. Evolving manufacturing

C. Responsive manufacturing

D. Informed manufacturing

**Answer: D**

9.5 The four essential elements of manufacturing that converge to create an advanced production environment are which of the following?

A. People, processes, infrastructure, mitigation

B. People, processes, infrastructure, analysis

C. Products, people, processes, infrastructure

D. Collection, correlation, processes, infrastructure

**Answer: C**

9.6 Which of the following are in the Open Web Application Security Project (OWASP) "Internet of Things Top Ten" security categories?

A. Lack of Transport Encryption

B. Insecure Web Interface

C. Insecure Cloud Interface

D. All of the above

**Answer: D**

9.7 The concept that companies should limit the data they collect and retain, and dispose of it once they no longer need it, is known as?

A. Information protection

B. Information preservation

C. Data minimization

D. Data conservation

**Answer: C**

9.8 Big Data refers to which one of the following?

A. A collection of data sets so large or complex that it exceeds the capability of traditional data processing technologies

B. A collection of large data sets that exceeds the traditional risk management paradigms

C. A collection of data sets so complex that it exceeds the capability of traditional security analytics

D. A collection of data sets so large or complex that it exceeds the capability of traditional risk analysis

**Answer: A**

9.9 The process of analyzing and mining Big Data, which can produce valuable information that can be used for a variety of purposes, is which of the following?

A. Big Data minimization

B. Big Data analytics

C. Big Data execution

D. Big Data solutions

**Answer: B**

9.10 Which of the following is NOT one of the "Vs" of Big Data?

A. Volume

B. Variety

C. Visibility

D. Velocity

**Answer: C**

9.11 Volume 4, *Security and Privacy*, of the *NIST Big Data Interoperability Framework* identifies a number of ways that security and privacy in Big Data projects can be different from traditional implementations. Which one of the following is NOT one of these differences?

A. The use of multiple Big Data sources not originally intended to be used together can compromise privacy, security, or both.

B. Big Data projects increasingly involve one or more streamed data sources that are used in conjunction with data at rest, creating unique security and privacy scenarios.

C. An increased reliance on sensor streams can create vulnerabilities that were more easily managed before they were amassed to Big Data scale.

D. Big Data projects often encompass homogeneous components in which a single security scheme has not been designed from the outset.

**Answer: D**

9.12 The additional characteristic of Big Data that is sometimes referred to as provenance and that addresses the problem of understanding the data's original source is known as which one of the following?

A. Veracity

B. Validity

C. Volatility

D. Volume

**Answer: A**

9.13 Which of the following describe differences between what is new for Big Data and the requirements of previous big system security and privacy, as discussed in volume 4 of the *NIST Big Data Interoperability* Framework?

A. Big Data may be gathered from diverse end points and from more sources than traditional providers and consumers.

B. Big Data systems may be operational outside formal, readily understood frameworks.

C. Data search and selection can lead to privacy or security policy concerns.

D. All of the above

**Answer: D**

9.14 The use of sensors, advanced analytics, and intelligent decision-making that forms the basis of significant improvements in the methods that field assets communicate with enterprise activities is known as which of the following?

A. Informed Internet of Things (IIoT)

B. Industrial Internet of Things (IIoT)

C. Information Internet of Things (IIoT)

D. Inclusive Internet of Things (IIoT)

**Answer: B**

9.15 Strategies that focus on pushing data into a centralized cloud platform that can bridge the gap between industrial and IT domains is known as which of the following?

A. Informed Internet of Things (IIoT)

B. Industrial Internet of Things (IIoT)

C. Information Internet of Things (IIoT)

D. Inclusive Internet of Things (IIoT)

**Answer: B**

9.16 The data sent to the cloud from industrial networks originates from devices that have their own particular characteristics and operating parameters. In this scenario, which of the following emerging standards are relevant?

A. Message Queueing Telemetry Transport (MQTT)

B. Asynchronous Messaging Queuing Protocol (AMQP)

C. Constrained Application Protocol (COAP)

D. All of the above

**Answer: D**

9.17 Big Data technologies can be divided into which of the following two groups?

A. Batch processing and stream processing

B. Distributed processing and stream processing

C. Multi-core processing and stream processing

D. Batch processing and distributed processing

**Answer: A**

9.18 Which of the following statements is FALSE in regards to how Big Data technologies will transform security analytics?

A. By collecting data at a massive scale from many internal enterprise sources and external sources such as vulnerability databases

B. By using nonintrusive methods when conducting penetration testing

C. By providing a consolidated view of security-related information

D. By achieving real-time analysis of streaming data

**Answer: B**

9.19 Targeted attacks against a high-value asset or a physical system that can be extended over a long period of time without the knowledge of the infected organization is known as which of the following?

A. Black box threats

B. Advanced persistent threats

C. Zero day threats

D. Automatic threats

**Answer: B**

9.20 Authenticity and integrity of data used for analytics is known as which of the following?

A. Data provenance

B. Data enumeration

C. Data optimization

D. Data resolution

**Answer: A**

9.21 Systems that integrate computation, communication, sensing, and actuation with physical systems to fulfill time-sensitive functions with varying degrees of interaction with the environment, including human interaction, are defined as which of the following?

A. Analog control systems

B. Remote control systems

C. Cyber-physical systems

D. Enabled control systems

**Answer: C**

9.22 The NIST *Framework for Cyber-Physical Systems* comprises which of the following key elements?

A. Domains

B. Aspects

C. Facets

D. All of the above

**Answer: D**

9.23 A facet in the NIST *Framework for Cyber-Physical Systems* comprises Views encompassing identified responsibilities in the systems engineering process. Each facet contains a set of well-defined activities and artifacts (outputs) for addressing concerns. The three identified facets in the View model are which of the following?

A. Conceptualization, realization, and assurance

B. Conceptualization, realization, and confirmation

C. Conceptualization, identification, and assurance

D. Evaluation, realization, and assurance

**Answer: A**

9.24 Depending on the application, a CPS can be categorized into which of the following complexities?

A. Critical tightly-coupled

B. Loosely-coupled

C. System of Systems (SoS)

D. All of the above

**Answer: D**

9.25 On 12 February 2013, Executive Order (EO) 13636, "Improving Critical Infrastructure Cybersecurity," established that "it is the policy of the United States to enhance the security and resilience of the Nation's critical infrastructure" and called for the development of a voluntary risk-based Cybersecurity Framework. The goal of the Framework, developed by NIST, is to provide a common taxonomy and mechanism for organizations to do which of the following?

A. Describe their current cybersecurity posture

B. Describe their target state for cybersecurity

C. Assess progress toward the target state

D. All of the above

**Answer: D**

9.26 The Cybersecurity Framework comprises which of the following three parts?

A. The Framework Core, Framework Elements, and a Framework Profile

B. The Framework Architecture, Framework Implementation Tiers, and a Framework Profile

C. The Framework Core, Framework Implementation Tiers, and a Framework Profile

D. The Framework Core, Framework Implementation Tiers, and the Framework Security Posture

**Answer: C**

9.27 The NIST Cybersecurity Framework describes how it can be applied by organizations in which of the following ways?

A. Basic review of cybersecurity practices

B. Establishing or improving a cybersecurity program

C. Communicating cybersecurity requirements with stakeholders

D. All of the above

**Answer: D**

9.28 The means that can be used to program the interconnections of switches, routers, firewalls, and network security appliances is known as which of the following?

A. Software-implemented networking

B. Software-defined networking

C. Elemental networking

D. Software-defined specification

**Answer: B**

# Appendix B
# ICS Supplemental Guidance for NIST SP 800-53 Security Controls

Figure B.1 uses the AU-4 control (the Audit Storage Capacity) as an example of the format and content of the detailed overlay control specifications.

❶ Control number and title.

❷ Column for control and control enhancement number.

❸ Column for control and control enhancement name.

❹ Columns for baselines. If the baselines have been supplemented, then "Supplemented" displays.

❺ A row for each control or control enhancement.

❻ Columns for low, moderate, and high baselines.

❼ "Selected" indicates the control is selected in NIST SP 800-53, Revision 4. "Added" indicates the control is added to a baseline in the ICS overlay. A blank cell indicates the control is not selected. "Removed" indicates the control is removed from the baseline.

❽ The ICS Supplemental Guidance. If there is none, that is stated.

❾ The Control Enhancement ICS Supplemental Guidance. If there is none, that is stated.

❿ The rationale for changing the presence of a control or control enhancement in the baseline.

❶ AU-4 AUDIT STORAGE CAPACITY

| ❷ CNTL NO. | ❸ CONTROL NAME *Control Enhancement Name* | ❹ SUPPLEMENTED CONTROL BASELINES | | | |
|---|---|---|---|---|---|
| | | LOW | MOD | HIGH | ❻ |
| AU-4 | Audit Storage Capacity | Selected | Selected | Selected | ❼ |
| ❺ AU-4 (1) | *AUDIT STORAGE CAPACITY \| TRANSFER TO ALTERNATE STORAGE* | Added | Added | Added | |

❽ ICS Supplemental Guidance: None.

❾ Control Enhancement: (1) ICS Supplemental Guidance: Legacy ICSs are typically configured with remote storage on a separate information system (e.g., the historian in the DMZ accumulates historical operational ICS data and is backed up for storage at a different site). ICSs are currently using online backup services and increasingly moving to cloud-based and virtualized services. Retention of some data (e.g., SCADA telemetry) may be required by regulatory authorities.

❿ Rationale for Adding Control to Baseline: Legacy ICS components typically do not have capacity to store or analyze audit data. The retention periods for some data, particularly compliance data, may require large volumes of storage.

**Figure B-1. Detailed Overlay Control Specifications Illustrated**

NIST SP 800-53, Revision 4, Appendix F, contains Supplemental Guidance for all Controls and Control Enhancements [22]. ICS Supplemental Guidance in this overlay provides organizations with additional information on the application of the security controls and control enhancements in NIST SP 800-53, Revision 4, Appendix F, to ICS and the environments in which these specialized systems operate. The ICS Supplemental Guidance also provides information as to why a particular security control or control enhancement may not be applicable in some ICS environments and may be a candidate for tailoring (i.e., the application of scoping guidance and/or compensating controls).

# Access Control – AC

## Tailoring Considerations for the Access Control Family

Before implementing controls in the AC family, consider the tradeoffs among security, privacy, latency, performance, throughput, and reliability. For example, the organization considers whether latency induced from the use of confidentiality and integrity mechanisms employing cryptographic mechanisms would adversely impact the operational performance of the ICS.

In situations where the ICS cannot support the specific access control requirements of a control, the organization employs compensating controls in accordance with the general tailoring guidance. Examples of compensating controls are given with each control, as appropriate.

## Supplemental Guidance

Supplemental Guidance for all Controls and Control Enhancements in NIST SP 800-53, Revision 4, Appendix F, should be used in conjunction with the ICS Supplemental Guidance in this overlay, if any.

### *AC-1 Access Control Policy and Procedures*

| CONTROL NUMBER | CONTROL NAME<br>*Control Enhancement Name* | CONTROL BASELINES | | |
|---|---|---|---|---|
| | | LOW | MOD | HIGH |
| **AC-1** | **Access Control Policy and Procedures** | Selected | Selected | Selected |

ICS Supplemental Guidance: The policy specifically addresses the unique properties and requirements of an ICS and its relationship to non-ICS systems. ICS access by vendors and maintenance staff can occur over a very large facility footprint or geographic area and into unobserved spaces, such as mechanical/electrical rooms, ceilings, floors, field substations, switch and valve vaults, and pump stations.

### *AC-2 Account Management*

| CONTROL NUMBER | CONTROL NAME<br>*Control Enhancement Name* | CONTROL BASELINES | | |
|---|---|---|---|---|
| | | LOW | MOD | HIGH |
| **AC-2** | **Account Management** | Selected | Selected | Selected |
| AC-2 (1) | *ACCOUNT MANAGEMENT \| AUTOMATED SYSTEM ACCOUNT MANAGEMENT* | | Selected | Selected |
| AC-2 (2) | *ACCOUNT MANAGEMENT \| REMOVAL OF TEMPORARY /EMERGENCY ACCOUNTS* | | Selected | Selected |
| AC-2 (3) | *ACCOUNT MANAGEMENT \|DISABLE INACTIVE ACCOUNTS* | | Selected | Selected |
| AC-2 (4) | *ACCOUNT MANAGEMENT \| AUTOMATED AUDIT ACTIONS* | | Selected | Selected |
| AC-2 (5) | *ACCOUNT MANAGEMENT \| INACTIVITY LOGOUT / TYPICAL USAGE MONITORING* | | | Selected |
| AC-2 (11) | *ACCOUNT MANAGEMENT \| USAGE CONDITIONS* | | | Selected |
| AC-2 (12) | *ACCOUNT MANAGEMENT \| ACCOUNT MONITORING / ATYPICAL USAGE* | | | Selected |
| AC-2 (13) | *ACCOUNT MANAGEMENT \|ACCOUNT REVIEWS* | | | Selected |

ICS Supplemental Guidance: Example compensating controls include providing increased physical security, personnel security, intrusion detection, and auditing measures.

Control Enhancement: (1, 3, 4) ICS Supplemental Guidance: Example compensating controls include employing nonautomated mechanisms or procedures.

Control Enhancement: (2) ICS Supplemental Guidance: In situations where the ICS (e.g., field devices) cannot support temporary or emergency accounts, this enhancement does not apply. Example compensating controls include employing nonautomated mechanisms or procedures.

Control Enhancement: (5) ICS Supplemental Guidance: Example compensating controls include employing nonautomated mechanisms or procedures.

Control Enhancement: (11, 12, 13) ICS Supplemental Guidance: None.

### *AC-3 Access Enforcement*

| CONTROL NUMBER | CONTROL NAME *Control Enhancement Name* | CONTROL BASELINES | | |
|---|---|---|---|---|
| | | LOW | MOD | HIGH |
| **AC-3** | **Access Enforcement** | Selected | Selected | Selected |

ICS Supplemental Guidance: The organization ensures that access enforcement mechanisms do not adversely impact the operational performance of the ICS. An example of a compensating control would be to include encapsulation. Policy for logical access control to nonaddressable and nonroutable system resources and the associated information is made explicit. Access control mechanisms include hardware, firmware, and software that control or have device access, such as device drivers and communications controllers. Physical access control may serve as a compensating control for logical access control; however, it may not provide sufficient granularity in situations where users require access to different functions. Logical access enforcement may be implemented in encapsulating hardware and software.

## AC-4 Information Flow Enforcement

| CONTROL NUMBER | CONTROL NAME *Control Enhancement Name* | CONTROL BASELINES | | |
|---|---|---|---|---|
| | | LOW | MOD | HIGH |
| AC-4 | Information Flow Enforcement | | Selected | Selected |

ICS Supplemental Guidance: Physical addresses (e.g., a serial port) may be implicitly or explicitly associated with labels or attributes (e.g., hardware I/O address). Manual methods are typically static. Label or attribute policy mechanisms may be implemented in hardware, firmware, and software that controls or has device access, such as device drivers and communications controllers. Information flow policy may be supported by labeling or coloring physical connectors as an aid to manual hookup. Inspection of message content may enforce information flow policy. For example, a message containing a command to an actuator may not be permitted to flow between the control network and any other network.

## AC-5 Separation of Duties

| CONTROL NUMBER | CONTROL NAME *Control Enhancement Name* | CONTROL BASELINES | | |
|---|---|---|---|---|
| | | LOW | MOD | HIGH |
| AC-5 | Separation of Duties | | Selected | Selected |

ICS Supplemental Guidance: Example compensating controls include providing increased personnel security and auditing. The organization carefully considers the appropriateness of a single individual performing multiple critical roles.

### *AC-6 Least Privilege*

| CONTROL NUMBER | CONTROL NAME<br>*Control Enhancement Name* | CONTROL BASELINES | | |
|---|---|---|---|---|
| | | LOW | MOD | HIGH |
| **AC-6** | **Least Privilege** | | Selected | Selected |
| AC-6 (1) | *LEAST PRIVILEGE \| AUTHORIZE ACCESS TO SECURITY FUNCTIONS* | | Selected | Selected |
| AC-6 (2) | *LEAST PRIVILEGE \| NONPRIVILEGED ACCESS FOR NONSECURITY FUNCTIONS* | | Selected | Selected |
| AC-6 (3) | *LEAST PRIVILEGE \| NETWORK ACCESS TO PRIVILEGED COMMANDS* | | | Selected |
| AC-6 (5) | *LEAST PRIVILEGE \| PRIVILEGED ACCOUNTS* | | Selected | Selected |
| AC-6 (9) | *LEAST PRIVILEGE \| AUDITING USE OF PRIVILEGED FUNCTIONS* | | Selected | Selected |
| AC-6 (10) | *LEAST PRIVILEGE \| PROHIBIT NONPRIVILEGED USERS FROM EXECUTING PRIVILEGED FUNCTIONS* | | Selected | Selected |

ICS Supplemental Guidance: Example compensating controls include providing increased personnel security and auditing. The organization carefully considers the appropriateness ofa single individual having multiple critical privileges. System privilege models may be tailored to enforce integrity and availability (e.g., lower privileges include read access and higher privileges include write access).

Control Enhancement: (1) ICS Supplemental Guidance: In situations where the ICS cannot support access control to security functions, the organization employs nonautomated mechanisms or procedures as compensating controls in accordance with the general tailoring guidance.

Control Enhancement: (2) ICS Supplemental Guidance: In situations where the ICS cannot support access control to nonsecurity functions, the organization employs nonautomated mechanisms or procedures as compensating controls in accordance with the general tailoring guidance.

Control Enhancement: (3) ICS Supplemental Guidance: In situations where the ICS cannot support network access control to privileged commands, the organization employs nonautomated mechanisms or procedures as compensating controls in accordance with the general tailoring guidance.

Control Enhancement: (5) ICS Supplemental Guidance: In situations where the ICS cannot support access control to privileged accounts, the organization employs nonautomated mechanisms or procedures as compensating controls in accordance with the general tailoring guidance.

Control Enhancement: (9) ICS Supplemental Guidance: In general, audit record processing is not performed on the ICS, but on a separate information system. An example of a compensating control would be providing an auditing capability on a separate information system.

Control Enhancement: (10) ICS Supplemental Guidance: An example of a compensating control would be enhanced auditing.

### *AC-7 Unsuccessful Login Attempts*

| CONTROL NUMBER | CONTROL NAME *Control Enhancement Name* | CONTROL BASELINES | | |
|---|---|---|---|---|
| | | LOW | MOD | HIGH |
| AC-7 | **Unsuccessful Login Attempts** | Selected | Selected | Selected |

ICS Supplemental Guidance: Many ICS must remain continuously on and operators remain logged onto the system at all times. A "log-over" capability may be employed. Example compensating controls include logging or recording all unsuccessful login attempts and alerting ICS security personnel though alarms or other means when the number of organization-defined, consecutive, invalid access attempts is exceeded.

### *AC-8 System Use Notification*

| CONTROL NUMBER | CONTROL NAME *Control Enhancement Name* | CONTROL BASELINES | | |
|---|---|---|---|---|
| | | LOW | MOD | HIGH |
| AC-8 | **System Use Notification** | Selected | Selected | Selected |

ICS Supplemental Guidance: Many ICS must remain continuously on and system use notification may not be supported or effective. An example of a compensating control would be posting physical notices in ICS facilities.

### *AC-10 Concurrent Session Control*

| CONTROL NUMBER | CONTROL NAME *Control Enhancement Name* | CONTROL BASELINES | | |
|---|---|---|---|---|
| | | LOW | MOD | HIGH |
| **AC-10** | **Concurrent Session Control** | | | Selected |

ICS Supplemental Guidance: The number, account type, and privileges of concurrent sessions takes into account the roles and responsibilities of the affected individuals. An example of compensating controls is providing increased auditing measures.

### *AC-11 Session Lock*

| CONTROL NUMBER | CONTROL NAME *Control Enhancement Name* | CONTROL BASELINES | | |
|---|---|---|---|---|
| | | LOW | MOD | HIGH |
| **AC-11** | **Session Lock** | | Selected | Selected |
| AC-11 (1) | *SESSION LOCK \| PATTERN-HIDING DISPLAYS* | | Selected | Selected |

ICS Supplemental Guidance: This control assumes a staffed environment where users interact with information system displays. When this assumption does not apply, the organization tailors the control appropriately (e.g., the ICS may be physically protected by placement in a locked enclosure). The control may also be tailored for ICS that are not configured with displays, but which have the capability to support displays (e.g., ICS to which a maintenance technician may attach a display). In some cases, session lock for ICS operator workstations/nodes is not advised (e.g., when immediate operator responses are required in emergency situations). An example of a compensating control would be locating the display in an area with physical access controls that limit access to individuals with permission and need-to-know for the displayed information.

Control Enhancement: (1) ICS Supplemental Guidance: ICS may employ physical protection to prevent access to a display or to prevent attachment of a display. In situations where the ICS cannot conceal displayed information, the organization employs nonautomated mechanisms or procedures as compensating controls in accordance with the general tailoring guidance.

## AC-12 Session Termination

| CONTROL NUMBER | CONTROL NAME *Control Enhancement Name* | CONTROL BASELINES | | |
|---|---|---|---|---|
| | | LOW | MOD | HIGH |
| **AC-12** | **Session Termination** | | Selected | Selected |

ICS Supplemental Guidance: Example compensating controls include providing increased auditing measures or limiting remote access privileges to key personnel.

## AC-14 Permitted Actions without Identification or Authentication

| CONTROL NUMBER | CONTROL NAME *Control Enhancement Name* | CONTROL BASELINES | | |
|---|---|---|---|---|
| | | LOW | MOD | HIGH |
| **AC-14** | **Permitted Actions without Identification or Authentication** | Selected | Selected | Selected |

ICS Supplemental Guidance: None.

## AC-17 Remote Access

| CONTROL NUMBER | CONTROL NAME *Control Enhancement Name* | CONTROL BASELINES | | |
|---|---|---|---|---|
| | | LOW | MOD | HIGH |
| **AC-17** | **Remote Access** | Selected | Selected | Selected |
| AC-17 (1) | *REMOTE ACCESS \| AUTOMATED MONITORING / CONTROL* | | Selected | Selected |
| AC-17 (2) | *REMOTE ACCESS \| PROTECTION OF CONFIDENTIALITY / INTEGRITY USING ENCRYPTION* | | Selected | Selected |
| AC-17 (3) | *REMOTE ACCESS \| MANAGED ACCESS CONTROL POINTS* | | Selected | Selected |
| AC-17 (4) | *REMOTE ACCESS \| PRIVILEGED COMMANDS / ACCESS* | | Selected | Selected |

ICS Supplemental Guidance: In situations where the ICS cannot implement any or all of the components of this control, the organization employs other mechanisms or procedures as compensating controls in accordance with the general tailoring guidance.

Control Enhancement: (1) ICS Supplemental Guidance: Example compensating controls include employing nonautomated mechanisms or procedures as compensating controls (e.g., following manual authentication [see IA- 2], dial-in remote access may be enabled for a specified period of time or a call may be placed from the ICS site to the authenticated remote entity).

Control Enhancement: (2) ICS Supplemental Guidance: ICS security objectives often rank confidentiality below availability and integrity. The organization explores all possible cryptographic mechanisms (e.g., encryption, digital signature, hash function). Each mechanism has a different delay impact. Example compensating controls include providing increased auditing for remote sessions or limiting remote access privileges to key personnel.

Control Enhancement: (3) ICS Supplemental Guidance: An example of a compensating control would be connection-specific manual authentication of the remote entity.

Control Enhancement: (4) ICS Supplemental Guidance: None.

ICS Supplemental Guidance: Example compensating controls include employing nonautomated mechanisms or procedures as compensating controls in accordance with the general tailoring guidance.

### *AC-18 Wireless Access*

| CONTROL NUMBER | CONTROL NAME<br>*Control Enhancement Name* | CONTROL BASELINES | | |
|---|---|---|---|---|
| | | LOW | MOD | HIGH |
| **AC-18** | **Wireless Access** | Selected | Selected | Selected |
| AC-18 (1) | *WIRELESS ACCESS \| AUTHENTICATION AND ENCRYPTION* | | Selected | Selected |
| AC-18 (4) | *WIRELESS ACCESS \| RESTRICT CONFIGURATIONS BY USERS* | | | Selected |
| AC-18 (5) | *WIRELESS ACCESS \| CONFINE WIRELESS COMMUNICATIONS* | | | Selected |

ICS Supplemental Guidance: In situations where the ICS cannot implement any or all of the components of this control, the organization employs other mechanisms or procedures as compensating controls in accordance with the general tailoring guidance.

Control Enhancement: (1) ICS Supplemental Guidance: See AC-17 Control Enhancement: (1) ICS Supplemental Guidance. Example compensating controls include providing increased auditing for wireless access or limiting wireless access privileges to key personnel.

Control Enhancement: (4) (5) ICS Supplemental Guidance: None.

### AC-19 Access Control for Mobile Devices

| CONTROL NUMBER | CONTROL NAME *Control Enhancement Name* | CONTROL BASELINES | | |
|---|---|---|---|---|
| | | LOW | MOD | HIGH |
| **AC-19** | **Access Control for Mobile Devices** | Selected | Selected | Selected |
| AC-19 (5) | *ACCESS CONTROL FOR MOBILE DEVICES \| FULL DEVICE / CONTAINER-BASED ENCRYPTION* | | Selected | Selected |

ICS Supplemental Guidance: None.

### AC-20 Use of External Information Systems

| CONTROL NUMBER | CONTROL NAME *Control Enhancement Name* | CONTROL BASELINES | | |
|---|---|---|---|---|
| | | LOW | MOD | HIGH |
| **AC-20** | **Use of External Information Systems** | Selected | Selected | Selected |
| AC-20 (1) | *USE OF EXTERNAL INFORMATION SYSTEMS \| LIMITS ON AUTHORIZED USE* | | Selected | Selected |
| AC-20 (2) | *USE OF EXTERNAL INFORMATION SYSTEMS \| PORTABLE STORAGE MEDIA* | | Selected | Selected |

ICS Supplemental Guidance: Organizations refine the definition of "external" to reflect lines of authority and responsibility; granularity of organization entity; and their relationships. An organization may consider a system to be external if that system performs different functions, implements different policies, comes under different managers, or does not provide sufficient visibility into the implementation of security controls to allow the establishment of a satisfactory trust relationship. For example, a process control system and a business data processing system would typically be considered external to each other. Access to an ICS for support by a business partner, such as a vendor or support contractor, is another common example. The definition and trustworthiness of external information systems is reexamined with respect to ICS functions, purposes, technology, and limitations to establish a clear, documented technical or business case for use and an acceptance of the risk inherent in the use of an external information system.

Control Enhancement: (1, 2) ICS Supplemental Guidance: None.

### AC-21 Information Sharing

| CONTROL NUMBER | CONTROL NAME *Control Enhancement Name* | CONTROL BASELINES | | |
|---|---|---|---|---|
| | | LOW | MOD | HIGH |
| AC-21 | **Collaboration and Information Sharing** | Added | Selected | Selected |

ICS Supplemental Guidance: The organization should collaborate and share information about potential incidents on a timely basis. The DHS National Cybersecurity & Communications Integration Center (NCCIC)[1] serves as a centralized location where operational elements involved in cybersecurity and communications reliance are coordinated and integrated. The Industrial Control Systems Cyber Emergency Response Team (ICS-CERT)[2] collaborates with international and private sector Computer Emergency Response Teams (CERTs) to share control systems-related security incidents and mitigation measures. Organizations should consider having the capabilities to share both unclassified and classified information.

Rationale for Adding AC-21 to Low Baseline: ICS systems provide essential services and control functions and are often connected to other ICS systems or business systems that can be vectors of attack. It is therefore necessary to provide a uniform defense encompassing all baselines.

### AC-22 Publicly Accessible Content

| CONTROL NUMBER | CONTROL NAME *Control Enhancement Name* | CONTROL BASELINES | | |
|---|---|---|---|---|
| | | LOW | MOD | HIGH |
| AC-22 | **Publicly Accessible Content** | Selected | Selected | Selected |

ICS Supplemental Guidance: Generally, public access to ICS is not permitted. Selected information may be transferred to a publicly accessible information system, possibly with added controls (e.g., introduction of fuzziness or delay).

# Awareness and Training – AT

## Supplemental Guidance

Supplemental Guidance for all Controls and Control Enhancements in NIST SP 800-53, Revision 4, Appendix F, should be used in conjunction with the ICS Supplemental Guidance in this overlay, if any.

### *AT-1 Security Awareness and Training Policy and Procedures*

| CONTROL NUMBER | CONTROL NAME *Control Enhancement Name* | CONTROL BASELINES | | |
|---|---|---|---|---|
| | | LOW | MOD | HIGH |
| **AT-1** | **Security Awareness and Training Policy and Procedures** | Selected | Selected | Selected |

ICS Supplemental Guidance: The policy specifically addresses the unique properties and requirements of ICS and the relationship to non-ICS systems.

### *AT-2 Security Awareness Training*

| CONTROL NUMBER | CONTROL NAME *Control Enhancement Name* | CONTROL BASELINES | | |
|---|---|---|---|---|
| | | LOW | MOD | HIGH |
| **AT-2** | **Security Awareness** | Selected | Selected | Selected |
| AT-2 (2) | *SECURITY AWARENESS \| INSIDER THREAT* | | Selected | Selected |

ICS Supplemental Guidance: Security awareness training includes initial and periodic review of ICS-specific policies, standard operating procedures, security trends, and vulnerabilities. The ICS security awareness program is consistent with the requirements of the security awareness and training policy established by the organization.

Control Enhancement: (2) ICS Supplemental Guidance: None.

### *AT-3 Role-Based Security Training*

| CONTROL NUMBER | CONTROL NAME *Control Enhancement Name* | CONTROL BASELINES | | |
|---|---|---|---|---|
| | | LOW | MOD | HIGH |
| AT-3 | Role-Based Security Training | Selected | Selected | Selected |

ICS Supplemental Guidance: Security training includes initial and periodic review of ICS-specific policies, standard operating procedures, security trends, and vulnerabilities. The ICS security training program is consistent with the requirements of the security awareness and training policy established by the organization.

### *AT-4 Security Training Records*

| CONTROL NUMBER | CONTROL NAME *Control Enhancement Name* | CONTROL BASELINES | | |
|---|---|---|---|---|
| | | LOW | MOD | HIGH |
| AT-4 | Security Training Records | Selected | Selected | Selected |

ICS Supplemental Guidance: None.

## Auditing and Accountability – AU

### Tailoring Considerations for the Audit Family

In general, audit information and audit tools are not present on legacy ICS, but on a separate information system (e.g., the historian). In situations where the ICS cannot support the specific Audit and Accountability requirements of a control, the organization employs compensating controls in accordance with the general tailoring guidance. Examples of compensating controls are given with each control, as appropriate.

### Supplemental Guidance

Supplemental Guidance for all Controls and Control Enhancements in NIST SP 800-53, Revision 4, Appendix F, should be used in conjunction with the ICS Supplemental Guidance in this overlay, if any.

## *AU-1 Audit and Accountability Policy and Procedures*

| CONTROL NUMBER | CONTROL NAME *Control Enhancement Name* | CONTROL BASELINES | | |
|---|---|---|---|---|
| | | LOW | MOD | HIGH |
| **AU-1** | **Audit and Accountability Policy and Procedures** | Selected | Selected | Selected |

ICS Supplemental Guidance: The policy specifically addresses the unique properties and requirements of ICS and the relationship to non-ICS systems.

## *AU-2 Audit Events*

| CONTROL NUMBER | CONTROL NAME *Control Enhancement Name* | CONTROL BASELINES | | |
|---|---|---|---|---|
| | | LOW | MOD | HIGH |
| **AU-2** | **Auditable Events** | Selected | Selected | Selected |
| AU-2 (3) | *AUDITABLE EVENTS \| REVIEWS AND UPDATES* | | Selected | Selected |

ICS Supplemental Guidance: The organization may designate ICS events as audit events, requiring that ICS data and/or telemetry be recorded as audit data.

Control Enhancement: (3) ICS Supplemental Guidance: None.

## *AU-3 Content of Audit Records*

| CONTROL NUMBER | CONTROL NAME *Control Enhancement Name* | CONTROL BASELINES | | |
|---|---|---|---|---|
| | | LOW | MOD | HIGH |
| **AU-3** | **Content of Audit Records** | Selected | Selected | Selected |
| AU-3 (1) | *CONTENT OF AUDIT RECORDS \| ADDITIONAL AUDIT INFORMATION* | | Selected | Selected |
| AU-3 (2) | *CONTENT OF AUDIT RECORDS \| CENTRALIZED MANAGEMENT OF PLANNED AUDIT RECORD CONTENT* | | | Selected |

ICS Supplemental Guidance: An example of a compensating control would be providing an auditing capability on a separate information system.

Control Enhancement: (1, 2) ICS Supplemental Guidance: None.

### AU-4 Audit Storage Capacity

| CONTROL NUMBER | CONTROL NAME *Control Enhancement Name* | SUPPLEMENTED CONTROL BASELINES | | |
|---|---|---|---|---|
| | | LOW | MOD | HIGH |
| **AU-4** | **Audit Storage Capacity** | Selected | Selected | Selected |
| AU-4 (1) | *AUDIT STORAGE CAPACITY \| TRANSFER TO ALTERNATE STORAGE* | Added | Added | Added |

ICS Supplemental Guidance: None.

Control Enhancement: (1) ICS Supplemental Guidance: Legacy ICS are typically configured with remote storage on a separate information system (e.g., the historian accumulates historical operational ICS data and is backed up for storage at a different site). ICS are currently using online backup services and increasingly moving to cloud-based and virtualized services. Retention of some data (e.g., SCADA telemetry) may be required by regulatory authorities.

Rationale for Adding AU-4 (1) to All Baselines: Legacy ICS components typically do not have the capacity to store or analyze audit data. The retention periods for some data, particularly compliance data, may require large volumes of storage.

### AU-5 Response to Audit Processing Failures

| CONTROL NUMBER | CONTROL NAME *Control Enhancement Name* | CONTROL BASELINES | | |
|---|---|---|---|---|
| | | LOW | MOD | HIGH |
| **AU-5** | **Response to Audit Processing Failures** | Selected | Selected | Selected |
| AU-5 (1) | *RESPONSE TO AUDIT PROCESSING FAILURES \| AUDIT STORAGE CAPACITY* | | | Selected |
| AU-5 (2) | *RESPONSE TO AUDIT PROCESSING FAILURES \| REAL-TIME ALERTS* | | | Selected |

ICS Supplemental Guidance: None.

## AU-6 Audit Review, Analysis, and Reporting

| CONTROL NUMBER | CONTROL NAME *Control Enhancement Name* | CONTROL BASELINES | | |
|---|---|---|---|---|
| | | LOW | MOD | HIGH |
| **AU-6** | **Audit Review, Analysis, and Reporting** | Selected | Selected | Selected |
| AU-6 (1) | *AUDIT REVIEW, ANALYSIS, AND REPORTING \| PROCESS INTEGRATION* | | Selected | Selected |
| AU-6 (3) | *AUDIT REVIEW, ANALYSIS, AND REPORTING \| CORRELATE AUDIT REPOSITORIES* | | Selected | Selected |
| AU-6 (5) | *AUDIT REVIEW, ANALYSIS, AND REPORTING \| INTEGRATION / SCANNING AND MONITORING CAPABILITIES* | | | Selected |
| AU-6 (6) | *AUDIT REVIEW, ANALYSIS, AND REPORTING \| CORRELATION WITH PHYSICAL MONITORING* | | | Selected |

ICS Supplemental Guidance: None.

Control Enhancement: (1) ICS Supplemental Guidance: Example compensating controls include manual mechanisms or procedures.

Control Enhancement: (3, 5, 6) ICS Supplemental Guidance: None.

## AU-7 Audit Reduction and Report Generation

| CONTROL NUMBER | CONTROL NAME *Control Enhancement Name* | CONTROL BASELINES | | |
|---|---|---|---|---|
| | | LOW | MOD | HIGH |
| **AU-7** | **Audit Reduction and Report Generation** | | Selected | Selected |
| AU-7 (1) | *AUDIT REDUCTION AND REPORT GENERATION \| AUTOMATIC PROCESSING* | | Selected | Selected |

ICS Supplemental Guidance: None.

Control Enhancement: (1) ICS Supplemental Guidance: None.

### *AU-8 Time Stamps*

| CONTROL NUMBER | CONTROL NAME *Control Enhancement Name* | CONTROL BASELINES | | |
|---|---|---|---|---|
| | | LOW | MOD | HIGH |
| **AU-8** | **Time Stamps** | Selected | Selected | Selected |
| AU-8 (1) | *TIME STAMPS \| SYNCHRONIZATION WITH AUTHORITATIVE TIME SOURCE* | | Selected | Selected |

ICS Supplemental Guidance: An example of a compensating control would be using a separate information system designated as an authoritative time source.

Control Enhancement: (1) ICS Supplemental Guidance: ICS employ suitable mechanisms (e.g., GPS, IEEE 1588) for time stamps.

### *AU-9 Protection of Audit Information*

| CONTROL NUMBER | CONTROL NAME *Control Enhancement Name* | CONTROL BASELINES | | |
|---|---|---|---|---|
| | | LOW | MOD | HIGH |
| **AU-9** | **Protection of Audit Information** | Selected | Selected | Selected |
| AU-9 (2) | *PROTECTION OF AUDIT INFORMATION \| AUDIT BACKUP ON SEPARATE PHYSICAL SYSTEMS / COMPONENTS* | | | Selected |
| AU-9 (3) | *PROTECTION OF AUDIT INFORMATION \| CRYPTOGRAPHIC PROTECTION* | | | Selected |
| AU-9 (4) | *PROTECTION OF AUDIT INFORMATION \| ACCESS BY SUBSET OF PRIVILEGED USERS* | | Selected | Selected |

ICS Supplemental Guidance: None.

### *AU-10 Nonrepudiation*

| CONTROL NUMBER | CONTROL NAME *Control Enhancement Name* | CONTROL BASELINES | | |
|---|---|---|---|---|
| | | LOW | MOD | HIGH |
| **AU-10** | **Nonrepudiation** | | | Selected |

ICS Supplemental Guidance: An example of a compensating control would be providing nonrepudiation on a separate information system.

### AU-11 Audit Record Retention

| CONTROL NUMBER | CONTROL NAME<br>*Control Enhancement Name* | CONTROL BASELINES | | |
|---|---|---|---|---|
| | | LOW | MOD | HIGH |
| AU-11 | **Audit Record Retention** | Selected | Selected | Selected |

ICS Supplemental Guidance: None.

### AU-12 Audit Generation

| CONTROL NUMBER | CONTROL NAME<br>*Control Enhancement Name* | CONTROL BASELINES | | |
|---|---|---|---|---|
| | | LOW | MOD | HIGH |
| AU-12 | **Audit Generation** | Selected | Selected | Selected |
| AU-12 (1) | *AUDIT GENERATION \| SYSTEM-WIDE / TIME-CORRELATED AUDIT TRAIL* | | | Selected |
| AU-12 (3) | *AUDIT GENERATION \| CHANGES BY AUTHORIZED INDIVIDUALS* | | | Selected |

ICS Supplemental Guidance: None.

Control Enhancement: (1) ICS Supplemental Guidance: An example of a compensating control would be providing time-correlated audit records on a separate information system.

Control Enhancement: (3) ICS Supplemental Guidance: Example compensating controls include employing nonautomated mechanisms or procedures.

# Security Assessment and Authorization – CA

## Tailoring Considerations for the Security Assessment and Authorization Family

In situations where the ICS cannot support the specific security assessment and authorization requirements of a control, the organization employs compensating controls in accordance with the general tailoring guidance. Examples of compensating controls are given with each control, as appropriate.

## Supplemental Guidance

Supplemental Guidance for all Controls and Control Enhancements in NIST SP 800-53, Revision 4, Appendix F, should be used in conjunction with the ICS Supplemental Guidance in this overlay, if any.

### *CA-1 Security Assessment and Authorization Policy and Procedures*

| CONTROL NUMBER | CONTROL NAME *Control Enhancement Name* | CONTROL BASELINES | | |
|---|---|---|---|---|
| | | LOW | MOD | HIGH |
| CA-1 | **Security Assessment and Authorization Policy and Procedures** | Selected | Selected | Selected |

ICS Supplemental Guidance: The policy specifically addresses the unique properties and requirements of ICS and the relationship to non-ICS systems.

### *CA-2 Security Assessments*

| CONTROL NUMBER | CONTROL NAME *Control Enhancement Name* | CONTROL BASELINES | | |
|---|---|---|---|---|
| | | LOW | MOD | HIGH |
| CA-2 | **Security Assessments** | Selected | Selected | Selected |
| CA-2 (1) | *SECURITY ASSESSMENTS \| INDEPENDENT ASSESSORS* | | Selected | Selected |
| CA-2 (2) | *SECURITY ASSESSMENTS \| TYPES OF ASSESSMENTS* | | | Selected |

ICS Supplemental Guidance: Assessments are performed and documented by qualified assessors (i.e., experienced in assessing ICS) authorized by the organization. The organization ensures that assessments do not interfere with ICS functions. The individual/group conducting the assessment fully understands the organizational information security policies and procedures, the ICS security policies and procedures, and the specific health, safety, and environmental risks associated with a particular facility and/or process. The organization ensures that the assessment does not affect system operation or result in unintentional system modification. If assessment activities must be performed on the production ICS, it may need to be taken off-line before an assessment can be conducted. If an ICS must be taken off-line to conduct an assessment, the assessment is scheduled to occur during planned ICS outages whenever possible.

Control Enhancement: (1) ICS Supplemental Guidance: None.

Control Enhancement: (2) ICS Supplemental Guidance: The organization conducts risk analysis to support the selection of assessment targets (e.g., the live system, an off-line replica, a simulation).

### CA-3 System Interconnections

| CONTROL NUMBER | CONTROL NAME *Control Enhancement Name* | CONTROL BASELINES | | |
|---|---|---|---|---|
| | | LOW | MOD | HIGH |
| **CA-3** | **Information System Connections** | Selected | Selected | Selected |
| CA-3 (5) | *SYSTEM INTERCONNECTIONS \| RESTRICTIONS ON EXTERNAL SYSTEM CONNECTIONS* | | Selected | Selected |

ICS Supplemental Guidance: Organizations perform risk-benefit analysis to support determining whether an ICS should be connected to other information system(s). The authorizing official fully understands the organizational information security policies and procedures; the ICS security policies and procedures; the risks to organizational operations and assets, individuals, other organizations and the nation associated with the connection to other information system(s); and the specific health, safety, and environmental risks associated with a particular interconnection. The AO documents risk acceptance in the ICS system security plan.

Control Enhancement: (5) ICS Supplemental Guidance: None.

### CA-5 Plan of Action and Milestones

| CONTROL NUMBER | CONTROL NAME *Control Enhancement Name* | CONTROL BASELINES | | |
|---|---|---|---|---|
| | | LOW | MOD | HIGH |
| **CA-5** | **Plan of Action and Milestones** | Selected | Selected | Selected |

ICS Supplemental Guidance: None.

### *CA-6 Security Authorization*

| CONTROL NUMBER | CONTROL NAME<br>*Control Enhancement Name* | CONTROL BASELINES | | |
|---|---|---|---|---|
| | | LOW | MOD | HIGH |
| **CA-6** | **Security Authorization** | Selected | Selected | Selected |

ICS Supplemental Guidance: None.

### *CA-7 Continuous Monitoring*

| CONTROL NUMBER | CONTROL NAME<br>*Control Enhancement Name* | CONTROL BASELINES | | |
|---|---|---|---|---|
| | | LOW | MOD | HIGH |
| **CA-7** | **Continuous Monitoring** | Selected | Selected | Selected |
| CA-7 (1) | *CONTINUOUS MONITORING \| INDEPENDENT ASSESSMENT* | | Selected | Selected |

ICS Supplemental Guidance: Continuous monitoring programs for ICS are designed, documented, and implemented by qualified personnel (i.e., experienced with ICS) selected by the organization. The organization ensures that continuous monitoring does not interfere with ICS functions. The individual/group designing and conducting the continuous monitoring fully understands the organizational information security policies and procedures, the ICS security policies and procedures, and the specific health, safety, and environmental risks associated with a particular facility and/or process. The organization ensures that continuous monitoring does not affect system operation or result in intentional or unintentional system modification. An example of a compensating control would be external monitoring.

Control Enhancement: (1) ICS Supplemental Guidance: None.

## CA-8 Penetration Testing

| CONTROL NUMBER | CONTROL NAME *Control Enhancement Name* | CONTROL BASELINES | | |
|---|---|---|---|---|
| | | LOW | MOD | HIGH |
| CA-8 | Penetration Testing | | | Selected |

ICS Supplemental Guidance: Penetration testing is used with care on ICS networks to ensure that ICS functions are not adversely impacted by the testing process. In general, ICS are highly sensitive to timing constraints and have limited resources. Example compensating controls include replicated, virtualized, or simulated systems that are employed to conduct penetration testing. Production ICS may need to be taken off-line before testing can be conducted. If ICS are taken off-line for testing, tests are scheduled to occur during planned ICS outages whenever possible. If penetration testing is performed on non-ICS networks, extra care is taken to ensure that tests do not propagate into the ICS network.

## CA-9 Internal System Connections

| CONTROL NUMBER | CONTROL NAME *Control Enhancement Name* | CONTROL BASELINES | | |
|---|---|---|---|---|
| | | LOW | MOD | HIGH |
| CA-9 | Internal System Connections | Selected | Selected | Selected |

ICS Supplemental Guidance: Organizations perform risk-benefit analysis to support determining whether an ICS should be connected to other internal information system(s) and (separate) constituent system components. The authorizing official fully understands the organizational information security policies and procedures; the ICS security policies and procedures; the risks to organizational operations and assets, individuals, other organizations, and the nation associated with the connection to other information system(s) and (separate) constituent system components, whether by authorizing each individual internal connection or authorizing internal connections for a class of components with common characteristics and/or configurations; and the specific health, safety, and environmental risks associated with a particular interconnection. The AO documents risk acceptance in the ICS system security plan.

# Configuration Management – CM

## Tailoring Considerations for the Configuration Management Family

In situations where the ICS cannot be configured to restrict the use of unnecessary functions or cannot support the use of automated mechanisms to implement configuration management functions, the organization employs nonautomated mechanisms or procedures as compensating controls in accordance with the general tailoring guidance. Examples of compensating controls are given with each control, as appropriate.

## Supplemental Guidance

Supplemental Guidance for all Controls and Control Enhancements in NIST SP 800-53, Revision 4, Appendix F, should be used in conjunction with the ICS Supplemental Guidance in this overlay, if any.

### *CM-1 Configuration Management Policy and Procedures*

| CONTROL NUMBER | CONTROL NAME<br>*Control Enhancement Name* | CONTROL BASELINES | | |
|---|---|---|---|---|
| | | LOW | MOD | HIGH |
| CM-1 | **Configuration Management Policy and Procedures** | Selected | Selected | Selected |

ICS Supplemental Guidance: The policy specifically addresses the unique properties and requirements of ICS and the relationship to non-ICS systems.

## CM-2 Baseline Configuration

| CONTROL NUMBER | CONTROL NAME *Control Enhancement Name* | CONTROL BASELINES | | |
|---|---|---|---|---|
| | | LOW | MOD | HIGH |
| **CM-2** | **Baseline Configuration** | Selected | Selected | Selected |
| CM-2 (1) | *BASELINE CONFIGURATION \| REVIEWS AND UPDATES* | | Selected | Selected |
| CM-2 (2) | *BASELINE CONFIGURATION \| AUTOMATION SUPPORT FOR ACCURACY / CURRENCY* | | | Selected |
| CM-2 (3) | *BASELINE CONFIGURATION \| RETENTION OF PREVIOUS CONFIGURATIONS* | | Selected | Selected |
| CM-2 (7) | *BASELINE CONFIGURATION \| CONFIGURE SYSTEMS, COMPONENTS, OR DEVICES FOR HIGH-RISKAREAS* | | Selected | Selected |

ICS Supplemental Guidance: None.

## CM-3 Configuration Change Control

| CONTROL NUMBER | CONTROL NAME *Control Enhancement Name* | CONTROL BASELINES | | |
|---|---|---|---|---|
| | | LOW | MOD | HIGH |
| **CM-3** | **Configuration Change Control** | | Selected | Selected |
| CM-3 (1) | *CONFIGURATION CHANGE CONTROL \| AUTOMATED DOCUMENT / NOTIFICATION / PROHIBITION OF CHANGES* | | | Selected |
| CM-3 (2) | *CONFIGURATION CHANGE CONTROL \| TEST / VALIDATE / DOCUMENT CHANGES* | | Selected | Selected |

ICS Supplemental Guidance: None.

## CM-4 Security Impact Analysis

| CONTROL NUMBER | CONTROL NAME *Control Enhancement Name* | CONTROL BASELINES | | |
|---|---|---|---|---|
| | | LOW | MOD | HIGH |
| **CM-4** | **Security Impact Analysis** | Selected | Selected | Selected |
| CM-4 (1) | *SECURITY IMPACT ANALYSIS \| SEPARATE TEST ENVIRONMENTS* | | | Selected |

ICS Supplemental Guidance: The organization considers ICS safety and security interdependencies.

Control Enhancement: (1) ICS Supplemental Guidance: None.

## CM-5 Access Restrictions for Change

| CONTROL NUMBER | CONTROL NAME *Control Enhancement Name* | CONTROL BASELINES | | |
|---|---|---|---|---|
| | | LOW | MOD | HIGH |
| **CM-5** | **Access Restrictions for Change** | | Selected | Selected |
| CM-5 (1) | *ACCESS RESTRICTIONS FOR CHANGE \| AUTOMATED ACCESS ENFORCEMENT / AUDITING* | | | Selected |
| CM-5 (2) | *ACCESS RESTRICTIONS FOR CHANGE \| AUDIT SYSTEM CHANGES* | | | Selected |
| CM-5 (3) | *ACCESS RESTRICTIONS FOR CHANGE \| SIGNED COMPONENTS* | | | Selected |

ICS Supplemental Guidance: None.

## CM-6 Configuration Settings

| CONTROL NUMBER | CONTROL NAME *Control Enhancement Name* | CONTROL BASELINES | | |
|---|---|---|---|---|
| | | LOW | MOD | HIGH |
| **CM-6** | **Configuration Settings** | Selected | Selected | Selected |
| CM-6 (1) | *CONFIGURATION SETTINGS \| AUTOMATED CENTRAL MANAGEMENT / APPLICATION / VERIFICATION* | | | Selected |
| CM-6 (2) | *CONFIGURATION SETTINGS \| RESPOND TO UNAUTHORIZED CHANGES* | | | Selected |

ICS Supplemental Guidance: None.

## *CM-7 Least Functionality*

| CONTROL NUMBER | CONTROL NAME *Control Enhancement Name* | CONTROL BASELINES | | |
|---|---|---|---|---|
| | | LOW | MOD | HIGH |
| **CM-7** | **Least Functionality** | Selected | Selected | Selected |
| CM-7 (1) | *LEAST FUNCTIONALITY \| PERIODIC REVIEW* | Added | Selected | Selected |
| CM-7 (2) | *LEAST FUNCTIONALITY \| PREVENT PROGRAM EXECUTION* | | Selected | Selected |
| CM-7 (4) | *LEAST FUNCTIONALITY \| UNAUTHORIZED SOFTWARE* | | Removed | |
| CM-7 (5) | *LEAST FUNCTIONALITY \| AUTHORIZED SOFTWARE* | | Added | Selected |

ICS Supplemental Guidance: Ports, as used in NIST SP 800-53, Revision 4, are part of the address space in network protocols and are often associated with specific protocols or functions. As such, ports are not relevant to nonroutable protocols and devices. When dealing with nonroutable and nonaddressable protocols and devices, prohibiting or restricting the use of specified functions, protocols, and/or services must be implemented for the (sub)system granularity that is available (e.g., at a low level, interrupts could be disabled; at a high level, set points could be made read-only except for privileged users). Example compensating controls include employing nonautomated mechanisms or procedures.

Control Enhancement: (1, 2, 5) ICS Supplemental Guidance: None.

Control Baseline Supplement Rationale: (1) Periodic review and removal of unnecessary and/or nonsecure functions, ports, protocols, and services are added to the low baseline because many of the low impact ICS components could adversely affect the systems to which they are connected.

Control Baseline Supplement Rationale: (4, 5) Whitelisting (CE 5) is more effective than blacklisting (CE 4). The set of applications that run in ICS is essentially static, making whitelisting practical. ICS-CERT recommends deploying application whitelisting on ICS.[3]

### CM-8 Information System Component Inventory

| CONTROL NUMBER | CONTROL NAME<br>*Control Enhancement Name* | CONTROL BASELINES | | |
|---|---|---|---|---|
| | | LOW | MOD | HIGH |
| **CM-8** | **Information System Component Inventory** | Selected | Selected | Selected |
| CM-8 (1) | *INFORMATION SYSTEM COMPONENT INVENTORY \| UPDATES DURING INSTALLATIONS / REMOVALS* | | Selected | Selected |
| CM-8 (2) | *INFORMATION SYSTEM COMPONENT INVENTORY \| AUTOMATED MAINTENANCE* | | | Selected |
| CM-8 (3) | *INFORMATION SYSTEM COMPONENT INVENTORY \| AUTOMATED UNAUTHORIZED COMPONENT DETECTION* | | Selected | Selected |
| CM-8 (4) | *INFORMATION SYSTEM COMPONENT INVENTORY \| PROPERTY ACCOUNTABILITY INFORMATION* | | | Selected |
| CM-8 (5) | *INFORMATION SYSTEM COMPONENT INVENTORY \| ALL COMPONENTS WITHIN AUTHORIZATION BOUNDARY* | | Selected | Selected |

ICS Supplemental Guidance: None.

### CM-9 Configuration Management Plan

| CONTROL NUMBER | CONTROL NAME<br>*Control Enhancement Name* | CONTROL BASELINES | | |
|---|---|---|---|---|
| | | LOW | MOD | HIGH |
| **CM-9** | **Configuration Management Plan** | | Selected | Selected |

ICS Supplemental Guidance: None.

### CM-10 Software Usage Restrictions

| CONTROL NUMBER | CONTROL NAME<br>*Control Enhancement Name* | CONTROL BASELINES | | |
|---|---|---|---|---|
| | | LOW | MOD | HIGH |
| **CM-10** | **Software Usage Restrictions** | Selected | Selected | Selected |

ICS Supplemental Guidance: None.

### *CM-11 User-Installed Software*

| CONTROL NUMBER | CONTROL NAME *Control Enhancement Name* | CONTROL BASELINES | | |
|---|---|---|---|---|
| | | LOW | MOD | HIGH |
| CM-11 | User-Installed Software | Selected | Selected | Selected |

ICS Supplemental Guidance: None.

# Contingency Planning – CP

## Tailoring Considerations for the Contingency Planning Family

ICS systems often contain a physical component at a fixed location. Such components may not be relocated logically. Some replacement components may not be readily available. Continuance of essential missions and business functions with little or no loss of operational continuity may not be possible. In situations where the organization cannot provide necessary essential services, support, or automated mechanisms during contingency operations, the organization provides nonautomated mechanisms or predetermined procedures as compensating controls in accordance with the general tailoring guidance. Examples of compensating controls are given with each control, as appropriate.

## Supplemental Guidance

Supplemental Guidance for all Controls and Control Enhancements in NIST SP 800-53, Revision 4, Appendix F, should be used in conjunction with the ICS Supplemental Guidance in this overlay, if any.

### *CP-1 Contingency Planning Policy and Procedures*

| CONTROL NUMBER | CONTROL NAME *Control Enhancement Name* | CONTROL BASELINES | | |
|---|---|---|---|---|
| | | LOW | MOD | HIGH |
| CP-1 | Contingency Planning Policy and Procedures | Selected | Selected | Selected |

ICS Supplemental Guidance: The policy specifically addresses the unique properties and requirements of ICS and the relationship to non-ICS systems.

### *CP-2 Contingency Plan*

| CONTROL NUMBER | CONTROL NAME<br>*Control Enhancement Name* | CONTROL BASELINES | | |
|---|---|---|---|---|
| | | LOW | MOD | HIGH |
| **CP-2** | **Contingency Plan** | Selected | Selected | Selected |
| CP-2 (1) | *CONTINGENCY PLAN \| COORDINATE WITH RELATED PLANS* | | Selected | Selected |
| CP-2 (2) | *CONTINGENCY PLAN \| CAPACITY PLANNING* | | | Selected |
| CP-2 (3) | *CONTINGENCY PLAN \| RESUME ESSENTIAL MISSIONS / BUSINESS FUNCTIONS* | | Selected | Selected |
| CP-2 (4) | *CONTINGENCY PLAN \| RESUME ALL MISSIONS / BUSINESS FUNCTIONS* | | | Selected |
| CP-2 (5) | *CONTINGENCY PLAN \| CONTINUE ESSENTIAL MISSIONS /BUSINESS FUNCTIONS* | | | Selected |
| CP-2 (8) | *CONTINGENCY PLAN \| IDENTIFY CRITICAL ASSETS* | | Selected | Selected |

ICS Supplemental Guidance: The organization defines contingency plans for categories of disruptions or failures. In the event of a loss of processing within the ICS or communication with operational facilities, the ICS executes predetermined procedures (e.g., alert the operator of the failure and then do nothing, alert the operator and then safely shut down the industrial process, alert the operator and then maintain the last operational setting prior to failure).

Control Enhancement: (1) ICS Supplemental Guidance: Organizational elements responsible for related plans may include suppliers such as electric power, fuel, fresh water and wastewater.

Control Enhancement: (2) ICS Supplemental Guidance: None.

Control Enhancement: (3, 4) ICS Supplemental Guidance: Plans for the resumption of essential missions and business functions, and for resumption of all missions and business functions take into account the effects of the disruption on the environment of operation. Restoration and resumption plans should include prioritization of efforts. Disruptions may affect the quality and quantity of resources in the environment, such as electric power, fuel, fresh water and wastewater, and the ability of these suppliers to also resume provision of essential mission and business functions. Contingency plans for widespread disruption may involve specialized organizations (e.g., FEMA, emergency services, regulatory authorities).[4]

Control Enhancement: (5, 8) ICS Supplemental Guidance: None.

## CP-3 Contingency Training

| CONTROL NUMBER | CONTROL NAME<br>*Control Enhancement Name* | CONTROL BASELINES | | |
|---|---|---|---|---|
| | | LOW | MOD | HIGH |
| **CP-3** | **Contingency Training** | Selected | Selected | Selected |
| CP-3 (1) | *CONTINGENCY TRAINING \| SIMULATED EVENTS* | | | Selected |

ICS Supplemental Guidance: None.

## CP-4 Contingency Plan Testing

| CONTROL NUMBER | CONTROL NAME<br>*Control Enhancement Name* | CONTROL BASELINES | | |
|---|---|---|---|---|
| | | LOW | MOD | HIGH |
| **CP-4** | **Contingency Plan Testing** | Selected | Selected | Selected |
| CP-4 (1) | *CONTINGENCY PLAN TESTING \| COORDINATE WITH RELATED PLANS* | | Selected | Selected |
| CP-4 (2) | *CONTINGENCY PLAN TESTING \| ALTERNATE PROCESSING SITE* | | | Selected |

ICS Supplemental Guidance: None.

## CP-6 Alternate Storage Site

| CONTROL NUMBER | CONTROL NAME<br>*Control Enhancement Name* | CONTROL BASELINES | | |
|---|---|---|---|---|
| | | LOW | MOD | HIGH |
| **CP-6** | **Alternate Storage Site** | | Selected | Selected |
| CP-6 (1) | *ALTERNATE STORAGE SITE \| SEPARATION FROM PRIMARY SITE* | | Selected | Selected |
| CP-6 (2) | *ALTERNATE STORAGE SITE \| RECOVERY TIME / POINT OBJECTIVES* | | | Selected |
| CP-6 (3) | *ALTERNATE STORAGE SITE \| ACCESSIBILITY* | | Selected | Selected |

ICS Supplemental Guidance: None.

### *CP-7 Alternate Processing Site*

| CONTROL NUMBER | CONTROL NAME *Control Enhancement Name* | CONTROL BASELINES | | |
|---|---|---|---|---|
| | | LOW | MOD | HIGH |
| **CP-7** | **Alternate Processing Site** | | Selected | Selected |
| CP-7 (1) | *ALTERNATE PROCESSING SITE \| SEPARATION FROM PRIMARY SITE* | | Selected | Selected |
| CP-7 (2) | *ALTERNATE PROCESSING SITE \|ACCESSIBILITY* | | Selected | Selected |
| CP-7 (3) | *ALTERNATE PROCESSING SITE \|PRIORITY OF SERVICE* | | Selected | Selected |
| CP-7 (4) | *ALTERNATE PROCESSING SITE \|CONFIGURATION FOR USE* | | | Selected |

ICS Supplemental Guidance: None.

### *CP-8 Telecommunications Services*

| CONTROL NUMBER | CONTROL NAME *Control Enhancement Name* | CONTROL BASELINES | | |
|---|---|---|---|---|
| | | LOW | MOD | HIGH |
| **CP-8** | **Telecommunications Services** | | Selected | Selected |
| CP-8 (1) | TELECOMMUNICATIONS SERVICES \| PRIORITY OF SERVICE PROVISIONS | | Selected | Selected |
| CP-8 (2) | TELECOMMUNICATIONS SERVICES \| SINGLE POINTS OF FAILURE | | Selected | Selected |
| CP-8 (3) | *TELECOMMUNICATIONS SERVICES \| SEPARATION OF PRIMARY /* ALTERNATE PROVIDERS | | | Selected |
| CP-8 (4) | TELECOMMUNICATIONS SERVICES \| PROVIDER CONTINGENCY PLAN | | | Selected |

ICS Supplemental Guidance: Quality of service factors for ICS include latency and throughput.

Control Enhancement: (1, 2, 3, 4) ICS Supplemental Guidance: None.

## CP-9 Information System Backup

| CONTROL NUMBER | CONTROL NAME *Control Enhancement Name* | CONTROL BASELINES | | |
|---|---|---|---|---|
| | | LOW | MOD | HIGH |
| **CP-9** | **Information System Backup** | Selected | Selected | Selected |
| CP-9 (1) | *INFORMATION SYSTEM BACKUP \| TESTING FOR RELIABILITY / INTEGRITY* | | Selected | Selected |
| CP-9 (2) | *INFORMATION SYSTEM BACKUP \| TEST RESTORATION USING SAMPLING* | | | Selected |
| CP-9 (3) | *INFORMATION SYSTEM BACKUP \| SEPARATE STORAGE FOR CRITICAL INFORMATION* | | | Selected |
| CP-9 (5) | *INFORMATION SYSTEM BACKUP \| TRANSFER TO ALTERNATE SITE* | | | Selected |

ICS Supplemental Guidance: None.

## CP-10 Information System Recovery and Reconstitution

| CONTROL NUMBER | CONTROL NAME *Control Enhancement Name* | CONTROL BASELINES | | |
|---|---|---|---|---|
| | | LOW | MOD | HIGH |
| **CP-10** | **Information System Recovery and Reconstitution** | Selected | Selected | Selected |
| CP-10 (2) | *INFORMATION SYSTEM RECOVERY AND RECONSTITUTION \| TRANSACTION RECOVERY* | | Selected | Selected |
| CP-10 (4) | *INFORMATION SYSTEM RECOVERY AND RECONSTITUTION \| RESTORE WITHIN TIME PERIOD* | | | Selected |

ICS Supplemental Guidance: Reconstitution of the ICS includes consideration whether system state variables should be restored to initial values or values before disruption (e.g., are valves restored to full open, full closed, or settings prior to disruption). Restoring system state variables may be disruptive to ongoing physical processes (e.g., valves initially closed may adversely affect system cooling).

Control Enhancement: (2, 4) ICS Supplemental Guidance: None.

### *CP-12 Safe Mode*

| CONTROL NUMBER | CONTROL NAME *Control Enhancement Name* | SUPPLEMENTED CONTROL BASELINES | | |
|---|---|---|---|---|
| | | LOW | MOD | HIGH |
| CP-12 | Safe Mode | Added | Added | Added |

ICS Supplemental Guidance: The organization-defined conditions and corresponding restrictions of safe mode of operation may vary among baselines. The same condition(s) may trigger different responses depending on the impact level. The conditions may be external to the ICS (e.g., electricity supply brown-out). Related controls: SI-17.

Rationale for Adding CP-12 to All Baselines: This control provides a framework for the organization to plan their policy and procedures for dealing with conditions beyond their control in the environment of operations. Creating a written record of the decision process for selecting incidents and the appropriate response is part of risk management in light of changing environment of operations.

## Identification and Authentication - IA

### Tailoring Considerations for the Identification and Authentication Family

Before implementing controls in the IA family, consider the tradeoffs among security, privacy, latency, performance, and throughput. For example, the organization considers whether latency induced from the use of authentication mechanisms employing cryptographic mechanisms would adversely impact the operational performance of the ICS.

In situations where the ICS cannot support the specific identification and authentication requirements of a control, the organization employs compensating controls in accordance with the general tailoring guidance. Examples of compensating controls are given with each control, as appropriate.

### Supplemental Guidance

Supplemental Guidance for all Controls and Control Enhancements in NIST SP 800-53, Revision 4, Appendix F, should be used in conjunction with the ICS Supplemental Guidance in this overlay, if any.

## IA-1 Identification and Authentication Policy and Procedures

| CONTROL NUMBER | CONTROL NAME *Control Enhancement Name* | CONTROL BASELINES | | |
|---|---|---|---|---|
| | | LOW | MOD | HIGH |
| IA-1 | **Security Identification and Authentication Policy and Procedures** | Selected | Selected | Selected |

ICS Supplemental Guidance: The policy specifically addresses the unique properties and requirements of ICS and the relationship to non-ICS systems.

## IA-2 User Identification and Authentication (Organizational Users)

| CONTROL NUMBER | CONTROL NAME *Control Enhancement Name* | CONTROL BASELINES | | |
|---|---|---|---|---|
| | | LOW | MOD | HIGH |
| IA-2 | **Identification and Authentication (Organizational Users)** | Selected | Selected | Selected |
| IA-2 (1) | *IDENTIFICATION AND AUTHENTICATION \| NETWORK ACCESS TO PRIVILEGED ACCOUNTS* | Selected | Selected | Selected |
| IA-2 (2) | *IDENTIFICATION AND AUTHENTICATION \| NETWORK ACCESS TO NONPRIVILEGED ACCOUNTS* | | Selected | Selected |
| IA-2 (3) | *IDENTIFICATION AND AUTHENTICATION \| LOCAL ACCESS TO PRIVILEGED ACCOUNTS* | | Selected | Selected |
| IA-2 (4) | *IDENTIFICATION AND AUTHENTICATION \| LOCAL ACCESS TO NONPRIVILEGED ACCOUNTS* | | | Selected |
| IA-2 (8) | *IDENTIFICATION AND AUTHENTICATION \| NETWORK ACCESS TO PRIVILEGED ACCOUNTS - REPLAY RESISTANT* | | Selected | Selected |
| IA-2 (9) | *IDENTIFICATION AND AUTHENTICATION \| NETWORK ACCESS TO NONPRIVILEGED ACCOUNTS - REPLAY RESISTANT* | | | Selected |
| IA-2 (11) | *IDENTIFICATION AND AUTHENTICATION \| REMOTE ACCESS - SEPARATE DEVICE* | | Selected | Selected |
| IA-2 (12) | *IDENTIFICATION AND AUTHENTICATION \| ACCEPTANCE OF PIV CREDENTIALS* | Selected | Selected | Selected |

ICS Supplemental Guidance: Where users function as a single group (e.g., control room operators), user identification and authentication may be role-based, group-based, or device-based. For certain ICS, the capability for immediate operator interaction is critical. Local emergency actions for ICS are not hampered by identification or authentication requirements. Access to these systems may be restricted by appropriate physical security controls.

Example compensating controls include providing increased physical security, personnel security, and auditing measures. For example, manual voice authentication of remote personnel and local, manual actions may be required in order to establish a remote access. See AC-17 ICS Supplemental Guidance. Local user access to ICS components is enabled only when necessary, approved, and authenticated.

Control Enhancement: (1, 2, 3, 4) ICS Supplemental Guidance: Example compensating controls include implemented physical security measures.

Control Enhancement: (8, 9) ICS Supplemental Guidance: An example of a compensating control would be providing replay-resistance in an external system.

Control Enhancement: (11) ICS Supplemental Guidance: None.

Control Enhancement: (12) ICS Supplemental Guidance: An example of a compensating control would be implementing support for PIV external to the ICS.

### *IA-3 Device Identification and Authentication*

| CONTROL NUMBER | CONTROL NAME *Control Enhancement Name* | SUPPLEMENTED CONTROL BASELINES | | |
|---|---|---|---|---|
| | | LOW | MOD | HIGH |
| **IA-3** | **Device Identification and Authentication** | Added | Selected | Selected |
| IA-3 (1) | *DEVICE IDENTIFICATION AND AUTHENTICATION \| CRYPTOGRAPHIC BIDIRECTIONAL AUTHENTICATION* | | Added | Added |
| IA-3 (4) | *DEVICE IDENTIFICATION AND AUTHENTICATION \| DEVICE ATTESTATION* | | Added | Added |

ICS Supplemental Guidance: The organization may permit connection of devices, also known as nonperson entities (NPE), belonging to and authorized by another organization (e.g., business partners) to their ICS. Especially when these devices are nonlocal, their identification and authentication can be vital. Organizations may perform risk and impact analysis to determine the required strength of authentication mechanisms. Example compensating controls for devices and protocols that do not provide authentication for remote network connections include implemented physical security measures.

Control Enhancement: (1, 4) ICS Supplemental Guidance: Configuration management for NPE identification and authentication customarily involves a human surrogate or representative for the NPE. Devices are provided with their identification and authentication credentials based on assertions by the human surrogate. The human surrogate also responds to events and anomalies (e.g., credential expiration). Credentials for software entities (e.g., autonomous processes not associated with a specific person) based on properties of that software (e.g., digital signatures) may change every time the software is changed or patched. Special purpose hardware (e.g., custom-integrated circuits and printed-circuit boards) may exhibit similar dependencies. The organization definition of parameters may be different among the impact levels.

Rationale (applies to control and control enhancements): ICS may exchange information with many external systems and devices. Identifying and authenticating the devices introduces situations that do not exist with humans. These controls include assignments that enable the organization to categorize devices by types, models, or other group characteristics. Assignments also enable the organizations to select appropriate controls for local, remote, and network connections.

### *IA-4 Identifier Management*

| CONTROL NUMBER | CONTROL NAME<br>*Control Enhancement Name* | CONTROL BASELINES | | |
|---|---|---|---|---|
| | | LOW | MOD | HIGH |
| IA-4 | **Identifier Management** | Selected | Selected | Selected |

ICS Supplemental Guidance: None.

### *IA-5 Authenticator Management*

| CONTROL NUMBER | CONTROL NAME *Control Enhancement Name* | CONTROL BASELINES | | |
|---|---|---|---|---|
| | | LOW | MOD | HIGH |
| **IA-5** | **Authenticator Management** | Selected | Selected | Selected |
| IA-5 (1) | *AUTHENTICATOR MANAGEMENT \| PASSWORD-BASED AUTHENTICATION* | Selected | Selected | Selected |
| IA-5 (2) | *AUTHENTICATOR MANAGEMENT \| PKI-BASED AUTHENTICATION* | | Selected | Selected |
| IA-5 (3) | *AUTHENTICATOR MANAGEMENT \| IN PERSON REGISTRATION* | | Selected | Selected |
| IA-5 (11) | *AUTHENTICATOR MANAGEMENT \| HARDWARE TOKEN-BASED AUTHENTICATION* | Selected | Selected | Selected |

ICS Supplemental Guidance: An example compensating control is physical access control, encapsulating the ICS to provide authentication external to the ICS.

Control Enhancement: (1, 2, 3, 11) ICS Supplemental Guidance: None.

### *IA-6 Authenticator Feedback*

| CONTROL NUMBER | CONTROL NAME *Control Enhancement Name* | CONTROL BASELINES | | |
|---|---|---|---|---|
| | | LOW | MOD | HIGH |
| **IA-6** | **Authenticator Feedback** | Selected | Selected | Selected |

ICS Supplemental Guidance: This control assumes a visual interface that provides feedback ofauthentication information during the authentication process. When ICS authentication uses an interface that does not support visual feedback (e.g., protocol-based authentication), this control may be tailored out.

### IA-7 Cryptographic Module Authentication

| CONTROL NUMBER | CONTROL NAME *Control Enhancement Name* | CONTROL BASELINES | | |
|---|---|---|---|---|
| | | LOW | MOD | HIGH |
| IA-7 | **Cryptographic Module Authentication** | Selected | Selected | Selected |

ICS Supplemental Guidance: None.

### IA-8 Identification and Authentication (Non-Organizational Users)

| CONTROL NUMBER | CONTROL NAME *Control Enhancement Name* | CONTROL BASELINES | | |
|---|---|---|---|---|
| | | LOW | MOD | HIGH |
| IA-8 | **Identification and Authentication (Non-Organizational Users)** | Selected | Selected | Selected |
| IA-8 (1) | *IDENTIFICATION AND AUTHENTICATION (NON-ORGANIZATIONAL USERS) \| ACCEPTANCE OF PIV CREDENTIALS FROM OTHER AGENCIES* | Selected | Selected | Selected |
| IA-8 (2) | *IDENTIFICATION AND AUTHENTICATION (NON-ORGANIZATIONAL USERS) \| ACCEPTANCE OF THIRD-PARTY CREDENTIALS* | Selected | Selected | Selected |
| IA-8 (3) | *IDENTIFICATION AND AUTHENTICATION (NON-ORGANIZATIONAL USERS) \| USE OF FICAM-APPROVED PRODUCTS* | Selected | Selected | Selected |
| IA-8 (4) | *IDENTIFICATION AND AUTHENTICATION (NON-ORGANIZATIONAL USERS) \| USE OF FICAM-ISSUED PROFILES* | Selected | Selected | Selected |

ICS Supplemental Guidance: The ICS Supplemental Guidance for IA-2, Identification and Authentication (Organizational Users), is applicable for Non-Organizational Users.

Control Enhancement: (1, 2, 3, 4) ICS Supplemental Guidance: Example compensating controls include implementing support external to the ICS and multifactor authentication.

## Incident Response – IR

### Tailoring Considerations for the Incident Response Family

The automated mechanisms used to support the tracking of security incidents are typically not part of, or connected to, the ICS.

## Supplemental Guidance

Supplemental Guidance for all Controls and Control Enhancements in NIST SP 800-53, Revision 4, Appendix F, should be used in conjunction with the ICS Supplemental Guidance in this overlay, if any.

### *IR-1 Incident Response Policy and Procedures*

| CONTROL NUMBER | CONTROL NAME *Control Enhancement Name* | CONTROL BASELINES | | |
|---|---|---|---|---|
| | | LOW | MOD | HIGH |
| IR-1 | **Incident Response Policy and Procedures** | Selected | Selected | Selected |

ICS Supplemental Guidance: The policy specifically addresses the unique properties and requirements of ICS and the relationship to non-ICS systems.

### *IR-2 Incident Response Training*

| CONTROL NUMBER | CONTROL NAME *Control Enhancement Name* | CONTROL BASELINES | | |
|---|---|---|---|---|
| | | LOW | MOD | HIGH |
| IR-2 | **Incident Response Training** | Selected | Selected | Selected |
| IR-2 (1) | *INCIDENT RESPONSE TRAINING \| SIMULATED EVENTS* | | | Selected |
| IR-2 (2) | *INCIDENT RESPONSE TRAINING \| AUTOMATED TRAINING ENVIRONMENTS* | | | Selected |

ICS Supplemental Guidance: None.

### *IR-3 Incident Response Testing*

| CONTROL NUMBER | CONTROL NAME *Control Enhancement Name* | CONTROL BASELINES | | |
|---|---|---|---|---|
| | | LOW | MOD | HIGH |
| IR-3 | **Incident Response Testing** | | Selected | Selected |
| IR-3 (2) | *INCIDENT RESPONSE TESTING \| COORDINATION WITH RELATED PLANS* | | Selected | Selected |

ICS Supplemental Guidance: None.

### *IR-4 Incident Handling*

| CONTROL NUMBER | CONTROL NAME *Control Enhancement Name* | CONTROL BASELINES | | |
|---|---|---|---|---|
| | | LOW | MOD | HIGH |
| **IR-4** | **Incident Handling** | Selected | Selected | Selected |
| IR-4 (1) | *INCIDENT HANDLING \| AUTOMATED INCIDENT HANDLING PROCESSES* | | Selected | Selected |
| IR-4 (4) | *INCIDENT HANDLING \| INFORMATION CORRELATION* | | | Selected |

ICS Supplemental Guidance: None.

### *IR-5 Incident Monitoring*

| CONTROL NUMBER | CONTROL NAME *Control Enhancement Name* | CONTROL BASELINES | | |
|---|---|---|---|---|
| | | LOW | MOD | HIGH |
| **IR-5** | **Incident Monitoring** | Selected | Selected | Selected |
| IR-5 (1) | *INCIDENT MONITORING \| AUTOMATED TRACKING / DATA COLLECTION / ANALYSIS* | | | Selected |

ICS Supplemental Guidance: None.

### *IR-6 Incident Reporting*

| CONTROL NUMBER | CONTROL NAME *Control Enhancement Name* | CONTROL BASELINES | | |
|---|---|---|---|---|
| | | LOW | MOD | HIGH |
| **IR-6** | **Incident Reporting** | Selected | Selected | Selected |
| IR-6 (1) | *INCIDENT REPORTING \| AUTOMATED REPORTING* | | Selected | Selected |

ICS Supplemental Guidance: The organization should report incidents on a timely basis. The DHS National Cybersecurity & Communications Integration Center (NCCIC)[5] serves as a centralized location where operational elements involved in cybersecurity and communications reliance are coordinated and integrated. The Industrial Control Systems Cyber Emergency Response Team (ICS-CERT)[6] collaborates with international and private sector Computer Emergency Response Teams (CERTs) to share control systems-related security incidents and mitigation measures.

Control Enhancement: (1) ICS Supplemental Guidance: The automated mechanisms used to support the incident reporting process are not necessarily part of, or connected to, the ICS.

### *IR-7 Incident Response Assistance*

| CONTROL NUMBER | CONTROL NAME *Control Enhancement Name* | CONTROL BASELINES | | |
|---|---|---|---|---|
| | | LOW | MOD | HIGH |
| IR-7 | **Incident Response Assistance** | Selected | Selected | Selected |
| IR-7 (1) | *INCIDENT RESPONSE ASSISTANCE \| AUTOMATION SUPPORT FOR AVAILABILITY OF INFORMATION / SUPPORT* | | Selected | Selected |

ICS Supplemental Guidance: None.

### *IR-8 Incident Response Plan*

| CONTROL NUMBER | CONTROL NAME *Control Enhancement Name* | CONTROL BASELINES | | |
|---|---|---|---|---|
| | | LOW | MOD | HIGH |
| IR-8 | **Incident Response Plan** | Selected | Selected | Selected |

ICS Supplemental Guidance: None.

## Maintenance – MA

### Tailoring Considerations for the Maintenance Family

The automated mechanisms used to schedule, conduct, and document maintenance and repairs are not necessarily part of, or connected to, the ICS.

In situations where the ICS cannot support the specific maintenance requirements of a control, the organization employs compensating controls in accordance with the general tailoring guidance. Examples of compensating controls are given with each control, as appropriate.

## Supplemental Guidance

Supplemental Guidance for all Controls and Control Enhancements in NISTSP 800-53, Revision 4, Appendix F, should be used in conjunction with the ICS Supplemental Guidance in this overlay, ifany.

### *MA-1 System Maintenance Policy and Procedures*

| CONTROL NUMBER | CONTROL NAME *Control Enhancement Name* | CONTROL BASELINES | | |
|---|---|---|---|---|
| | | LOW | MOD | HIGH |
| **MA-1** | **Maintenance Policy and Procedures** | Selected | Selected | Selected |

ICS Supplemental Guidance: The policy specifically addresses the unique properties and requirements ofICS and the relationship to non-ICS systems.

### *MA-2 Controlled Maintenance*

| CONTROL NUMBER | CONTROL NAME *Control Enhancement Name* | CONTROL BASELINES | | |
|---|---|---|---|---|
| | | LOW | MOD | HIGH |
| **MA-2** | **Controlled Maintenance** | Selected | Selected | Selected |
| MA-2 (2) | *CONTROLLED MAINTENANCE \| AUTOMATED MAINTENANCEACTIVITIES* | | | Selected |

ICS Supplemental Guidance: None.

### *MA-3 Maintenance Tools*

| CONTROL NUMBER | CONTROL NAME *Control Enhancement Name* | CONTROL BASELINES | | |
|---|---|---|---|---|
| | | LOW | MOD | HIGH |
| **MA-3** | **Maintenance Tools** | | Selected | Selected |
| MA-3 (1) | *MAINTENANCE TOOLS \| INSPECT TOOLS* | | Selected | Selected |
| MA-3 (2) | *MAINTENANCE TOOLS \| INSPECT MEDIA* | | Selected | Selected |
| MA-3 (3) | *MAINTENANCE TOOLS \| PREVENT UNAUTHORIZED REMOVAL* | | | Selected |

ICS Supplemental Guidance: None.

### MA-4 Nonlocal Maintenance

| CONTROL NUMBER | CONTROL NAME *Control Enhancement Name* | CONTROL BASELINES | | |
|---|---|---|---|---|
| | | LOW | MOD | HIGH |
| **MA-4** | **Nonlocal Maintenance** | Selected | Selected | Selected |
| MA-4 (2) | *NONLOCAL MAINTENANCE \| DOCUMENT NONLOCAL MAINTENANCE* | | Selected | Selected |
| MA-4 (3) | *NONLOCAL MAINTENANCE \| COMPARABLE SECURITY / SANITIZATION* | | | Selected |

ICS Supplemental Guidance: None.

Control Enhancement: (2) ICS Supplemental Guidance: None.

Control Enhancement: (3) ICS Supplemental Guidance: In crisis or emergency situations, the organization may need immediate access to nonlocal maintenance and diagnostic services in order to restore essential ICS operations or services. Example compensating controls include limiting the extent of the maintenance and diagnostic services to the minimum essential activities, and carefully monitoring and auditing the nonlocal maintenance and diagnostic activities.

### MA-5 Maintenance Personnel

| CONTROL NUMBER | CONTROL NAME *Control Enhancement Name* | CONTROL BASELINES | | |
|---|---|---|---|---|
| | | LOW | MOD | HIGH |
| **MA-5** | **Maintenance Personnel** | Selected | Selected | Selected |
| MA-5 (1) | *MAINTENANCE PERSONNEL \| INDIVIDUALS WITHOUT APPROPRIATE ACCESS* | | | Selected |

ICS Supplemental Guidance: None.

### MA-6 Timely Maintenance

| CONTROL NUMBER | CONTROL NAME<br>*Control Enhancement Name* | CONTROL BASELINES | | |
|---|---|---|---|---|
| | | LOW | MOD | HIGH |
| MA-6 | Timely Maintenance | | Selected | Selected |

<u>ICS Supplemental Guidance</u>: None.

# Media Protection – MP

## Supplemental Guidance

Supplemental Guidance for all Controls and Control Enhancements in NIST SP 800-53, Revision 4, Appendix F, should be used in conjunction with the ICS Supplemental Guidance in this overlay, if any.

### MP-1 Media Protection Policy and Procedures

| CONTROL NUMBER | CONTROL NAME<br>*Control Enhancement Name* | CONTROL BASELINES | | |
|---|---|---|---|---|
| | | LOW | MOD | HIGH |
| MP-1 | Media Protection Policy and Procedures | Selected | Selected | Selected |

<u>ICS Supplemental Guidance</u>: The policy specifically addresses the unique properties and requirements of ICS and the relationship to non-ICS systems.

### MP-2 Media Access

| CONTROL NUMBER | CONTROL NAME<br>*Control Enhancement Name* | CONTROL BASELINES | | |
|---|---|---|---|---|
| | | LOW | MOD | HIGH |
| MP-2 | Media Access | Selected | Selected | Selected |

<u>ICS Supplemental Guidance</u>: None.

### MP-3 Media Marking

| CONTROL NUMBER | CONTROL NAME *Control Enhancement Name* | CONTROL BASELINES | | |
|---|---|---|---|---|
| | | LOW | MOD | HIGH |
| **MP-3** | **Media Marking** | | Selected | Selected |

ICS Supplemental Guidance: None.

### MP-4 Media Storage

| CONTROL NUMBER | CONTROL NAME *Control Enhancement Name* | CONTROL BASELINES | | |
|---|---|---|---|---|
| | | LOW | MOD | HIGH |
| **MP-4** | **Media Storage** | | Selected | Selected |

ICS Supplemental Guidance: None.

### MP-5 Media Transport

| CONTROL NUMBER | CONTROL NAME *Control Enhancement Name* | CONTROL BASELINES | | |
|---|---|---|---|---|
| | | LOW | MOD | HIGH |
| **MP-5** | **Media Transport** | | Selected | Selected |
| MP-5 (4) | *MEDIA TRANSPORT \| CRYPTOGRAPHIC PROTECTION* | | Selected | Selected |

ICS Supplemental Guidance: None.

### MP-6 Media Sanitization

| CONTROL NUMBER | CONTROL NAME *Control Enhancement Name* | CONTROL BASELINES | | |
|---|---|---|---|---|
| | | LOW | MOD | HIGH |
| **MP-6** | **Media Sanitization** | Selected | Selected | Selected |
| MP-6 (1) | *MEDIA SANITIZATION \| TRACKING / DOCUMENTING / VERIFYING* | | | Selected |
| MP-6 (2) | *MEDIA SANITIZATION \| EQUIPMENT TESTING* | | | Selected |
| MP-6 (3) | *MEDIA SANITIZATION \| NONDESTRUCTIVE TECHNIQUES* | | | Selected |

ICS Supplemental Guidance: None.

### *MP-7 Media Use*

| CONTROL NUMBER | CONTROL NAME *Control Enhancement Name* | CONTROL BASELINES | | |
|---|---|---|---|---|
| | | LOW | MOD | HIGH |
| **MP-7** | **Media Use** | Selected | Selected | Selected |
| MP-7 (1) | *MEDIA USE \| ORGANIZATIONAL RESTRICTIONS* | | Selected | Selected |

ICS Supplemental Guidance: None.

# Physical and Environmental Protection – PE

## Supplemental Guidance

Supplemental Guidance for all Controls and Control Enhancements in NIST SP 800-53 Revision 4, Appendix F, should be used in conjunction with the ICS Supplemental Guidance in this overlay, if any.

### *PE-1 Physical and Environmental Protection Policy and Procedures*

| CONTROL NUMBER | CONTROL NAME *Control Enhancement Name* | CONTROL BASELINES | | |
|---|---|---|---|---|
| | | LOW | MOD | HIGH |
| **PE-1** | **Physical and Environmental Protection Policy and Procedures** | Selected | Selected | Selected |

ICS Supplemental Guidance: The policy specifically addresses the unique properties and requirements of ICS and the relationship to non-ICS systems. The ICS components can be distributed over a large facility footprint or geographic area and can be an entry point into the entire organizational network ICS. Regulatory controls may also apply.

### *PE-2 Physical Access Authorizations*

| CONTROL NUMBER | CONTROL NAME *Control Enhancement Name* | CONTROL BASELINES | | |
|---|---|---|---|---|
| | | LOW | MOD | HIGH |
| **PE-2** | **Physical Access Authorizations** | Selected | Selected | Selected |

ICS Supplemental Guidance: None.

### PE-3 Physical Access Control

| CONTROL NUMBER | CONTROL NAME *Control Enhancement Name* | CONTROL BASELINES | | |
|---|---|---|---|---|
| | | LOW | MOD | HIGH |
| PE-3 | **Physical Access Control** | Selected | Selected | Selected |
| PE-3 (1) | *PHYSICAL ACCESS CONTROL \| INFORMATION SYSTEM ACCESS* | | | Selected |

ICS Supplemental Guidance: The organization considers ICS safety and security interdependencies. The organization considers access requirements in emergency situations. During an emergency-related event, the organization may restrict access to ICS facilities and assets to authorized individuals only. ICS are oftenconstructed of devices that either do not have or cannot use comprehensive access control capabilities due to time-restrictive safety constraints. Physical access controls and defense-in-depth measures are used by the organization when necessary and possible to supplement ICS security when electronic mechanisms are unable to fulfill the security requirements of the organization's security plan. Primary nodes, distribution closets, and mechanical/electrical rooms should be locked, require key or electronic access control, and incorporate intrusion detection sensors.

Control Enhancement: (1) ICS Supplemental Guidance: None.

### PE-4 Access Control for Transmission Medium

| CONTROL NUMBER | CONTROL NAME *Control Enhancement Name* | CONTROL BASELINES | | |
|---|---|---|---|---|
| | | LOW | MOD | HIGH |
| PE-4 | **Access Control for Transmission Medium** | | Selected | Selected |

ICS Supplemental Guidance: None.

## PE-5 Access Control for Output Devices

| CONTROL NUMBER | CONTROL NAME *Control Enhancement Name* | CONTROL BASELINES | | |
|---|---|---|---|---|
| | | LOW | MOD | HIGH |
| PE-5 | **Access Control for Output Devices** | | Selected | Selected |

ICS Supplemental Guidance: None.

## PE-6 Monitoring Physical Access

| CONTROL NUMBER | CONTROL NAME *Control Enhancement Name* | SUPPLEMENTED CONTROL BASELINES | | |
|---|---|---|---|---|
| | | LOW | MOD | HIGH |
| PE-6 | **Monitoring Physical Access** | Selected | Selected | Selected |
| PE-6 (1) | *MONITORING PHYSICAL ACCESS \| INTRUSION ALARMS / SURVEILLANCE EQUIPMENT* | | Selected | Selected |
| PE-6 (4) | *MONITORING PHYSICAL ACCESS \| MONITORING PHYSICAL ACCESS TO INFORMATION SYSTEMS* | | Added | Selected |

ICS Supplemental Guidance: Physical access controls and defense-in-depth measures are used as compensating controls by the organization when necessary and possible to supplement ICS security when electronic mechanisms are unable to monitor, detect and alarm when an ICS has been accessed. These compensating controls are in addition to the PE-6 controls (e.g., employing PE-3(4) Lockable Casings and/or PE-3(5) Tamper Protection).

Control Enhancement: (1) ICS Supplemental Guidance: None.

Control Enhancement: (4) ICS Supplemental Guidance: The locations of ICS components (e.g., field devices, remote terminal units) can include various remote locations (e.g., substations, pumping stations).

Rationale (adding CE 4 to MODERATE baseline): Many of the ICS components are in remote geographical and dispersed locations with little capability to monitor all ICS components. Other components may be in ceilings, floors, or distribution closets with minimal physical barriers to detect, delay, or deny access to the devices and no electronic surveillance or guard forces response capability.

## *PE-8 Visitor Access Records*

| CONTROL NUMBER | CONTROL NAME *Control Enhancement Name* | CONTROL BASELINES | | |
|---|---|---|---|---|
| | | LOW | MOD | HIGH |
| **PE-8** | **Visitor Access Records** | Selected | Selected | Selected |
| PE-8 (1) | *VISITOR ACCESS RECORDS \| AUTOMATED RECORDS MAINTENANCE / REVIEW* | | | Selected |

ICS Supplemental Guidance: None.

## *PE-9 Power Equipment and Cabling*

| CONTROL NUMBER | CONTROL NAME *Control Enhancement Name* | SUPPLEMENTED CONTROL BASELINES | | |
|---|---|---|---|---|
| | | LOW | MOD | HIGH |
| **PE-9** | **Power Equipment and Cabling** | | Selected | Selected |
| PE-9 (1) | *POWER EQUIPMENT AND CABLING \| REDUNDANT CABLING* | | Added | Added |

ICS Supplemental Guidance: None.

Control Enhancement: (1) ICS Supplemental Guidance: None.

Rationale (for adding (1)): Continuity of ICS control and operation requires redundant power cabling.

## *PE-10 Emergency Shutoff*

| CONTROL NUMBER | CONTROL NAME *Control Enhancement Name* | CONTROL BASELINES | | |
|---|---|---|---|---|
| | | LOW | MOD | HIGH |
| **PE-10** | **Emergency Shutoff** | | Selected | Selected |

ICS Supplemental Guidance: It may not be possible or advisable to shutoff power to some ICS. Example compensating controls include fail in known state and emergency procedures.

## *PE-11 Emergency Power*

| CONTROL NUMBER | CONTROL NAME *Control Enhancement Name* | SUPPLEMENTED CONTROL BASELINES | | |
|---|---|---|---|---|
| | | LOW | MOD | HIGH |
| **PE-11** | **Emergency Power** | Added | Selected | Selected |
| PE-11 (1) | *EMERGENCY POWER \| LONG-TERM ALTERNATE POWER SUPPLY-MINIMAL OPERATIONAL CAPABILITY* | Added | Added | Selected |
| PE-11 (2) | *EMERGENCY POWER \| LONG-TERM ALTERNATE POWER SUPPLY-SELF-CONTAINED* | | | Added |

ICS Supplemental Guidance: Emergency power production, transmission, and distribution systems are a type of ICS that are required to meet extremely high performance specifications. The systems are governed by international, national, state, and local building codes; must be tested on a continual basis; and must be repaired and placed back into operations within a short period of time. Traditionally, emergency power has been provided by generators for short- to mid-term power (typically for fire and life safety systems, some IT load, and evacuation transport) and UPS battery packs in distribution closets and within work areas to allow some level of business continuity and for the orderly shutdown of nonessential IT and facility systems. Traditional emergency power systems typically are off-line until a loss of power occurs and are typically on a separate network and control system specific to the facility they support. New methods of energy generation and storage (e.g., solar voltaic, geothermal, flywheel, microgrid, distributed energy) that have a real-time demand and storage connection to local utilities or cross connected to multiple facilities should be carefully analyzed to ensure that the power can meet the load and signal quality without disruption of mission essential functions.

Control Enhancement: (1) ICS Supplemental Guidance: None.

Rationale for Adding Control to Baseline: ICS may support critical activities that will be needed for safety and reliability even in the absence of reliable power from the public grid.

### *PE-12 Emergency Lighting*

| CONTROL NUMBER | CONTROL NAME<br>*Control Enhancement Name* | CONTROL BASELINES | | |
|---|---|---|---|---|
| | | LOW | MOD | HIGH |
| **PE-12** | **Emergency Lighting** | Selected | Selected | Selected |

ICS Supplemental Guidance: None.

### *PE-13 Fire Protection*

| CONTROL NUMBER | CONTROL NAME<br>*Control Enhancement Name* | CONTROL BASELINES | | |
|---|---|---|---|---|
| | | LOW | MOD | HIGH |
| **PE-13** | **Fire Protection** | Selected | Selected | Selected |
| PE-13 (1) | *FIRE PROTECTION \| DETECTION DEVICES / SYSTEMS* | | | Selected |
| PE-13 (2) | *FIRE PROTECTION \| SUPPRESSION DEVICES / SYSTEMS* | | | Selected |
| PE-13 (3) | *FIRE PROTECTION \| AUTOMATIC FIRE SUPPRESSION* | | Selected | Selected |

ICS Supplemental Guidance: Fire suppression mechanisms should take the ICS environment into account (e.g., water sprinkler systems could be hazardous in specific environments).

Control Enhancement: (1, 2, 3) ICS Supplemental Guidance: None.

### *PE-14 Temperature and Humidity Controls*

| CONTROL NUMBER | CONTROL NAME<br>*Control Enhancement Name* | CONTROL BASELINES | | |
|---|---|---|---|---|
| | | LOW | MOD | HIGH |
| **PE-14** | **Temperature and Humidity Controls** | Selected | Selected | Selected |

ICS Supplemental Guidance: Temperature and humidity controls are typically components of other ICS systems, such as the HVAC, process, or lighting systems, or can be a standalone and unique ICS system. ICS can operate in extreme environments and in both interior and exterior locations. For a specific ICS, the temperature and humidity design and operational parameters dictate the performance specifications. As ICS and IS become interconnected and the network provides connectivity across the hybrid domain, power circuits, distribution closets, routers, and switches that support fire protection and life safety systems must be maintained at the proper temperature and humidity.

### *PE-15 Water Damage Protection*

| CONTROL NUMBER | CONTROL NAME<br>*Control Enhancement Name* | CONTROL BASELINES | | |
|---|---|---|---|---|
| | | LOW | MOD | HIGH |
| PE-15 | **Water Damage Protection** | Selected | Selected | Selected |
| PE-15 (1) | *WATER DAMAGE PROTECTION \| AUTOMATION SUPPORT* | | | Selected |

ICS Supplemental Guidance: Water damage protection and use of shutoff and isolation valves is both a procedural action and a specific type of ICS. ICS that are used in the manufacturing, hydropower, transportation/navigation, water and wastewater industries rely on the movement of water and are specifically designed to manage the quantity/flow and pressure of water. As ICS and IS become interconnected and the network provides connectivity across the hybrid domain, power circuits, distribution closets, routers and switches that support fire protection and life safety. Systems should ensure that water will not disable the system (e.g., a fire that activates the sprinkler system does not spray onto the fire control servers, router, and switches and short out the alarms, egress systems, emergency lighting, and suppression systems).

Control Enhancement: (1) ICS Supplemental Guidance: None.

### *PE-16 Delivery and Removal*

| CONTROL NUMBER | CONTROL NAME<br>*Control Enhancement Name* | CONTROL BASELINES | | |
|---|---|---|---|---|
| | | LOW | MOD | HIGH |
| PE-16 | **Delivery and Removal** | Selected | Selected | Selected |

ICS Supplemental Guidance: None.

### *PE-17 Alternate Work Site*

| CONTROL NUMBER | CONTROL NAME *Control Enhancement Name* | CONTROL BASELINES | | |
|---|---|---|---|---|
| | | LOW | MOD | HIGH |
| PE-17 | Alternate Work Site | | Selected | Selected |

ICS Supplemental Guidance: None.

### *PE-18 Location of Information System Components*

| CONTROL NUMBER | CONTROL NAME *Control Enhancement Name* | CONTROL BASELINES | | |
|---|---|---|---|---|
| | | LOW | MOD | HIGH |
| PE-18 | Location of Information System Components | | | Selected |

ICS Supplemental Guidance: None.

# Planning – PL

## Supplemental Guidance

Supplemental Guidance for all Controls and Control Enhancements in NIST SP 800-53 Revision 4, Appendix F, should be used in conjunction with the ICS Supplemental Guidance in this overlay, if any.

### *PL-1 Security Planning Policy and Procedures*

| CONTROL NUMBER | CONTROL NAME *Control Enhancement Name* | CONTROL BASELINES | | |
|---|---|---|---|---|
| | | LOW | MOD | HIGH |
| PL-1 | Security Planning Policy and Procedures | Selected | Selected | Selected |

ICS Supplemental Guidance: The policy specifically addresses the unique properties and requirements of ICS and the relationship to non-ICS systems.

### PL-2 System Security Plan

| CONTROL NUMBER | CONTROL NAME *Control Enhancement Name* | SUPPLEMENTED CONTROL BASELINES | | |
|---|---|---|---|---|
| | | LOW | MOD | HIGH |
| **PL-2** | **System Security Plan** | Selected | Selected | Selected |
| PL-2 (3) | *SYSTEM SECURITY PLAN \| PLAN / COORDINATE WITH OTHER ORGANIZATIONAL ENTITIES* | Added | Selected | Selected |

ICS Supplemental Guidance: None.

Control Enhancement: (3) ICS Supplemental Guidance: None.

Rationale for Adding PL-2 (3) to Low Baseline: When systems are highly interconnected, coordinated planning is essential. A low-impact system could adversely affect a higher-impact system.

### PL-4 Rules of Behavior

| CONTROL NUMBER | CONTROL NAME *Control Enhancement Name* | CONTROL BASELINES | | |
|---|---|---|---|---|
| | | LOW | MOD | HIGH |
| **PL-4** | **Rules of Behavior** | Selected | Selected | Selected |
| PL-4 (1) | *RULES OF BEHAVIOR \| SOCIAL MEDIA AND NETWORKING RESTRICTIONS* | | Selected | Selected |

ICS Supplemental Guidance: None.

### PL-7 Security Concept of Operations (CONOPS)

| CONTROL NUMBER | CONTROL NAME *Control Enhancement Name* | SUPPLEMENTED CONTROL BASELINES | | |
|---|---|---|---|---|
| | | LOW | MOD | HIGH |
| **PL-7** | **Security Concept of Operations** | | Added | Added |

ICS Supplemental Guidance: None.

Rationale for Adding PL-7 to Moderate and High Baselines: ICS are complex systems. Organizations typically employ a CONOPS to help define a system and share that understanding with personnel involved with that system and other systems with which it interacts. A CONOPS often helps identify information protection requirements.

### *PL-8 Information Security Architecture*

| CONTROL NUMBER | CONTROL NAME *Control Enhancement Name* | CONTROL BASELINES | | |
|---|---|---|---|---|
| | | LOW | MOD | HIGH |
| **PL-8** | **Information Security Architecture** | | Selected | Selected |

ICS Supplemental Guidance: None.

# Personnel Security – PS

## Supplemental Guidance

Supplemental Guidance for all Controls and Control Enhancements in NIST SP 800-53, Revision 4, Appendix F, should be used in conjunction with the ICS Supplemental Guidance in this overlay, if any.

### *PS-1 Personnel Security Policy and Procedures*

| CONTROL NUMBER | CONTROL NAME *Control Enhancement Name* | CONTROL BASELINES | | |
|---|---|---|---|---|
| | | LOW | MOD | HIGH |
| **PS-1** | **Personnel Security Policy and Procedures** | Selected | Selected | Selected |

ICS Supplemental Guidance: The policy specifically addresses the unique properties and requirements of ICS and the relationship to non-ICS systems.

### *PS-2 Position Risk Designation, Selected*

| CONTROL NUMBER | CONTROL NAME *Control Enhancement Name* | CONTROL BASELINES | | |
|---|---|---|---|---|
| | | LOW | MOD | HIGH |
| **PS-2** | **Position Risk Designation** | Selected | Selected | Selected |

ICS Supplemental Guidance: None.

## PS-3 Personnel Screening

| CONTROL NUMBER | CONTROL NAME *Control Enhancement Name* | CONTROL BASELINES | | |
|---|---|---|---|---|
| | | LOW | MOD | HIGH |
| PS-3 | Personnel Screening | Selected | Selected | Selected |

ICS Supplemental Guidance: None.

## PS-4 Personnel Termination

| CONTROL NUMBER | CONTROL NAME *Control Enhancement Name* | CONTROL BASELINES | | |
|---|---|---|---|---|
| | | LOW | MOD | HIGH |
| PS-4 | Personnel Termination | Selected | Selected | Selected |
| PS-4 (2) | *PERSONNEL TERMINATION \| AUTOMATED NOTIFICATION* | | | Selected |

ICS Supplemental Guidance: None.

## PS-5 Personnel Transfer

| CONTROL NUMBER | CONTROL NAME *Control Enhancement Name* | CONTROL BASELINES | | |
|---|---|---|---|---|
| | | LOW | MOD | HIGH |
| PS-5 | Personnel Transfer | Selected | Selected | Selected |

ICS Supplemental Guidance: None.

## PS-6 Access Agreements

| CONTROL NUMBER | CONTROL NAME *Control Enhancement Name* | CONTROL BASELINES | | |
|---|---|---|---|---|
| | | LOW | MOD | HIGH |
| PS-6 | Access Agreements | Selected | Selected | Selected |

ICS Supplemental Guidance: None.

### PS-7 Third-Party Personnel Security

| CONTROL NUMBER | CONTROL NAME *Control Enhancement Name* | CONTROL BASELINES | | |
|---|---|---|---|---|
| | | LOW | MOD | HIGH |
| PS-7 | Third-Party Personnel Security | Selected | Selected | Selected |

ICS Supplemental Guidance: None.

### PS-8 Personnel Sanctions

| CONTROL NUMBER | CONTROL NAME *Control Enhancement Name* | CONTROL BASELINES | | |
|---|---|---|---|---|
| | | LOW | MOD | HIGH |
| PS-8 | Personnel Sanctions | Selected | Selected | Selected |

ICS Supplemental Guidance: None.

## Risk Assessment – RA

### Supplemental Guidance

Supplemental Guidance for all Controls and Control Enhancements in NIST SP 800-53 Revision 4, Appendix F, should be used in conjunction with the ICS Supplemental Guidance in this overlay, if any.

### RA-1 Risk Assessment Policy and Procedures

| CONTROL NUMBER | CONTROL NAME *Control Enhancement Name* | CONTROL BASELINES | | |
|---|---|---|---|---|
| | | LOW | MOD | HIGH |
| RA-1 | Risk Assessment Policy and Procedures | Selected | Selected | Selected |

ICS Supplemental Guidance: The policy specifically addresses the unique properties and requirements of ICS and the relationship to non-ICS systems.

## RA-2 Security Categorization

| CONTROL NUMBER | CONTROL NAME<br>*Control Enhancement Name* | CONTROL BASELINES | | |
|---|---|---|---|---|
| | | LOW | MOD | HIGH |
| **RA-2** | **Security Categorization** | Selected | Selected | Selected |

ICS Supplemental Guidance: None.

## RA-3 Risk Assessment

| CONTROL NUMBER | CONTROL NAME<br>*Control Enhancement Name* | CONTROL BASELINES | | |
|---|---|---|---|---|
| | | LOW | MOD | HIGH |
| **RA-3** | **Risk Assessment** | Selected | Selected | Selected |

ICS Supplemental Guidance: None.

## RA-5 Vulnerability Scanning

| CONTROL NUMBER | CONTROL NAME<br>*Control Enhancement Name* | CONTROL BASELINES | | |
|---|---|---|---|---|
| | | LOW | MOD | HIGH |
| **RA-5** | **Vulnerability Scanning** | Selected | Selected | Selected |
| RA-5 (1) | *VULNERABILITY SCANNING \| UPDATE TOOL CAPABILITY* | | Selected | Selected |
| RA-5 (2) | *VULNERABILITY SCANNING \| UPDATE BY FREQUENCY / PRIOR TO NEW SCAN / WHEN IDENTIFIED* | | Selected | Selected |
| RA-5 (4) | *VULNERABILITY SCANNING \| DISCOVERABLE INFORMATION* | | | Selected |
| RA-5 (5) | *VULNERABILITY SCANNING \| PRIVILEGED ACCESS* | | Selected | Selected |

ICS Supplemental Guidance: Active vulnerability scanning, which introduces network traffic, is used with care on ICS systems to ensure that ICS functions are not adversely impacted by the scanning process. The organization makes a risk-based determination whether to employ active scanning. Passive monitoring/sniffing may be used as part of a compensating control. Example compensating controls include replicated, virtualized, or simulated systems that are used to conduct scanning. Production ICS may need to be taken off-line before scanning can be conducted. If ICS are taken off-line for scanning, scans are scheduled to occur during planned ICS outages whenever possible. If vulnerability scanning tools are used on non-ICS networks, extra care is taken to ensure that they do not scan the ICS network. Network scanning is not applicable to nonaddressable communications. Vulnerability examination may be performed using other mechanisms than scanning to identify the objects being examined. Host-based vulnerability examination is an example of compensating control.

Control Enhancement: (1, 2, 4, 5) ICS Supplemental Guidance: None.

## System and Services Acquisition – SA

### Tailoring Considerations for the System and Services Acquisition Family

In situations where the ICS cannot support the specific system and services acquisition requirements of a control, the organization employs compensating controls in accordance with the general tailoring guidance. Examples of compensating controls are given with each control, as appropriate.

### Supplemental Guidance

Supplemental Guidance for all controls and control enhancements in NIST SP 800-53, Revision 4, Appendix F, should be used in conjunction with the ICS Supplemental Guidance in this overlay, if any.

### *SA-1 System and Services Acquisition Policy and Procedures*

| CONTROL NUMBER | CONTROL NAME *Control Enhancement Name* | CONTROL BASELINES | | |
|---|---|---|---|---|
| | | LOW | MOD | HIGH |
| SA-1 | **System and Services Acquisition Policy and Procedures** | Selected | Selected | Selected |

ICS Supplemental Guidance: The policy specifically addresses the unique properties and requirements of ICS and the relationship to non-ICS systems.

## *SA-2 Allocation of Resources*

| CONTROL NUMBER | CONTROL NAME<br>*Control Enhancement Name* | CONTROL BASELINES | | |
|---|---|---|---|---|
| | | LOW | MOD | HIGH |
| SA-2 | **Allocation of Resources** | Selected | Selected | Selected |

ICS Supplemental Guidance: None.

## *SA-3 System Development Life Cycle*

| CONTROL NUMBER | CONTROL NAME<br>*Control Enhancement Name* | CONTROL BASELINES | | |
|---|---|---|---|---|
| | | LOW | MOD | HIGH |
| SA-3 | **System Development Life Cycle** | Selected | Selected | Selected |

ICS Supplemental Guidance: None.

## *SA-4 Acquisition Process*

| CONTROL NUMBER | CONTROL NAME<br>*Control Enhancement Name* | CONTROL BASELINES | | |
|---|---|---|---|---|
| | | LOW | MOD | HIGH |
| **SA-4** | **Acquisition Process** | Selected | Selected | Selected |
| SA-4 (1) | *ACQUISITION PROCESS \| FUNCTIONAL PROPERTIES OF SECURITY CONTROLS* | | Selected | Selected |
| SA-4 (2) | *ACQUISITION PROCESS \| DESIGN / IMPLEMENTATION INFORMATION FOR SECURITY CONTROLS* | | Selected | Selected |
| SA-4 (9) | *ACQUISITION PROCESS \| FUNCTIONS / PORTS / PROTOCOLS / SERVICES IN USE* | | Selected | Selected |
| SA-4 (10) | *ACQUISITION PROCESS \| USE OF APPROVED PIV PRODUCTS* | Selected | Selected | Selected |

ICS Supplemental Guidance: Since ICS security has historically focused on physical protection and isolation, vendors and developers may be unfamiliar with cybersecurity. Organizations should anticipate a need to engage with ICS suppliers to raise awareness of cybersecurity needs. The SCADA/Control Systems Procurement Project provides example cybersecurity procurement language for ICS.[7]

Control Enhancements: (1, 2, 9) ICS Supplemental Guidance: Developers may not have access to required information.

Control Enhancement: (10) ICS Supplemental Guidance: Examples of compensating controls would be external products on the FIPS 201 approved products list that are employed for Personal Identity Verification (PIV) capability in conjunction with ICS products.

### SA-5 Information System Documentation

| CONTROL NUMBER | CONTROL NAME *Control Enhancement Name* | CONTROL BASELINES | | |
|---|---|---|---|---|
| | | LOW | MOD | HIGH |
| SA-5 | **Information System Documentation** | Selected | Selected | Selected |

ICS Supplemental Guidance: None.

### SA-8 Security Engineering Principles

| CONTROL NUMBER | CONTROL NAME *Control Enhancement Name* | CONTROL BASELINES | | |
|---|---|---|---|---|
| | | LOW | MOD | HIGH |
| SA-8 | **Security Engineering Principles** | | Selected | Selected |

ICS Supplemental Guidance: None.

### SA-9 External Information System Services

| CONTROL NUMBER | CONTROL NAME *Control Enhancement Name* | CONTROL BASELINES | | |
|---|---|---|---|---|
| | | LOW | MOD | HIGH |
| SA-9 | **External Information System Services** | Selected | Selected | Selected |
| SA-9 (2) | *EXTERNAL INFORMATION SYSTEMS \| IDENTIFICATION OF FUNCTIONS / PORTS / PROTOCOLS / SERVICES* | | Selected | Selected |

ICS Supplemental Guidance: None.

## SA-10 Developer Configuration Management

| CONTROL NUMBER | CONTROL NAME *Control Enhancement Name* | CONTROL BASELINES | | |
|---|---|---|---|---|
| | | LOW | MOD | HIGH |
| SA-10 | Developer Configuration Management | | Selected | Selected |

ICS Supplemental Guidance: None.

## SA-11 Developer Security Testing and Evaluation

| CONTROL NUMBER | CONTROL NAME *Control Enhancement Name* | CONTROL BASELINES | | |
|---|---|---|---|---|
| | | LOW | MOD | HIGH |
| SA-11 | Developer Security Testing and Evaluation | | Selected | Selected |

ICS Supplemental Guidance: None.

## SA-12 Supply Chain Protection

| CONTROL NUMBER | CONTROL NAME *Control Enhancement Name* | CONTROL BASELINES | | |
|---|---|---|---|---|
| | | LOW | MOD | HIGH |
| SA-12 | Supply Chain Protection | | | Selected |

ICS Supplemental Guidance: None.

## SA-15 Development Process, Standards, and Tools

| CONTROL NUMBER | CONTROL NAME *Control Enhancement Name* | CONTROL BASELINES | | |
|---|---|---|---|---|
| | | LOW | MOD | HIGH |
| SA-15 | Development Process, Standards, and Tools | | | Selected |

ICS Supplemental Guidance: None.

### *SA-16 Developer-Provided Training*

| CONTROL NUMBER | CONTROL NAME *Control Enhancement Name* | CONTROL BASELINES | | |
|---|---|---|---|---|
| | | LOW | MOD | HIGH |
| SA-16 | Developer-Provided Training | | | Selected |

ICS Supplemental Guidance: None.

### *SA-17 Developer Security Architecture and Design*

| CONTROL NUMBER | CONTROL NAME *Control Enhancement Name* | CONTROL BASELINES | | |
|---|---|---|---|---|
| | | LOW | MOD | HIGH |
| SA-17 | Developer Security Architecture and Design | | | Selected |

ICS Supplemental Guidance: None.

## System and Communications Protection – SC

### Tailoring Considerations for the System and Communications Protection Family

The use of cryptography is determined after careful consideration of the security needs and the potential ramifications on system performance. For example, the organization considers whether latency induced from the use of cryptography would adversely impact the operational performance of the ICS. While the legacy devices commonly found within ICS often lack direct support of cryptographic functions, compensating controls (e.g., encapsulations) may be used to meet the intent of the control.

In situations where the ICS cannot support the specific system and communications protection requirements of a control, the organization employs compensating controls in accordance with the general tailoring guidance. Examples of compensating controls are given with each control, as appropriate.

## Supplemental Guidance

Supplemental Guidance for all Controls and Control Enhancements in NIST SP 800-53, Revision 4, Appendix F, should be used in conjunction with the ICS Supplemental Guidance in this overlay, if any.

### *SC-1 System and Communications Protection Policy and Procedures*

| CONTROL NUMBER | CONTROL NAME *Control Enhancement Name* | CONTROL BASELINES | | |
|---|---|---|---|---|
| | | LOW | MOD | HIGH |
| SC-1 | System and Communications Protection Policy and Procedures | Selected | Selected | Selected |

ICS Supplemental Guidance: The policy specifically addresses the unique properties and requirements of ICS and the relationship to non-ICS systems.

### *SC-2 Application Partitioning*

| CONTROL NUMBER | CONTROL NAME *Control Enhancement Name* | CONTROL BASELINES | | |
|---|---|---|---|---|
| | | LOW | MOD | HIGH |
| SC-2 | Application Partitioning | | Selected | Selected |

ICS Supplemental Guidance: Systems used to manage the ICS should be separate from the operational ICS components. An example of compensating controls is providing increased auditing measures.

### *SC-3 Security Function Isolation*

| CONTROL NUMBER | CONTROL NAME *Control Enhancement Name* | CONTROL BASELINES | | |
|---|---|---|---|---|
| | | LOW | MOD | HIGH |
| SC-3 | Security Function Isolation | | | Selected |

ICS Supplemental Guidance: Example compensating controls include providing increased auditing measures, limiting network connectivity, and architectural allocation.

### SC-4 Information in Shared Resources

| CONTROL NUMBER | CONTROL NAME *Control Enhancement Name* | CONTROL BASELINES | | |
|---|---|---|---|---|
| | | LOW | MOD | HIGH |
| **SC-4** | **Information in Shared Resources** | | Selected | Selected |

ICS Supplemental Guidance: An example of a compensating control would be architecting the use of the ICS to prevent sharing system resources.

### SC-5 Denial-of-Service Protection

| CONTROL NUMBER | CONTROL NAME *Control Enhancement Name* | CONTROL BASELINES | | |
|---|---|---|---|---|
| | | LOW | MOD | HIGH |
| **SC-5** | **Denial-of-Service Protection** | Selected | Selected | Selected |

ICS Supplemental Guidance: An example of a compensating control would be ensuring a loss of communication results in the ICS operating in nominal or safe mode. Risk-based analysis informs the establishment of policy and procedure.

### SC-7 Boundary Protection

| CONTROL NUMBER | CONTROL NAME *Control Enhancement Name* | SUPPLEMENTED CONTROL BASELINES | | |
|---|---|---|---|---|
| | | LOW | MOD | HIGH |
| **SC-7** | **Boundary Protection** | Selected | Selected | Selected |
| SC-7 (3) | *BOUNDARY PROTECTION \| ACCESS POINTS* | | Selected | Selected |
| SC-7 (4) | *BOUNDARY PROTECTION \| EXTERNAL TELECOMMUNICATIONS SERVICES* | | Selected | Selected |
| SC-7 (5) | *BOUNDARY PROTECTION \| DENY BY DEFAULT / ALLOW BY EXCEPTION* | | Selected | Selected |
| SC-7 (7) | *BOUNDARY PROTECTION \| PREVENT SPLIT TUNNELING FOR REMOTE DEVICES* | | Selected | Selected |
| SC-7 (8) | *BOUNDARY PROTECTION \| ROUTE TRAFFIC TO AUTHENTICATED PROXY SERVERS* | | | Selected |
| SC-7 (18) | *BOUNDARY PROTECTION \| FAIL SECURE* | | Added | Selected |
| SC-7 (21) | *BOUNDARY PROTECTION \| ISOLATION OF INFORMATION SYSTEM COMPONENTS* | | | Selected |

ICS Supplemental Guidance: None.

Control Enhancement: (3, 4, 5, 7, 8, 21) ICS Supplemental Guidance: None.

Control Enhancement: (18) ICS Supplemental Guidance: The organization selects an appropriate failure mode (e.g., permit or block all communications).

Rationale for Adding SC-7 (18) to Moderate Baseline: As part of the architecture and design of the ICS, the organization selects an appropriate failure mode in accordance with the function performed by the ICS and the operational environment. The ability to choose the failure mode for the physical part of the ICS differentiates the ICS from other IT systems. This choice may be a significant influence in mitigating the impact of a failure.

### *SC-8 Transmission Confidentiality and Integrity*

| CONTROL NUMBER | CONTROL NAME<br>*Control Enhancement Name* | CONTROL BASELINES | | |
|---|---|---|---|---|
| | | LOW | MOD | HIGH |
| **SC-8** | **Transmission Confidentiality and Integrity** | | Selected | Selected |
| SC-8 (1) | Transmission Confidentiality and Integrity \| Cryptographic or Alternate Physical Protection | | Selected | Selected |

ICS Supplemental Guidance: None.

Control Enhancement: (1) ICS Supplemental Guidance: The organization explores all possible cryptographic integrity mechanisms (e.g., digital signature, hash function). Each mechanism has a different delay impact.

### *SC-10 Network Disconnect*

| CONTROL NUMBER | CONTROL NAME<br>*Control Enhancement Name* | CONTROL BASELINES | | |
|---|---|---|---|---|
| | | LOW | MOD | HIGH |
| **SC-10** | **Network Disconnect** | | Selected | Selected |

ICS Supplemental Guidance: Example compensating controls include providing increased auditing measures or limiting remote access privileges to key personnel.

### SC-12 Cryptographic Key Establishment and Management

| CONTROL NUMBER | CONTROL NAME *Control Enhancement Name* | CONTROL BASELINES | | |
|---|---|---|---|---|
| | | LOW | MOD | HIGH |
| **SC-12** | **Cryptographic Key Establishment and Management** | Selected | Selected | Selected |
| SC-12 (1) | *CRYPTOGRAPHIC KEY ESTABLISHMENT AND MANAGEMENT \| AVAILABILITY* | | | Selected |

ICS Supplemental Guidance: The use of cryptographic key management in ICS is intended to supportinternal nonpublic use.

Control Enhancement: (1) ICS Supplemental Guidance: None.

### SC-13 Cryptographic Protection

| CONTROL NUMBER | CONTROL NAME *Control Enhancement Name* | CONTROL BASELINES | | |
|---|---|---|---|---|
| | | LOW | MOD | HIGH |
| **SC-13** | **Cryptographic Protection** | Selected | Selected | Selected |

ICS Supplemental Guidance: None.

### SC-15 Collaborative Computing Devices

| CONTROL NUMBER | CONTROL NAME *Control Enhancement Name* | CONTROL BASELINES | | |
|---|---|---|---|---|
| | | LOW | MOD | HIGH |
| **SC-15** | **Collaborative Computing Devices** | Selected | Selected | Selected |

ICS Supplemental Guidance: None.

### SC-17 Public Key Infrastructure Certificates

| CONTROL NUMBER | CONTROL NAME *Control Enhancement Name* | CONTROL BASELINES | | |
|---|---|---|---|---|
| | | LOW | MOD | HIGH |
| **SC-17** | **Public Key Infrastructure Certificates** | | Selected | Selected |

ICS Supplemental Guidance: None.

### SC-18 Mobile Code

| CONTROL NUMBER | CONTROL NAME *Control Enhancement Name* | CONTROL BASELINES | | |
|---|---|---|---|---|
| | | LOW | MOD | HIGH |
| SC-18 | Mobile Code | | Selected | Selected |

ICS Supplemental Guidance: None.

### SC-19 Voice Over Internet Protocol (VoIP)

| CONTROL NUMBER | CONTROL NAME *Control Enhancement Name* | CONTROL BASELINES | | |
|---|---|---|---|---|
| | | LOW | MOD | HIGH |
| SC-19 | Voice Over Internet Protocol | | Selected | Selected |

ICS Supplemental Guidance: The use of VoIP technologies is determined after careful consideration and verification that it does not adversely impact the operational performance of the ICS.

### SC-20 Secure Name/Address Resolution Service (Authoritative Source)

| CONTROL NUMBER | CONTROL NAME *Control Enhancement Name* | CONTROL BASELINES | | |
|---|---|---|---|---|
| | | LOW | MOD | HIGH |
| SC-20 | Secure Name/Address Resolution Service (Authoritative Source) | Selected | Selected | Selected |

ICS Supplemental Guidance: The use of secure name/address resolution services is determined after careful consideration and verification that it does not adversely impact the operation of the ICS.

### *SC-21 Secure Name/Address Resolution Service (Recursive or Caching Resolver)*

| CONTROL NUMBER | CONTROL NAME<br>*Control Enhancement Name* | CONTROL BASELINES | | |
|---|---|---|---|---|
| | | LOW | MOD | HIGH |
| SC-21 | **Secure Name/Address Resolution Service (Recursive or Caching Resolver)** | Selected | Selected | Selected |

ICS Supplemental Guidance: The use of secure name/address resolution services is determined after careful consideration and verification that it does not adversely impact the operation of the ICS.

### *SC-22 Architecture and Provisioning for Name/Address Resolution Service*

| CONTROL NUMBER | CONTROL NAME<br>*Control Enhancement Name* | CONTROL BASELINES | | |
|---|---|---|---|---|
| | | LOW | MOD | HIGH |
| SC-22 | **Architecture and Provisioning for Name/Address Resolution Service** | Selected | Selected | Selected |

ICS Supplemental Guidance: The use of secure name/address resolution services is determined after careful consideration and verification that it does not adversely impact the operational performance of the ICS.

### *SC-23 Session Authenticity*

| CONTROL NUMBER | CONTROL NAME<br>*Control Enhancement Name* | CONTROL BASELINES | | |
|---|---|---|---|---|
| | | LOW | MOD | HIGH |
| SC-23 | **Session Authenticity** | | Selected | Selected |

ICS Supplemental Guidance: Example compensating controls include auditing measures.

### SC-24 Fail in Known State

| CONTROL NUMBER | CONTROL NAME *Control Enhancement Name* | SUPPLEMENTED CONTROL BASELINES | | |
|---|---|---|---|---|
| | | LOW | MOD | HIGH |
| SC-24 | Fail in Known State | | Added | Selected |

ICS Supplemental Guidance: The organization selects an appropriate failure state. Preserving ICS state information includes consistency among ICS state variables and the physical state which the ICS represents (e.g., whether valves are open or closed, or communication is permitted or blocked, operations continue).

Rationale for Adding SC-24 to Moderate Baseline: As part of the architecture and design of the ICS, the organization selects an appropriate failure state of an ICS in accordance with the function performed by the ICS and the operational environment. The ability to choose the failure mode for the physical part of the ICS differentiates the ICS from other IT systems. This choice may be a significant influence in mitigating the impact of a failure, since it may be disruptive to ongoing physical processes (e.g., valves failing in closed position may adversely affect system cooling).

### SC-28 Protection of Information at Rest

| CONTROL NUMBER | CONTROL NAME *Control Enhancement Name* | CONTROL BASELINES | | |
|---|---|---|---|---|
| | | LOW | MOD | HIGH |
| SC-28 | Protection of Information at Rest | | Selected | Selected |

ICS Supplemental Guidance: The use of cryptographic mechanisms is determined after careful consideration and after verification that it does not adversely impact the operational performance of the ICS.

### *SC-39 Process Isolation*

| CONTROL NUMBER | CONTROL NAME *Control Enhancement Name* | CONTROL BASELINES | | |
|---|---|---|---|---|
| | | LOW | MOD | HIGH |
| SC-39 | Process Isolation | Selected | Selected | Selected |

ICS Supplemental Guidance: Example compensating controls include partition processes used to separate platforms.

### *SC-41 Port and I/O Device Access*

| CONTROL NUMBER | CONTROL NAME *Control Enhancement Name* | SUPPLEMENTED CONTROL BASELINES | | |
|---|---|---|---|---|
| | | LOW | MOD | HIGH |
| SC-41 | Port and I/O Device Access | Added | Added | Added |

ICS Supplemental Guidance: None.

Rationale for Adding SC-24 to All Baselines: The function of ICS can be readily determined in advance, making it easier to identify ports and I/O devices that are unnecessary. Disabling or removing ports reinforces air-gap policy.

## System and Information Integrity – SI

### Tailoring Considerations for the System and Information Integrity Family

In situations where the ICS cannot support the specific system and information integrity requirements of a control, the organization employs compensating controls in accordance with the general tailoring guidance. Examples of compensating controls are given with each control, as appropriate.

### Supplemental Guidance

Supplemental Guidance for all Controls and Control Enhancements in NIST SP 800-53, Revision 4, Appendix F, should be used in conjunction with the ICS Supplemental Guidance in this overlay, if any.

## SI-1 System and Information Integrity Policy and Procedures

| CONTROL NUMBER | CONTROL NAME *Control Enhancement Name* | CONTROL BASELINES | | |
|---|---|---|---|---|
| | | LOW | MOD | HIGH |
| SI-1 | **System and Information Integrity Policy and Procedures** | Selected | Selected | Selected |

ICS Supplemental Guidance: The policy specifically addresses the unique properties and requirements of ICS and the relationship to non-ICS systems.

## SI-2 Flaw Remediation

| CONTROL NUMBER | CONTROL NAME *Control Enhancement Name* | CONTROL BASELINES | | |
|---|---|---|---|---|
| | | LOW | MOD | HIGH |
| SI-2 | **Flaw Remediation** | Selected | Selected | Selected |
| SI-2 (1) | *FLAW REMEDIATION \| CENTRAL MANAGEMENT* | | | Selected |
| SI-2 (2) | *FLAW REMEDIATION \| AUTOMATED FLAW REMEDIATION STATUS* | | Selected | Selected |

ICS Supplemental Guidance: Flaw remediation is complicated since many ICS employ operating systems and other software that is not current, is no longer being maintained by the vendors, and is not resistant to current threats. ICS operators are often dependent on product vendors to validate the operability of a patch and also sometimes to perform the installation. Often flaws cannot be remediated based on circumstances outside of the ICS operator's control (e.g., lack of a vendor patch). Sometimes the organization has no choice but to accept additional risk. In these situations, compensating controls should be implemented (e.g., limit the exposure of the vulnerable system). Other compensating controls that do not decrease the residual risk but increase the ability to respond may be desirable (e.g., provide a timely response in case of an incident; devise a plan to ensure the ICS can identify the exploitation of the flaw). Testing flaw remediation in an ICS may require more resources than the organization can commit.

Control Enhancement: (1) ICS Supplemental Guidance: None.

Control Enhancement: (2) ICS Supplemental Guidance: In situations where the ICS cannot support the use of automated mechanisms to conduct and report on the status of flaw remediation, the organization employs nonautomated mechanisms or procedures which incorporate methods to apply, track, and verify mitigation efforts as compensating controls in accordance with the general tailoring guidance.

### *SI-3 Malicious Code Protection*

| CONTROL NUMBER | CONTROL NAME<br>*Control Enhancement Name* | CONTROL BASELINES | | |
|---|---|---|---|---|
| | | LOW | MOD | HIGH |
| **SI-3** | **Malicious Code Protection** | Selected | Selected | Selected |
| SI-3 (1) | *MALICIOUS CODE PROTECTION \| CENTRAL MANAGEMENT* | | Selected | Selected |
| SI-3 (2) | *MALICIOUS CODE PROTECTION \| AUTOMATIC UPDATES* | | Selected | Selected |

ICS Supplemental Guidance: The use and deployment of malicious code protection is determined after careful consideration and after verification that it does not adversely impact the operation of the ICS. Malicious code protection tools should be configured to minimize their potential impact on the ICS (e.g., employ notification rather than quarantine). Example compensating controls include increased traffic monitoring and auditing.

Control Enhancement: (1) ICS Supplemental Guidance: The organization implements central management of malicious code protection with consideration of the impact on operation of the ICS. An example of a compensating control would be increased auditing.

Control Enhancement: (2) ICS Supplemental Guidance: The organization implements automatic updates of malicious code protection with consideration of the impact on the operation of the ICS. In situations where the ICS cannot support the use of automatic update of malicious code protection, the organization employs nonautomated procedures as compensating controls in accordance with the general tailoring guidance.

## SI-4 Information System Monitoring

| CONTROL NUMBER | CONTROL NAME *Control Enhancement Name* | CONTROL BASELINES | | |
|---|---|---|---|---|
| | | LOW | MOD | HIGH |
| **SI-4** | **Information System Monitoring** | Selected | Selected | Selected |
| SI-4 (2) | *INFORMATION SYSTEM MONITORING \|AUTOMATED TOOLS FOR REAL-TIME ANALYSIS* | | Selected | Selected |
| SI-4 (4) | *INFORMATION SYSTEM MONITORING \| INBOUND AND OUTBOUND COMMUNICATIONS TRAFFIC* | | Selected | Selected |
| SI-4 (5) | *INFORMATION SYSTEM MONITORING \| SYSTEM-GENERATED ALERTS* | | Selected | Selected |

ICS Supplemental Guidance: The organization ensures that the use of monitoring tools and techniques does not adversely impact the operational performance of the ICS. An example of a compensating control would be deploying sufficient network monitoring.

Control Enhancement: (2) ICS Supplemental Guidance: In situations where the ICS cannot support the use of automated tools to support near real-time analysis of events, the organization employs compensating controls (e.g., providing an auditing capability on a separate system, nonautomated mechanisms, or procedures) in accordance with the general tailoring guidance.

Control Enhancement: (4) ICS Supplemental Guidance: In situations where the ICS cannot monitor inbound and outbound communications traffic, the organization employs compensating controls, which includes providing a monitoring capability on a separate information system.

Control Enhancement: (5) ICS Supplemental Guidance: Example compensating controls include manual methods of generating alerts.

## SI-5 Security Alerts, Advisories, and Directives

| CONTROL NUMBER | CONTROL NAME *Control Enhancement Name* | CONTROL BASELINES | | |
|---|---|---|---|---|
| | | LOW | MOD | HIGH |
| **SI-5** | **Security Alerts, Advisories, and Directives** | Selected | Selected | Selected |
| SI-5 (1) | *SECURITY ALERTS, ADVISORIES, AND DIRECTIVES \| AUTOMATED ALERTS AND ADVISORIES* | | | Selected |

ICS Supplemental Guidance: The DHS Industrial Control Systems Cyber Emergency Response Team (ICS- CERT) generates security alerts and advisories relative to ICS.[8]

Control Enhancement: (1) ICS Supplemental Guidance: None.

## SI-6 Security Functionality Verification

| CONTROL NUMBER | CONTROL NAME *Control Enhancement Name* | CONTROL BASELINES | | |
|---|---|---|---|---|
| | | LOW | MOD | HIGH |
| **SI-6** | **Security Function Verification** | | | Selected |

ICS Supplemental Guidance: The shutting down and restarting of the ICS may not always be feasible upon the identification of an anomaly; these actions should be scheduled according to ICS operational requirements.

## SI-7 Software and Information Integrity

| CONTROL NUMBER | CONTROL NAME *Control Enhancement Name* | CONTROL BASELINES | | |
|---|---|---|---|---|
| | | LOW | MOD | HIGH |
| **SI-7** | **Software, Firmware, and Information Integrity** | | Selected | Selected |
| SI-7 (1) | *SOFTWARE, FIRMWARE, AND INFORMATION INTEGRITY \| INTEGRITY CHECKS* | | Selected | Selected |
| SI-7 (2) | *SOFTWARE, FIRMWARE, AND INFORMATION INTEGRITY \| AUTOMATED NOTIFICATIONS OF INTEGRITY VIOLATIONS* | | | Selected |
| SI-7 (5) | *SOFTWARE, FIRMWARE, AND INFORMATION INTEGRITY \| AUTOMATED RESPONSE TO INTEGRITY VIOLATIONS* | | | Selected |
| SI-7 (7) | *SOFTWARE, FIRMWARE, AND INFORMATION INTEGRITY \|INTEGRATION OF DETECTION AND RESPONSE* | | Selected | Selected |
| SI-7 (14) | *SOFTWARE, FIRMWARE, AND INFORMATION INTEGRITY \| BINARY OR MACHINE EXECUTABLE CODE* | | | Selected |

ICS Supplemental Guidance: The organization determines whether the use of integrity verification applications would adversely impact the operation of the ICS and employs compensating controls (e.g., manual integrity verifications that do not affect performance).

Control Enhancements: (1) ICS Supplemental Guidance: The organization ensures that the use of integrity verification applications does not adversely impact the operational performance of the ICS.

Control Enhancement: (2) ICS Supplemental Guidance: In situations where the organization cannot employ automated tools that provide notification of integrity discrepancies, the organization employs nonautomated mechanisms or procedures. An example of a compensating control would be performing scheduled manual inspections for integrity violations.

Control Enhancement: (5) ICS Supplemental Guidance: The shutting down and restarting of the ICS may not always be feasible upon the identification of an anomaly; these actions should be scheduled according to ICS operational requirements.

Control Enhancement: (7) ICS Supplemental Guidance: In situations where the ICS cannot detect unauthorized security-relevant changes, the organization employs compensating controls (e.g., manual procedures) in accordance with the general tailoring guidance.

Control Enhancement: (14) ICS Supplemental Guidance: None.

### *SI-8 Spam Protection*

| CONTROL NUMBER | CONTROL NAME<br>*Control Enhancement Name* | SUPPLEMENTED CONTROL BASELINES | | |
|---|---|---|---|---|
| | | LOW | MOD | HIGH |
| **SI-8** | **Spam Protection** | | Selected | Selected |
| SI-8 (1) | *SPAM PROTECTION \| CENTRAL MANAGEMENT OF PROTECTION MECHANISMS* | | Selected | Selected |
| SI-8 (2) | *SPAM PROTECTION \| AUTOMATIC UPDATES* | | Selected | Selected |

ICS Supplemental Guidance: ICS spam protection may be implemented by removing spam transport mechanisms, functions, and services (e.g., electronic mail, Internet access) from the ICS. If any spam transport mechanisms, functions, and services are present in the ICS, spam protection in ICS takes into account operational characteristics of ICS that differ from general purpose information systems (e.g., unusual traffic flow that may be misinterpreted and detected as spam). Example compensating controls include whitelist mail transfer agents (MTA), digitally signed messages, acceptable sources, and acceptable message types.

Control Enhancement: (1) ICS Supplemental Guidance: Example compensating controls include employing local mechanisms or procedures.

Control Enhancement: (2) ICS Supplemental Guidance: None.

### *SI-10 Information Input Validation*

| CONTROL NUMBER | CONTROL NAME<br>*Control Enhancement Name* | CONTROL BASELINES | | |
|---|---|---|---|---|
| | | LOW | MOD | HIGH |
| **SI-10** | **Information Input Validation** | | Selected | Selected |

ICS Supplemental Guidance: None.

## SI-11 Error Handling

| CONTROL NUMBER | CONTROL NAME *Control Enhancement Name* | CONTROL BASELINES | | |
|---|---|---|---|---|
| | | LOW | MOD | HIGH |
| SI-11 | Error Handling | | Selected | Selected |

ICS Supplemental Guidance: None.

## SI-12 Information Handling and Retention

| CONTROL NUMBER | CONTROL NAME *Control Enhancement Name* | CONTROL BASELINES | | |
|---|---|---|---|---|
| | | LOW | MOD | HIGH |
| SI-12 | Information Handling and Retention | Selected | Selected | Selected |

ICS Supplemental Guidance: None.

## SI-13 Predictable Failure Prevention

| CONTROL NUMBER | CONTROL NAME *Control Enhancement Name* | SUPPLEMENTED CONTROL BASELINES | | |
|---|---|---|---|---|
| | | LOW | MOD | HIGH |
| SI-13 | Predictable Failure Prevention | | | Added |

ICS Supplemental Guidance: Failures in ICS can be stochastic or deterministic. Stochastic failures can be analyzed using probability theory, while analysis of deterministic failures is based on nonrandom properties of the system. Known ICS failure modes and causes are considered. The calculation and use ofstatistical descriptors, such as mean time to failure (MTTF), should incorporate additional analysis to determine how those failures manifest within the cyber and physical domains. Knowledge of these possible manifestations may be necessary to detect whether a failure has occurred within the ICS, as failures of the information systems may not be easily identifiable. Emergent properties, which may arise both within the information systems and physical processes and can potentially cause system failures, should be incorporated into the analysis. For example, cumulative effects of resource exhaustion (e.g., memory leakage) or errors (e.g., rounding and truncation) can occur when ICS processes execute for unexpectedly long periods. Deterministic failures (e.g., integer counter overflow), once identified, are preventable.

Often substitute components may not be available or may not be sufficient to protect against faults occurring before predicted failure. Nonautomated mechanisms or physical safeguards should be in place in order to protect against these failures.

In addition to information concerning newly discovered vulnerabilities (i.e., latent flaws) potentially affecting the system/applications that are discovered by forensic studies, new vulnerabilities may be identified by organizations with the responsibility for disseminating vulnerability information (e.g., ICS-CERT) based upon an analysis of a similar pattern of incidents reported to them or vulnerabilities reported by other researchers.

Related Controls: IR-5, IR-6, RA-5, SI-2, SI-5, and SI-11.

Rationale for Adding Control to Baseline: ICS are designed and built with certain boundary conditions, design parameters, and assumptions about their environment and mode of operation. ICS may run much longer than conventional systems, allowing latent flaws to become effective that are not manifest in other environments. For example, integer overflow might never occur in systems that are re-initialized more frequently than the occurrence of the overflow. Experience and forensic studies of anomalies and incidents in ICS can lead to identification of emergent properties that were previously unknown, unexpected, or unanticipated. Preventative and restorative actions (e.g., restarting the system or application) are prudent but may not be acceptable for operational reasons in ICS.

### *SI-16 Memory Protection*

| CONTROL NUMBER | CONTROL NAME<br>*Control Enhancement Name* | CONTROL BASELINES | | |
|---|---|---|---|---|
| | | LOW | MOD | HIGH |
| SI-16 | **Memory Protection** | | Selected | Selected |

ICS Supplemental Guidance: None.

### *SI-17 Fail-Safe Procedures*

| CONTROL NUMBER | CONTROL NAME *Control Enhancement Name* | SUPPLEMENTED CONTROL BASELINES | | |
|---|---|---|---|---|
| | | LOW | MOD | HIGH |
| SI-17 | Fail-Safe Procedures | Added | Added | Added |

ICS Supplemental Guidance: The selected failure conditions and corresponding procedures may vary among baselines. The same failure event may trigger a different response depending on the impact level. Mechanical and analog systems can be used to provide mechanisms to ensure fail-safe procedures. Fail-safe states should incorporate potential impacts to human safety, physical systems, and the environment. Related controls: CP-6.

Rationale for Adding SI-17 to All Baselines: This control provides a structure for the organization to identify their policy and procedures for dealing with failures and other incidents. Creating a written record of the decision process for selecting incidents and appropriate responses is part of risk management in light of the changing environment of operations.

## Organization-Wide Information Security Program Management Controls – PM

### Characteristics of Organization-Wide Information Security Program Management Control Family

Organization-Wide Information Security Program Management Controls are deployed with organization-wide support of the information security program. They are not associated with security control baselines and are independent of any system impact level.

### Supplemental Guidance

Supplemental Guidance for all Controls and Control Enhancements in NIST SP 800-53, Revision 4, Appendix F, should be used in conjunction with the ICS Supplemental Guidance in this overlay, if any.

### PM-1 Information Security Program Plan

| CONTROL NUMBER | CONTROL NAME<br>*Control Enhancement Name* |
|---|---|
| PM-1 | Information Security Program Plan Policy and Procedures |

ICS Supplemental Guidance: The policy specifically addresses the unique properties and requirements of ICS, the relationship to non-ICS systems, and the relationship to other programs concerned with operational characteristics of ICS (e.g., safety, efficiency, reliability, resilience).

### PM-2 Senior Information Security Officer

| CONTROL NUMBER | CONTROL NAME<br>*Control Enhancement Name* |
|---|---|
| PM-2 | Senior Information Security Officer |

ICS Supplemental Guidance: None.

### PM-3 Information Security Resources

| CONTROL NUMBER | CONTROL NAME<br>*Control Enhancement Name* |
|---|---|
| PM-3 | Information Security Resources |

ICS Supplemental Guidance: Capital planning and investment decisions address all the relevant technologies and all phases of the life cycle and needs to be informed by ICS experts as well as other subject matter experts (e.g., information security). Marshaling interdisciplinary working teams to advise capital planning and investment decisions can help tradeoff and balance among conflicting equities, objectives, and responsibilities, such as capability, adaptability, resilience, safety, security, usability, and efficiency.

### PM-4 Plan of Action and Milestones Process

| CONTROL NUMBER | CONTROL NAME<br>*Control Enhancement Name* |
|---|---|
| PM-4 | Plan of Action and Milestones Process |

ICS Supplemental Guidance: The plan of action and milestones includes both computational and physical ICS components. Records of observed shortcomings and appropriate remedial action may be maintained in a single document or in multiple coordinated documents (e.g., future engineering plans).

### PM-5 Information System Inventory

| CONTROL NUMBER | CONTROL NAME<br>*Control Enhancement Name* |
|---|---|
| PM-5 | Information System Inventory |

ICS Supplemental Guidance: None.

### PM-6 Information Security Measures of Performance

| CONTROL NUMBER | CONTROL NAME<br>*Control Enhancement Name* |
|---|---|
| PM-6 | Information Security Measures of Performance |

ICS Supplemental Guidance: None.

### PM-7 Enterprise Architecture

| CONTROL NUMBER | CONTROL NAME<br>*Control Enhancement Name* |
|---|---|
| PM-7 | Enterprise Architecture |

ICS Supplemental Guidance: None.

### PM-8 Critical Infrastructure Plan

| CONTROL NUMBER | CONTROL NAME<br>*Control Enhancement Name* |
|---|---|
| PM-8 | Critical Infrastructure Plan |

ICS Supplemental Guidance: None.

Reference: Executive Order 13636 – Improving Critical Infrastructure Cybersecurity, February 12,2013[9].

### PM-9 Risk Management Strategy

| CONTROL NUMBER | CONTROL NAME<br>*Control Enhancement Name* |
|---|---|
| PM-9 | Risk Management Strategy |

ICS Supplemental Guidance: Risk management of ICS is considered along with other organizational risks affecting mission/business success from an organization-wide perspective. Organization-wide risk management strategy includes sector-specific guidance as appropriate.

### PM-10 Security Authorization Process

| CONTROL NUMBER | CONTROL NAME<br>*Control Enhancement Name* |
|---|---|
| PM-10 | Security Authorization Process |

ICS Supplemental Guidance: The authorization to operate processes for ICS involves multiple disciplines that have existing approval and risk management process (e.g., physical security, safety). Organization-wide risk management requires harmonization among these disciplines.

### *PM-11 Mission/Business Process Definition*

| CONTROL NUMBER | CONTROL NAME<br>*Control Enhancement Name* |
|---|---|
| PM-11 | Mission/Business Process Definition |

ICS Supplemental Guidance: Mission/business processes refinement requires protection of physical assets from damage originating in the cyber domain. These needs are derived from the mission/business needs defined by the organization, the mission/business processes selected to meet the stated needs, and the organizational risk management strategy.

### *PM-12 Insider Threat Program*

| CONTROL NUMBER | CONTROL NAME<br>*Control Enhancement Name* |
|---|---|
| PM-12 | Insider Threat Program |

ICS Supplemental Guidance: None.

### *PM-13 Information Security Workforce*

| CONTROL NUMBER | CONTROL NAME<br>*Control Enhancement Name* |
|---|---|
| PM-13 | Information Security Workforce |

ICS Supplemental Guidance: All aspects of information security workforce development and improvement programs include knowledge and skill levels in both computational and physical ICS components.

### *PM-14 Testing, Training, and Monitoring*

| CONTROL NUMBER | CONTROL NAME<br>*Control Enhancement Name* |
|---|---|
| PM-14 | Testing, Training, and Monitoring |

ICS Supplemental Guidance: None.

### PM-15 Contacts with Security Groups and Associations

| CONTROL NUMBER | CONTROL NAME *Control Enhancement Name* |
|---|---|
| PM-15 | Contacts with Security Groups and Associations |

ICS Supplemental Guidance: None.

### PM-16 Threat Awareness Program

| CONTROL NUMBER | CONTROL NAME *Control Enhancement Name* |
|---|---|
| PM-16 | Threat Awareness Program |

ICS Supplemental Guidance: The organization should collaborate and share information about potential incidents on a timely basis. The DHS National Cybersecurity & Communications Integration Center (NCCIC)[10] serves as a centralized location where operational elements involved in cybersecurity and communications reliance are coordinated and integrated. The Industrial Control Systems Cyber Emergency Response Team (ICS-CERT)[11] collaborates with international and private sector Computer Emergency Response Teams (CERTs) to share control systems-related security incidents and mitigation measures. Organizations should consider having the capabilities to share both unclassified and classified information

## References

1. NCCIC (National Cybersecurity & Communications Integration Center). http://www.dhs.gov/about-national-cybersecurity-communications-integration-center.

2. ICS-CERT (Industrial Control Systems Cyber Emergency Response Team). http://ics-cert.us-cert.gov/ics-cert.

3. ICS-CERT. *Recommendations on Deploying Application Whitelisting on ICS*. http://ics-cert.us- cert.gov/tips/ICS-TIP-12-146-01B.

4. NFPA 1600. *Standard on Disaster/Emergency Management and Business Continuity Programs*. Quincy, MA: NFPA (National Fire Protection Agency), 2012.

5. NCCIC (National Cybersecurity & Communications Integration Center). http://www.dhs.gov/about-national-cybersecurity-communications-integration-center.

6. ICS-CERT (Industrial Control Systems Cyber Emergency Response Team). http://ics-cert.us-cert.gov/ics-cert.

7. SCADA/Control Systems Procurement Project provides example cybersecurity procurement language for ICS. https://ics-cert.us-cert.gov/sites/default/files/documents/Procurement_Language_Rev4_100809.pdf.

8. The DHS Industrial Control Systems Cyber Emergency Response Team (ICS- CERT) generates security alerts and advisories relative to ICS, http://ics-cert.us-cert.gov.

9. Executive Order (EO) 13636. *Improving Critical Infrastructure Cybersecurity*. February 12, 2013.

10. NCCIC (National Cybersecurity & Communications Integration Center). http://www.dhs.gov/about-national-cybersecurity-communications-integration-center.

11. ICS-CERT (Industrial Control Systems Cyber Emergency Response Team). http://ics-cert.us-cert.gov/ics-cert.

# Glossary and Acronyms

**Abuse** – Malicious misuse, with the objective of intentional denial, alteration, or destruction. See *misuse*.

**Acceptance testing** – A type of testing used to determine whether or not software is acceptable to the actual users.

**Access** – A specific type of interaction between a subject and an object that results in the flow of information from one to the other.

**Access control mechanisms** – Hardware or software features, operating procedures, management procedures, and various combinations thereof that are designed to detect and prevent unauthorized access and to permit authorized access in an automated system.

**Access control** – The process of limiting access to system or software resources only to authorized users, programs, processes, or other systems (on a network). This term is synonymous with controlled access and limited access.

**Access list** – A list of users, programs, and/or processes and the specifications of the access categories to which each is assigned; a list of the users, programs, and/or processes that have access privileges to a particular resource, and the type of access assigned to each.

**Access point (AP)** – A wireless LAN transceiver interface between a wireless network and a wired network. Access points forward frames between wireless devices and hosts on the LAN.

**Access type** – The nature of an access right to a particular device, program, or file (e.g., read, write, execute, append, modify, delete, or create).

**Accountability** – A property that allows auditing of IT system activities to be traced to persons or processes that may then be held responsible for their actions. Accountability includes authenticity and nonrepudiation.

**Accreditation** – A formal declaration by the designated approving authority (DAA) that the automated information system (AIS) is approved to operate in a particular security mode by using a prescribed set of safeguards. Accreditation is the official management authorization for operation of an AIS and is based on the certification process as well as other management considerations. The accreditation statement affixes security responsibility with the DAA and shows that due care has been taken for security.

**Accreditation authority** – Synonymous with designated approving authority.

**Acquisition** – The act of procurement or purchase of a product or service under license, subscription, or contract.

**Advanced Encryption Standard (AES) (originally, Rijndael)** – A symmetric block cipher with a block size of 128 bits, in which the key can be 128, 192, or 256 bits. The Advanced Encryption Standard replaces the Data Encryption Standard (DES) and was announced on 26 November 2001 as Federal Information Processing Standards Publication 197 (FIPS PUB 197).

**AIS** – See *automated information system.*

**API** – Application programming interface.

**Application layer** – The top layer of the Open Systems Interconnection (OSI) model, which is concerned with application programs. It provides services such as file transfer and email to the network's end users.

**Application process** – An entity, either human or software, that uses the services offered by the application layer of the OSI reference model.

**Application programming interface (API)** – An interface to a library of software functions. An API is designed to allow software developers to call functions from the library and to make requests of an operating system or another software component.

**Application software** – Software that accomplishes functions such as database access, electronic mail, and menu prompts.

**Architecture** – When referring to a computer system, architecture describes the type of components, interfaces, and protocols the system uses and how they fit together. Architecture is the configuration of any equipment or interconnected system or subsystems of equipment that is used in the acquisition, storage, manipulation, management, movement, control, display, switching, interchange, transmission, or reception of data or information; it includes computers, ancillary equipment, and services ( including support services and related resources).

**Asset** – An object of value. In the context of this book, an asset could be a computer system, network, PLC, plant equipment, controllers, data acquisition systems, and so on.

**Assurance** – A measure of confidence that the security features and architecture of an automated information system (AIS) accurately mediate and enforce the security policy; the grounds for confidence that an IT product or system meets its security objectives.

**Asymmetric (public) key encryption** – A cryptographic system that employs two keys – a public key and a private key. The public key is made available to anyone wishing to send an encrypted message to an individual holding the corresponding private key of the public-private key pair. Any message encrypted with one of these keys can be decrypted with the other. The private key is always kept private. It should not be possible to derive the private key from the public key.

**Asynchronous transmission** – The transmission of data at a variable bit rate without the use of an external clock signal; the timing required to decode the data is contained within the communication stream.

**Attack** – The act of trying to bypass security controls on a system. An attack can be active, resulting (or potentially resulting) in data modification, or passive, resulting (or potentially resulting) in the release of data.

**NOTE** The fact that an attack is made does not necessarily mean that it will succeed. The degree of success depends on the vulnerability of the system or activity and the effectiveness of existing countermeasures.

**Audit trail** – A chronological record of system activities that is sufficient to enable the reconstruction, reviewing, and examination of the sequence of environments and activities surrounding or leading to an operation, a procedure, or an event in a transaction from its inception to its final result.

**Authenticate** – (1) To verify the identity of a user, device, or other entity in a computer system, often as a prerequisite to allowing access to system resources; (2) To verify the integrity of data that have been stored, transmitted, or otherwise exposed to possible unauthorized modification.

**Authentication** – The process of verifying "who" is at the other end of a transmission.

**Authentication device** – A device whose identity has been verified during the lifetime of the current link based on the authentication procedure.

**Authenticity** – A property that allows the ability to validate the claimed identity of a system entity.

**Authorization** – The granting of access rights to a user, program, or process.

**Automated information system (AIS)** – An assembly of computer hardware, software, and/or firmware that is configured to collect, create, communicate, compute, disseminate, process, store, and/or control data or information.

**Automated information system security** – Measures and controls that protect an automated information system (AIS) against denial of service (DoS) and unauthorized (accidental or intentional) disclosure, modification, or destruction of an AIS and data. AIS security includes consideration of all hardware and/or software functions, characteristics, and/or features; operational procedures, accountability procedures, and access controls at the central computer facility, remote computers and terminal facilities; management constraints; physical structures and devices; and personnel and communication controls that are needed to provide an acceptable level of risk for the AIS and for the data and information contained in the AIS. It includes the totality of security safeguards needed to provide an acceptable protection level for an AIS and for data handled by an AIS.

**Automated security monitoring** – The use of automated procedures to ensure that security controls are not circumvented.

**Availability** – Refers to the ability of an authorized person or process to have prompt and reliable access to computational resources when needed.

**Availability of data** – The condition in which data is in the place needed by the user, at the time the user needs it, and in the form needed by the user.

**Backbone network** – A network that interconnects other networks.

**Back door** – Synonymous with trap door.

**Backup plan** – Synonymous with contingency plan.

**Bandwidth** – Specification of the amount of the frequency spectrum that is usable for data transfer. In other words, bandwidth is the maximum data rate a signal can attain on the medium without encountering significant attenuation (loss of power). Also, the amount of information that may be sent through a connection.

**Bell–LaPadula model** – A formal state transition model of computer security policy that describes a set of access control rules. In this formal model, the entities in a computer system are divided into abstract sets of subjects and objects. The notion of a secure state is defined, and each state transition preserves security by moving from secure state to secure state, thereby inductively proving that the system is secure. A system state is defined to be secure only if the permitted access modes of subjects to objects are in accordance with a specific security policy. In order to determine whether a specific access mode is allowed, the clearance of a subject is compared to the classification of the object, and a determination is made as to whether the subject is authorized for the specific access mode. Also see *star property* (or * property) and *simple security property.*

**Between-the-lines entry** – Unauthorized access obtained by tapping the temporarily inactive terminal of a legitimate user. See *piggyback.*

**Binary digit** – See *bit.*

**Biometrics** – Access control method in which an individual's physiological or behavioral characteristics are used to determine that individual's access to a particular resource.

**BIOS** – Basic input/output system. The BIOS is the first program to run when the computer is turned on. BIOS initializes and tests the computer hardware, loads and runs the operating system, and manages setup for making changes in the computer.

**Bit** – Short for binary digit. A single digit number in binary (0 or 1).

**Bit decay** – The gradual loss of information stored in bits on storage media over time.

**Bit splitting** – Splitting bits into groups of bits and processing the groups such that after they are recombined the correct result is obtained. A countermeasure to a differential power analysis attack that does not expose the target bits internally to the processor so that the power trace is not affected.

**Black box test** – A test of system security in which an ethical hacking team has no knowledge of the target network.

**Black hat hacker** – A hacker who conducts unethical and illegal attacks against information systems to gain unauthorized access to sensitive information.

**Block cipher** – A symmetric key algorithm that operates on a fixed-length block of plaintext and transforms it into a fixed-length block of ciphertext. A block cipher is obtained by segregating plaintext into blocks of *n* characters or bits and applying the same encryption algorithm and key to each block.

**Browsing** – The act of searching through storage to locate or acquire information without necessarily knowing the existence or the format of the information being sought.

**BSI ISO/IEC 17799 – 2000, BS 7799-I – 2000. *Information Technology—Code of Practice for Information Security Management*** – A British Standards Institution (London, UK) standard intended to "provide a comprehensive set of controls comprising best practices in information security." ISO refers to the International Organization for Standardization and IEC is the International Electrotechnical Commission.

**Buffer overflow** – A condition in which more input is placed into a buffer or data holding area than the allowed or allocated capacity, overwriting other information. Such a condition is exploited by attackers to crash or gain control of a system.

**Bug** – An error, defect, mistake, vulnerability, failure, or fault in a computer system or in software.

**Byte** – A set of bits, usually eight, that represent a single character.

**C&A** – Certification and accreditation.

**CA** – Certificate authority/agent. See *certificate authority.*

**Calls** – The operations performed by an application to perform a task.

**Call graph** – A visual representation of a sequence of calls.

**Capability** – A protected identifier that identifies an object and specifies the access rights allowed to the accessor who possesses the capability. In a capability-based system, access to protected objects (such as files) is granted if the would-be accessor possesses a capability for the object.

**CAPI** – See *cryptographic application programming interface.*

**Category** – A restrictive label that has been applied to classified or unclassified data as a means of increasing the protection of the data and further restricting its access.

**CBC** – See *cipher-block chaining.*

**CC** – See *Common Criteria.*

**CDDI** – See *copper data distributed interface.*

**Cell** – See *lines, units, cells.*

**Central processing unit (CPU)** – The microprocessor unit or units responsible for interpreting and executing instructions in a computer system.

**Computer Emergency Response Team Coordination Center (CERT/CC)** – A unit of the Carnegie Mellon University Software Engineering Institute (SEI). SEI is a federally funded R&D center. CERT's mission is to alert the Internet community to vulnerabilities and attacks and to conduct research and training in the areas of computer security, including incident response.

**Certification** – The comprehensive evaluation of the technical and nontechnical security features of an automated information system (AIS) and other safeguards, made in support of the accreditation process, that establishes the extent to which a particular design and implementation meet a specified set of security requirements.

**Certificate authority (CA)** – The official responsible for performing the comprehensive evaluation of the technical and nontechnical security features of an IT system and other safeguards, made in support of the accreditation process, to establish the extent that a particular design and implementation meet a set of specified security requirements.

**Certificate revocation list (CRL)** – A list of certificates that have been revoked or are no longer valid and should not be relied on.

**Channel** – Specific communication link established within a communication conduit. See *conduit*.

**Cipher-block chaining (CBC)** – An encryption mode of the Data Encryption Standard (DES) that operates on plaintext blocks 64 bits in length. Each block of plaintext is Exclusive ORed (XORed) with the previous ciphertext block before being encrypted. See *Exclusive OR*.

**Cipher** – A cryptographic transformation that operates on characters or bits.

**Ciphertext or cryptogram** – An unintelligible encrypted message.

**Client** – A computer that accesses a server's resources.

**Client/server architecture** – A network system design in which a processor or computer designated as a file server or database server provides services to other client processors or computers. Applications are distributed between a host server and a remote client.

**Cluster** – A group of computers linked together over a fast local area network or other means. Clustered computers work closely together such that they act and appear like a single large computer. Clusters are typically created to improve availability, performance, or redundancy beyond that provided by a single computer.

**Collision** – An event in which simultaneous transmissions on a communications medium interfere with one another or "collide."

**Component Object Model (COM)** – A Microsoft technology that enables software components to communicate with each other.

**Common Criteria (CC)** – A standard for specifying and evaluating the features of computer products and systems.

**Common Object Request Broker Architecture (CORBA)** – A standard that uses the object request broker (ORB) to implement exchanges among objects in a heterogeneous, distributed environment.

**Communications security (COMSEC)** – Measures and controls taken to deny unauthorized persons information derived from telecommunications and to ensure the authenticity of such telecommunications. Communications security includes cryptosecurity, transmission security, emission security, and physical security of COMSEC material and information.

**Component** – A constituent element of a larger system. In software, a component is a functional part of a larger program.

**Compromise** – A violation of a system's security policy such that unauthorized disclosure of sensitive information might have occurred.

**Compromising emanations** – Unintentional data-related or intelligence-bearing signals that, when intercepted and analyzed, disclose the information transmission that is received, handled, or otherwise processed by any information processing equipment. See *TEMPEST.*

**Computer abuse** – The misuse, alteration, disruption, or destruction of data-processing resources. The key is that computer abuse is intentional and improper.

**Computer cryptography** – The use of a crypto-algorithm in a computer, microprocessor, or microcomputer to perform encryption or decryption in order to protect information or to authenticate users, sources, or information.

**Computer facility** – The physical structure housing data-processing operations.

**Computer forensics** – Collection of information from and about computer systems that is admissible in a court of law.

**Computer fraud** – Computer-related crimes involving deliberate misrepresentation, alteration, or disclosure of data in order to obtain something of value (usually monetary gain). A computer system must have been involved in the perpetration or cover-up of the act or series of acts. A computer system might have been involved through improper manipulation of input data, output or results, application programs, data files, computer operations, communications, computer hardware, systems software, or firmware.

**Computer security (COMPUSEC)** – Synonymous with automated information system security.

**Computing environment** – The total environment in which an automated information system, network, or component operates. The environment includes physical, administrative, and personnel procedures as well as communication and networking relationships with other information systems.

**COMSEC** – See *communications security.*

**Concealment** – Keeping a secret attribute of a program hidden to prevent it from being discovered by an attacker.

**Conduit** – The means by which something is transmitted or conveyed, such as information.

**Confidentiality** – Assurance that information is not disclosed to unauthorized persons, processes, or devices. The concept of holding sensitive data in confidence, limited to an appropriate set of individuals or organizations.

**Configuration control** – The process of controlling modifications to the system's hardware, firmware, software, and documentation in a way that provides sufficient assurance that the system is protected against the introduction of improper modifications prior to, during, and after system implementation. Compare with configuration management.

**Configuration management** – The management of security features and assurances through control of changes made to a system's hardware, software, firmware, documentation, test, test fixtures, and test documentation throughout the development and operational life of the system. Compare with configuration control.

**Confinement** – The prevention of the leaking of sensitive data from a program.

**Confinement property** – Synonymous with star property (* property).

**Conformance testing** – Planned activities to ensure that software processes and products conform to applicable requirements, standards, and procedures.

**Connection-oriented service** – Service that establishes a logical connection that provides flow control and error control between two stations that need to exchange data.

**Connectivity** – The condition of establishing and maintaining a connection between two or more points in a telecommunications system.

**Containment** – Preventing a successful attack on a software system from spreading to other parts of an organization's computing resources.

**Contamination** – The intermixing of data at different sensitivity and need-to-know levels. The lower-level data is said to be contaminated by the higher-level data, with the result that the contaminating (higher-level) data might not receive the required level of protection.

**Contingency management** – Establishing actions to be taken before, during, and after a threatening incident.

**Contingency plan** – A plan for emergency response, backup operations, and post-disaster recovery maintained by an activity as a part of its security program; this plan ensures the availability of critical resources and facilitates the continuity of operations in an emergency situation. Synonymous with backup plan, disaster plan, and emergency plan.

**Control center** – Central location used to operate a set of physical assets.

**NOTE** Infrastructure industries typically use one or more control centers to supervise or coordinate their operations. If there are multiple control centers (e.g., a backup center at a separate site), they are typically connected via a wide area network. The control center in such industries typically contains the SCADA host computers and associated operator display devices plus ancillary information systems such as a historian.

**Control equipment** – Class that includes distributed control systems, programmable logic controllers, SCADA systems, associated operator interface consoles, and field sensing and control devices used to manage and control the process.

**NOTE** The term also includes fieldbus networks where control logic and algorithms are executed on intelligent electronic devices that coordinate actions with each other, as well as systems used to monitor the process and the systems used to maintain the process.

**Control network** – Time-critical network that is typically connected to equipment that controls physical processes.

**NOTE** The control network can be subdivided into zones, and there can be multiple separate control networks within one company or site.

**Controlled access** – Synonymous with access control.

**Correctness** – If software performs all of its intended functions as specified, it is said to be correct, and exhibits the property of correctness.

**Continuity of operations** – Maintenance of essential functions and services under a broad range of circumstances.

**Controlled sharing** – The condition that exists when access control is applied to all users and components of a system. See *access control*.

**Copper Data Distributed Interface (CDDI)** – A version of FDDI specifying the use of unshielded twisted pair wiring. See *FDDI*.

**CORBA** – See *Common Object Request Broker Architecture*.

**Cost-risk analysis** – The assessment of the cost of providing data protection for a system versus the cost of losing or compromising the data.

**COTS** – Commercial off-the-shelf. Refers to commercial items that are sold in substantial quantities in the commercial marketplace.

**Countermeasure** – Any action, device, procedure, technique, or other reactive measure that reduces the vulnerability of or threat to a system.

**Covert channel** – A communications channel that enables two cooperating processes to transfer information in a manner that violates the system's security policy.

**Covert storage channel** – A covert channel that comprises the direct or indirect writing of a storage location by one process and the direct or indirect reading of the storage location by another process. Covert storage channels typically involve a finite resource (e.g., sectors on a disk) shared by two subjects at different security levels.

**Covert timing channel** – A covert channel in which one process signals information to another by modulating its own use of system resources (e.g., CPU time) in such a way that this manipulation affects the real response time observed by the second process.

**CPU** – See *central processing unit.*

**Criteria** – Standards by which something can be judged or decided. Also, see *DoD Trusted Computer System Evaluation Criteria (TCSEC).*

**CRC** – See *cyclic redundancy check.*

**CRL** – See *certificate revocation list.*

**Cryptanalysis** – The formal term for attacking cryptographic systems. In cryptanalysis, an attacker attempts to obtain the original, unencrypted message (plaintext) from the encrypted message (ciphertext).

**Crypto-algorithm** – A well-defined procedure, sequence of rules, or steps used to produce a keystream or ciphertext from plaintext, and vice versa. A step-by-step procedure that is used to encipher plaintext and decipher ciphertext. Also called a *cryptographic algorithm.*

**Cryptographic application programming interface (CAPI)** – An interface to a library of software functions that provides security and cryptography services. CAPI is designed for software developers to call functions from the library, which makes it easier to implement security services.

**Cryptography** – The principles, means, and methods for rendering information unintelligible and for restoring encrypted information to intelligible form. The word cryptography comes from the Greek kryptos, meaning "hidden," and graphein, "to write."

**Cryptosystem** – A set of transformations from a message space to a ciphertext space. This system includes all cryptovariables (keys), plaintexts, and ciphertexts associated with the transformation algorithm (crypto-algorithm).

**Cryptovariable** – See *key.*

**CSS** – See *eCSS.*

**Cyclic redundancy check (CRC)** – A common error-detection process used in control systems. A mathematical operation is applied to the data when it is transmitted. The result is appended to the core packet. Upon receipt, the same mathematical operation is performed and the result is checked against the CRC. A mismatch indicates a very high probability that an error has occurred during transmission.

**DAA** – See *designated approving authority.*

**Damage potential** – The level of harm the attacker can cause to a system in conducting an attack.

**DAC** – See *discretionary access control.*

**Data dictionary** – A database that comprises tools to support the analysis, design, and development of software and to support good software engineering practices.

**Data Encryption Standard (DES)** – A cryptographic algorithm for the protection of unclassified data, published in Federal Information Processing Standard (FIPS) 46. The DES, which was approved by the National Institute of Standards and Technology (NIST), is intended for public and government use.

**Data flow control** – See *information flow control.*

**Data integrity** – The attribute of data that is related to the preservation of its meaning and completeness, the consistency of its representation(s), and its correspondence to what it represents. Data is said to have integrity when it meets a prior expectation of quality.

**Data link layer** – The OSI level that performs the assembly and transmission of data packets, including error control.

**Data security** – The protection of data from unauthorized (whether accidental or intentional) modification, destruction, or disclosure.

**Database** – A persistent collection of data items that form relationships among each other.

**Datagram service** – A connectionless form of packet switching whereby the source does not need to establish a connection with the destination before sending data packets.

**DCOM** – See *Distributed Component Object Model.*

**Decipher** – To reverse the encipherment process in order to make an encrypted message human readable.

**Declassification of AIS storage media** – An administrative decision or procedure to remove or reduce the security classification of the subject media.

**DeCSS** – A program that bypasses the Content Scrambling System (CSS) software used to prevent the viewing of DVD movie disks on unlicensed platforms.

**Default** – A value or option that is automatically chosen when no other value or option is specified.

**Degauss** – To degauss a magnetic storage medium is to remove all the data stored on it by demagnetization. A degausser is a device used for this purpose.

**Degausser products list (DPL)** – A list of commercially produced degaussers that meet National Security Agency specifications. This list is included in the NSA Information Systems Security Products and Services Catalogue and is available through the Government Printing Office.

**Demilitarized zone (DMZ)** – A host or network segment inserted as a "neutral zone" between an organization's private network and the Internet.

**Denial of service (DoS)** – Any action (or series of actions) that prevents any part of a system from functioning in accordance with its intended purpose. This includes any action that causes unauthorized destruction or modification of data or delay of service.

**Dependability** – A desirable condition of software that can be attained with justifiable confidence when the software functions as intended.

**DES** – See *Data Encryption Standard.*

**Designated approving authority (DAA)** – The official who has the authority to decide whether to accept the security safeguards prescribed for an automated information system (AIS), or the official who is responsible for issuing an accreditation statement that records the decision to accept those safeguards.

**Developer** – The organization or individual that develops the information system.

**DHCP** – See *Dynamic Host Configuration Protocol.*

**Dial-up** – A service whereby a computer terminal can use a telephone to initiate and effect communication with a computer.

**Diffusion** – A method of obscuring redundancy in plaintext by dissipating the statistical structure of plaintext over the bulk of ciphertext.

**Digital Millennium Copyright Act (DMCA) of 1998** – Addresses copyright licensing and ownership and prohibits trading, manufacturing, or selling in any way that is intended to bypass copyright protection mechanisms.

**Direct-sequence spread spectrum (DSSS)** – A method of spread spectrum modulation for digital signal transmission over the airwaves. The transmission is split into 14 channels, each with a frequency range, by combining a data signal with a chipping sequence. The chipping sequence or chipping code, as it is sometimes known, divides the data according to a spreading ratio. The redundant chipping code helps the signal resist interference and also enables the original data to be recovered if data bits are damaged during transmission. Data rates of 1, 2, 5.5, and 11 Mbps are obtainable.

**Disaster** – A sudden, unexpected, calamitous event that produces great damage or loss; any event that creates an inability on an organization's part to provide critical business functions for some undetermined period of time.

**Disaster plan** – Synonymous with contingency plan.

**Disaster recovery plan** – Procedure for emergency response, extended backup operations, and post-disaster recovery when an organization suffers a loss of computer resources and physical facilities.

**Discretionary access control (DAC)** – A means of restricting access to objects based on the identity and need-to-know of the user, process, and/or groups to which they belong. The controls are discretionary in the sense that a subject that has certain access permissions is capable of passing that permission (perhaps indirectly) on to any other subject. Compare with mandatory access control.

**Disk image backup** – The process of conducting a bit-level copy of a disk, sector by sector, which provides the capability to examine slack space, undeleted clusters, and (possibly) deleted files.

**Distributed control system (DCS)** – A type of control system in which the system elements are dispersed but are operated in a coupled manner.

**NOTE** Distributed control systems may have shorter coupling time constants than those typically found in SCADA systems.

**NOTE** Distributed control systems are commonly associated with continuous processes such as electric power generation, oil and gas refining, chemical, and paper manufacturing as well as discrete processes, such as automobile and other goods manufacturing, packaging, and warehousing.

**Distributed Component Object Model (DCOM)** – A distributed object model that is similar to the Common Object Request Broker Architecture (CORBA). DCOM is the distributed version of COM that supports remote objects as if the objects reside in the client's address space. A COM client can access a COM object through the use of a pointer to one of the object's interfaces and then invoke methods through that pointer.

**DITSCAP** – See *DoD Information Technology Security Certification and Accreditation Process.*

**DMCA** – See *Digital Millennium Copyright Act of 1998.*

**DNS enumeration** – Gathering information on domain name service (DNS) servers.

**DoD** – U.S. Department of Defense.

**DoD Information Technology Security Certification and Accreditation Process (DITSCAP)** – Establishes for the defense entities a standard process, set of activities, general task descriptions, and management structure to certify and accredit IT systems that will maintain the required security posture.

**DoD Trusted Computer System Evaluation Criteria (TCSEC)** – A document published by the National Computer Security Center containing a uniform set of basic requirements and evaluation classes for assessing degrees of assurance in the effectiveness of hardware and software security controls built into systems. These criteria are intended for use in the design and evaluation of systems that process and/or store sensitive or classified data. This document is Government Standard DoD 5200.28-STD and is frequently referred to as *the Criteria* (see *criteria*) or *the Orange Book.*

**Domain** – The unique context (e.g., access control parameters) in which a program is operating; in effect, the set of objects that a subject has the ability to access. See *process and subject.*

**DoS** – See *denial of service.*

**DoS attack** – Denial-of-service attack.

**DPL** – See *degausser products list.*

**DSSS** – See *direct-sequence spread spectrum.*

**Due care** – The care which an ordinary prudent person would have exercised under the same or similar circumstances. The terms due care and reasonable care are used interchangeably.

**Dynamic Host Configuration Protocol (DHCP)** – A protocol that issues Internet Protocol (IP) addresses automatically within a specified range to devices such as PCs when they are first powered on. The device retains the use of the IP address for a specific license period that the system administrator can define.

**EAL** – See *evaluation assurance level.*

**Electronics Industry Association (EIA)** – A U.S. standards organization that represents a large number of electronics firms.

**Emanations** – See *compromising emanations.*

**Embedded software** – Software that is an element of a larger system and performs some of the requirements of that system, such as controlling, measuring, or monitoring the actions of the system's physical components.

**Embedded system** – A system that performs or controls a function, either in whole or in part, as an integral element of a larger system or subsystem.

**Emergency plan** – Synonymous with contingency plan.

**Emission(s) security (EMSEC)** – The protection resulting from all measures taken to deny unauthorized persons information of value derived from the intercept and analysis of compromising emanations from crypto-equipment or an IT system.

**Encipher** – To make a message unintelligible to all but the intended recipients.

**End-to-end encryption** – Encrypted information sent from the point of origin to the final destination. In symmetric key encryption, this process requires the sender and the receiver to have identical keys for the session.

**Enumeration** – Gathering detailed information about a target information system.

**Environment** – The aggregate of external procedures, conditions, and objects that affect the development, operation, and maintenance of a system.

**Equipment under control** – Equipment, machinery, apparatus, or plant used for manufacturing, process, transportation, medical, or other activities.

**Erasure** – A process by which a signal recorded on magnetic media is removed. Erasure is accomplished in two ways – (1) by alternating current erasure, by which the information is destroyed when an alternating high and low magnetic field is applied to the media; (2) by direct current erasure, in which the media is saturated by applying a unidirectional magnetic field. See *degauss*.

**Error** – An error occurs (1) if an individual interacts with a software system and creates an error, for example, a coding error or an operational failure; (2) when the value actually produced by the software is different than the correct value; when one of the software's states changes from correct to incorrect.

**Ethernet switch** – Ethernet switches establish a data link in which a circuit or a channel is connected to an Ethernet network. Switches are used to interconnect different LANs. A switch operates in the data link layer of the ISO/OSI Reference model.

**Ethical hacker** – Trusted individual who performs penetration tests without malicious intent.

**Evaluation** – Assessment of an IT or IACS against defined security functional and assurance criteria, performed by a combination of testing and analytic techniques.

**Event** – Something that occurs, such as a specific situation or an activity. Within a software system, an event handler is a subroutine that handles input received from the software.

**Execution environment** – The entities in software's operational environment such as servers, middleware, and network devices that support, affect, or influence its execution.

**Exploit** – To exploit means to take advantage of a security weakness in software in order to compromise the software to, for example, gain control of a system. An exploit also refers to the portion of code, data, or sequence of commands used to conduct an attack.

**Evaluation assurance level (EAL)** – In the Common Criteria (CC), the degree of examination of the product to be tested. EALs range from EA1 (functional testing) to EA7 (detailed testing and formal design verification). Each numbered package represents a point on the CC's predefined assurance scale. An EAL can be considered a level of confidence in the security functions of an IT product or system.

**Exclusive OR** – A Boolean function where the output is a logical 0 when two input bits are identical (both 1s or both 0s) and the output is a logical 1 when the two input bits are not identical (one input is a 0 and the other input is a 1).

**Executive state** – One of several states in which a system can operate and the only one in which certain privileged instructions can be executed. Such instructions cannot be executed when the system is operating in other (e.g., user) states. Synonymous with supervisor state.

**Exploitable channel** – Any information channel that is usable or detectable by subjects that are external to the trusted computing base whose purpose is to violate the security policy of the system. See *covert channel*.

**Exposure** – An instance of being exposed to losses from a threat.

**External dependencies** – The network and architectural components that a network interacts with but does not control.

**Fail-over** – A condition in which operations automatically switch over to a backup system when one system/application fails.

**Fail-safe** – If a failure occurs, the system should fail in a secure manner (i.e., security controls and settings remain in effect and are enforced). It is usually better to lose functionality rather than security.

**Fail-secure** – A system design principle that preserves a secure state during and after the occurrence of failures.

**Failure** – A condition that occurs when software or hardware are unable to perform their intended functions within the operational parameters specified for those functions.

**False negative** – A condition that happens when a security tool does not report a weakness where one is present.

**False positive** – A condition that happens when a security tool reports a weakness where no weakness is present.

**False positive rate** – The number of false positives divided by the sum of the number of false positives and the number of true positives.

**Fault** – A condition that causes a device or system component to fail to perform in the required manner.

**FCC** – Federal Communications Commission.

**FDMA** – See *frequency division multiple access.*

**FDX** – See *full-duplex.*

**Federal Intelligence Surveillance Act (FISA) of 1978** – An act that limited wiretapping for national security purposes as a result of the Nixon administration's history of using illegal wiretaps.

**Fetch protection** – A system-provided restriction to prevent a program from accessing data in another user's segment of storage.

**Fiber-Distributed Data Interface (FDDI)** – A set of ANSI and ISO standards for data transmission on fiber optic lines in a local area network (LAN) that can extend in range up to 200 km (124 miles). The FDDI protocol is based on the token ring protocol. In addition to being large geographically, an FDDI local area network can support thousands of users. FDDI is frequently used on the backbone for a wide area network (WAN).

**Fiestel cipher** – An iterated block cipher that encrypts by breaking a plaintext block into two halves and, with a subkey, applies a "round" transformation to one of the halves. The output of this transformation is then XORed with the remaining half. The round is completed by swapping the two halves.

**FIFO** – See *first in, first out.*

**File server –** A computer that provides network stations with controlled access to sharable resources. The network operating system (NOS) is loaded on the file server and most sharable devices, including disk subsystems and printers, are attached to it.

**File protection –** The aggregate of all processes and procedures in a system designed to inhibit unauthorized access, contamination, or elimination of a file.

**File security –** The means by which access to computer files is limited to authorized users only.

**File Transfer Protocol (FTP) –** A TCP/IP protocol for file transfer.

**FIPS –** Federal Information Processing Standards. A set of standards that describes document processing, encryption algorithms, and other information technology standards for use within nonmilitary government agencies and by government contractors and vendors who work with the agencies.

**Firewall –** A network device that shields a trusted network from unauthorized users in an untrusted network by blocking specific types of traffic. Many types of firewalls exist, including packet filtering and stateful inspection.

**Firmware –** Executable programs stored in nonvolatile memory.

**First in, first out (FIFO) –** In computing, a method of processing elements in a queue on a first-come, first-served basis.

**Flaw –** A shortcoming in software's requirements, architecture, or design specification that results in a weak design or errors in the implementation. A flaw may or may not be a vulnerability.

**Flow-sensitive analysis –** Analysis of a computer program that takes into account the flow of control.

**Flow control –** See *information flow control.*

**Footprinting –** Gathering information in both active and passive modes.

**Formal method** – A technique used to verify, through the use of mathematical proofs, that software is consistent with its specified requirements, architecture, design, or security policy.

**Formal proof** – A complete and convincing mathematical argument presenting the full logical justification for each proof step for the truth of a theorem or set of theorems.

**Formal security policy model** – A mathematically precise statement of a security policy. To be adequately precise, such a model must represent the initial state of a system, the way in which the system progresses from one state to another, and a definition of a secure state of the system. To be acceptable as a basis for a trusted computing base (TCB), the model must be supported by a formal proof that if the initial state of the system satisfies the definition of a secure state and if all assumptions required by the model hold, then all future states of the system will be secure. Some formal modeling techniques include state transition models, denotational semantics models, and algebraic specification models. See *Bell–LaPadula model*.

**Formal verification** – The process of using formal proofs to demonstrate the consistency between a formal specification of a system and a formal security policy model (design verification) or between the formal specification and its high-level program implementation (implementation verification).

**Frequency division multiple access (FDMA)** – A spectrum-sharing technique whereby the available spectrum is divided into a number of individual radio channels.

**FTP** – See *File Transfer Protocol*.

**Full-duplex (FDX)** – Communications method where information can be transmitted in both directions simultaneously.

**Functional testing** – The segment of security testing in which the advertised security mechanisms of the system are tested under operational conditions for correct operation.

**Gateway** – A network component that provides interconnectivity at higher network layers.

**Gigabyte (GB, Gbyte)** – A unit of measure for memory or disk storage capacity; usually 1,073,741,824 bytes.

**Gigahertz (GHz)** – A measure of frequency; one billion hertz.

**GOTS** – Government off-the-shelf software.

**Gramm–Leach–Bliley (GLB) Act of November 1999** – An act that removed Depression-era restrictions on banks that limited certain business activities, mergers, and affiliations. It repealed the restrictions on banks affiliating with securities firms contained in sections 20 and 32 of the Glass–Steagall Act. GLB became effective on November 13, 2001. GLB also requires health plans and insurers to protect member and subscriber data in electronic and other formats. These health plans and insurers will fall under new state laws and regulations that are being passed to implement GLB because GLB explicitly assigns enforcement of the health plan and insurer regulations to state insurance authorities (15 U.S.C. §6805). Some of the privacy and security requirements of Gramm–Leach–Bliley are similar to those of the Health Insurance Portability and Accountability Act (HIPAA).

**Granularity** – An expression of the relative size of a data object; for example, protection at the file level is considered coarse granularity, whereas protection at the field level is considered to be of a finer granularity.

**Gray box test** – A test in which the ethical hacking team has partial knowledge of the target information system.

**Gray hat hacker** – A hacker who normally performs ethical hacking but sometimes reverts to malicious, black hat hacking.

**Guard** – A processor that provides a filter between two disparate systems operating at different security levels or between a user terminal and a database in order to filter out data that the user is not authorized to access.

**Hamming weight** – The Hamming weight of a string is the number of symbols that are different from the zero-symbol of the alphabet used.

**Handshaking procedure** – A dialogue between two entities (e.g., a user and a computer, a computer and another computer, or a program and another program) for the purpose of identifying and authenticating the entities to one another.

**Hertz (Hz)** – A unit of frequency measurement; one cycle of a periodic event per second. Used to measure frequency.

**HIPAA (Health Insurance Portability and Accountability Act)** – See *Kennedy–Kassebaum Act of 1996*.

**Host** – A time-sharing computer accessed via terminals or terminal emulation; a computer to which an expansion device attaches.

**HTML (hypertext markup language)** – A standard used on the Internet for defining hypertext links between documents.

**HTTP** – Hypertext Transfer Protocol. An application layer protocol designed within the framework of the Internet Protocol suite that is the fundamental protocol of the World Wide Web.

**I&A** – Identification and authentication. Identification is the process of presenting the identity of a user, process, or device, usually as a prerequisite for granting access to resources in an IT system. Authentication is verifying the identity of a user, process, or device, often as a prerequisite to allowing access to resources in an information system.

**IA** – Information assurance. Measures that protect and defend information and information systems by ensuring their availability, integrity, authentication, confidentiality, and nonrepudiation. These measures include providing for restoration of information systems by incorporating protection, detection, and reaction capabilities.

**IAC** – See *inquiry access code*.

**ICV** – Integrity check value. In Wired Equivalent Privacy (WEP) encryption, the frame is run through an integrity algorithm, and the generated ICV is placed at the end of the encrypted data in the frame. Then the receiving station runs the data through its integrity algorithm and compares it to the ICV received in the frame. If it matches, the unencrypted frame is passed to the higher layers. If it does not match, the frame is discarded.

**ID –** Common abbreviation for "identifier" or "identity."

**Identification –** The process that enables a system to recognize an entity, generally by the use of unique machine-readable user names.

**Identity –** The label applied to the type of service provided by a software component, for example, NETWORK_SERVICE.

**IDS –** See *Intrusion detection system.*

**IEEE –** See *Institute of Electrical and Electronic Engineers.*

**IETF –** Internet Engineering Task Force. The Internet Engineering Task Force is a volunteer body that develops and promotes Internet standards. The mission of the IETF is to improve the Internet by producing high-quality, relevant technical documents that influence the way the Internet is designed, used, and managed.

**IKE –** See *Internet Key Exchange.*

**Impersonating –** Synonymous with spoofing.

**Implementation –** The phase in the software life cycle when the software is coded from specifications and/or integrated from components.

**Incomplete parameter checking –** A method used to detect system design flaws that result when all parameters have not been fully examined for accuracy and consistency, thus making the system vulnerable to penetration.

**Individual accountability –** The ability to identify an individual and to hold that individual responsible for his or her actions.

**Industrial automation and control system –** Collection of personnel, hardware, and software that can affect or influence the safe, secure, and reliable operation of an industrial process.

**NOTE** These systems include, but are not limited to:

- Industrial control systems, including distributed control systems (DCSs), programmable logic controllers (PLCs), remote terminal units (RTUs), intelligent electronic devices, supervisory control and data acquisition (SCADA), networked electronic sensing and control, and monitoring and diagnostic systems. (In this context, process control systems include basic process control system and safety instrumented system [SIS] functions, whether they are physically separate or integrated.)
- Associated information systems such as advanced or multivariable control, online optimizers, dedicated equipment monitors, graphical interfaces, process historians, manufacturing execution systems, and plant information management systems.
- Associated internal, human, network, or machine interfaces used to provide control, safety, and manufacturing operations functionality to continuous, batch, discrete, and other processes.

**Industrial, scientific, and medicine (ISM) bands** – Radio frequency bands authorized by the Federal Communications Commission (FCC) for wireless LANs used for industrial, scientific, and medicine-related purposes. The ISM bands are located at 902 MHz, 2.400 GHz, and 5.7 GHz. The transmitted power is commonly less than 600mW, and no FCC license is required.

**Information flow control** – A procedure undertaken to ensure that information transfers within a system are not made from a higher security level object to a lower security level object. See *covert channel, simple security property, and star property (* property)*. Synonymous with data flow control and flow control.

**Information security policy** – The aggregate of public law, directives, regulations, and rules that regulate how an organization manages, protects, and distributes information. For example, the information security policy for financial data processed on DoD systems may be in U.S. Code (USC), Executive Order (EO), and DoD directives and in local regulations. The information security policy lists all the security requirements applicable to specific information.

**Information system (IS)** – Any telecommunications or computer-related equipment or interconnected systems or subsystems of equipment used in the acquisition, storage, manipulation, management, movement, control, display, switching, interchange, transmission, or reception of voice and/or data; includes software, firmware, and hardware.

**Information system security engineering/engineer (ISSE)** – Process of capturing and refining information protection requirements to ensure their integration into information systems acquisition and information systems development through purposeful security design or configuration. Individual assigned responsibility for conducting information system security engineering activities.

**Information system security officer (ISSO)** – (1) The person who is responsible to the designated approving authority (DAA) for ensuring that security is provided for and implemented throughout the life cycle of an automated information system (AIS), from the beginning of the concept development plan through its design, development, operation, maintenance, and secure disposal. (2) In certification and accreditation (C&A), the person responsible to the DAA for ensuring that the security of an IT system is approved, operated, and maintained throughout its life cycle in accordance with the SSAA.

**Information technology (IT)** – The hardware, firmware, and software used as part of the information system to perform information-related functions. This definition includes computers, telecommunications, automated information systems, and automatic data processing equipment. IT includes any assembly of computer hardware, software, and/or firmware configured to collect, create, communicate, compute, disseminate, process, store, and/or control information.

**Information technology security (ITSEC)** – (1) Protection of information technology against unauthorized access to or modification of information, whether in storage, processing, or transit, and against the denial of service to authorized users, including those measures necessary to detect, document, and counter such threats; (2) protection and maintenance of confidentiality, integrity, availability, and accountability.

**Infrared (IR) light** – Light waves that range in length from about 0.75 to 1,000 microns; IR light operates at a lower frequency than the spectral colors but at a higher frequency than radio waves.

**Inheritance (in object-oriented programming) –** When all the methods of one class, called a *superclass*, are inherited by a subclass. Thus, all messages understood by the superclass are understood by the subclass.

**Input validation –** Checking the data that is input to a software application for acceptable parameters such as data type, length, and range.

**Inquiry access code (IAC) –** Inquiry access code; used in inquiry procedures. The IAC can be one of two types – a dedicated IAC for specific devices or a generic IAC for all devices.

**Institute of Electrical and Electronic Engineers (IEEE) –** A U.S.-based standards organization participating in the development of standards for data transmission systems. The IEEE has made significant progress in the establishment of standards for LANs, namely the IEEE 802 series.

**Integration testing –** A testing process used to verify the interface among components such as hardware or software modules as the components are installed. The testing personnel should integrate components into the network one by one and perform integration testing when necessary to ensure proper gradual integration of components.

**Integrator –** An organization or individual that unites, combines, or otherwise incorporates information system components with another system(s).

**Integrity –** (1) The quality of being in sound, unimpaired, or perfect condition.

A quality of an IT system reflecting the logical correctness and reliability of the operating system, the logical completeness of the hardware and software implementing the protection mechanisms, and the consistency of the data structures and occurrence of the stored data. It is composed of data integrity and system integrity. (3) Integrity is concerned with unauthorized modification of data, whether by an authorized or an unauthorized individual or process. If a message is intercepted, its contents are changed, and it is then forwarded to a receiver, integrity has been violated.

**Integrity check value –** See *ICV*.

**Interdiction –** See *denial of service.*

**Inter-file analysis –** Analysis of code resulting in different files that have procedural, data, or other interdependencies.

**Internal security controls –** Hardware, firmware, and software features within a system that restrict access to resources (hardware, software, and data) to authorized subjects only (persons, programs, or devices).

**International Standards Organization (ISO) –** A nontreaty standards organization active in the development of international standards, such as the Open System Interconnection (OSI) network architecture.

**International Telecommunications Union (ITU) –** An intergovernmental agency of the United States responsible for making recommendations and developing standards regarding telephone and data communications systems for public and private telecommunication organizations and for providing coordination for the development of international standards.

**Internet –** The largest network in the world. The successor to the Advanced Research Projects Agency Network (ARPANET), the Internet includes other large subnetworks. The Internet uses the TCP/IP protocol suite and connects universities, government agencies, business, industry, and individuals around the world.

**Internet Key Exchange (IKE)** – The protocol used to set up a security association (SA) in the IPsec protocol suite in order to establish secure communications.

**Internet Protocol (IP) –** The Internet standard protocol that defines the Internet datagram as the information unit passed across the Internet. IP provides the basis of a best-effort packet delivery service. The Internet Protocol suite is often referred to as TCP/IP because IP is one of the two fundamental protocols, the other being the Transfer Control Protocol.

**Internet Service Provider (ISP)** – A commercial organization that provides user connections to the Internet for a fee.

**Internetwork packet exchange (IPX) –** Novell NetWare operating system protocol for the exchange of message packets on an internetwork. IPX passes application requests for network services to the network drives and then to other workstations, servers, or devices on the internetwork.

**Intrusion detection system –** See *IDS*. Hardware or software product that gathers and analyzes information from various areas within a computer or a network to identify possible security breaches, which include both intrusions (attacks from outside the organizations) and misuse (attacks from within the organizations).

**Ionization –** The act of converting an atom or molecule into an ion by adding or removing charged particles such as electrons or other ions.

**Ionize –** To convert an atom or molecule into an ion by adding or removing charged particles such as electrons or other ions.

**IP –** See *Internet Protocol.*

**IPX –** See *internetwork packet exchange.*

**IS –** See *information system.*

**ISM bands –** See *industrial, scientific, and medicine bands.*

**Isolation –** The containment of subjects and objects in a system in such a way that they are separated from one another as well as from the protection controls of the operating system.

**ISO –** See *International Standards Organization.*

**ISP –** See *Internet service provider.*

**ISSE –** See *information system security engineering/engineer.*

**ISSO –** See *information system security officer.*

**IT –** See *information technology.*

**ITSEC** – See *information technology security.*

**ITU** – See *International Telecommunications Union.*

**IV** – An initialization vector (IV) is a random or pseudorandom character sequence that is used with a secret key for data encryption. The use of an IV makes it more difficult for an attacker to decipher the encrypted message.

**Justifiable confidence** – A high level of certainty, achieved with actions, arguments, and evidence.

**Kennedy–Kassebaum Health Insurance Portability and Accountability Act (HIPAA) of 1996** – A set of regulations that mandates the use of standards in health care record keeping and electronic transactions. The act requires that health care plans, providers, insurers, and clearinghouses do the following:

- Provide for restricted access by the patient to personal health care information
- Implement administrative simplification standards
- Enable the portability of health insurance
- Establish strong penalties for health care fraud

**Key** – Information or sequence that controls the enciphering and deciphering of messages. Also known as a cryptovariable. Used with a particular algorithm to encipher the plaintext message or decipher the encrypted message.

**Keystream** – A sequence of random or pseudorandom characters that are combined with a plaintext message to produce an encrypted message (the ciphertext).

**Kilobyte (KB, Kbyte)** – A unit of measurement of memory or disk storage capacity; a data unit of 210 (1,024) bytes.

**Kilohertz (kHz)** – A unit of frequency measurement; equivalent to 1,000 Hz.

**Knowledge base** – The rules and facts of the particular problem domain in an expert system. An expert system is a type of artificial intelligence element that makes decisions based on a knowledge base and rules derived from interviews with domain experts.

**LAN** – See *local area network.*

**Least privilege** – In the principle of least privilege, a user or process is given the minimum amount of privileges, authorization, and so on for the least amount of time that will permit him or her to accomplish assigned tasks and complete a mission. Least privilege reduces the prospect of an individual or process having unauthorized access to sensitive information.

**Light-emitting diode (LED)** – Used in conjunction with optical fiber, an LED emits light when current is passed through it. Its advantages include low cost and long life, and it is capable of transmitting data in the million bits per second (Mbps) range.

**Limited access** – Synonymous with access control.

**Lines, units, cells** – Lower-level control system elements (entities) that perform manufacturing, field device control, or vehicle functions.

**NOTE** Entities at this level may be connected together by an area control network and may contain information systems related to the operations performed in that entity.

**Local area network (LAN)** – A network that interconnects devices in the same office, floor, or building, or in adjacent or nearby buildings.

**MAC** – See *mandatory access control* if used in the context of a type of access control. MAC also refers to the media access control address assigned to a network interface card on an Ethernet network.

**Magnetic remanence** – A measure of the magnetic flux density that remains after removal of an applied magnetic force. Refers to any data remaining on magnetic storage media after removal of the power.

**Mail gateway** – A type of gateway that interconnects dissimilar email systems.

**Maintainer** – The organization or individual that maintains the information system.

**Maintenance organization** – The organization that keeps an IT or industrial automation and control system (IACS) operating in accordance with prescribed requirements, laws, policies, procedures, and regulations.

**Malicious code** – Software or firmware that is intentionally included in a system for an unauthorized purpose (e.g., a Trojan horse).

**Malicious logic** – Synonymous with malicious code.

**Malware** – A merging of the words malicious and software. Malware is inserted into a system, usually covertly, with the intention of compromising the confidentiality, availability, or integrity of the system's data, applications, and operating system. It may come to the attention of a user by inhibiting the operational abilities of the system. Often referred to as malicious code.

**Mandatory access control (MAC)** – A means of restricting access to objects based on the sensitivity (as represented by a label) of the information contained in the objects and the formal authorization (in other words, clearance) of subjects to access information of such sensitivity. Compare with discretionary access control.

**Manufacturing operations** – Collection of production, maintenance, and quality assurance operations and their relationship to other activities of a production facility. Manufacturing operations include:

- Manufacturing or processing facility activities that coordinate the personnel, equipment, and material involved in the conversion of raw materials or parts into products
- Functions that may be performed by physical equipment, human effort, and information systems
- Managing information about the schedules, use, capability, definition, history, and status of all resources (personnel, equipment, and material) within the manufacturing facility

**Masquerading** – Synonymous with spoofing.

**Media access control (MAC)** – An IEEE 802 standards sublayer used to control access to a network medium, such as a wireless LAN. Also deals with collision detection. Each computer has its own unique MAC address.

**Megabits per second (Mbps)** – One million bits per second, typically referring to a data transmission rate.

**Megabyte (MB, Mbyte)** – A unit of measurement for memory or disk storage capacity, usually 1,048,576 bytes.

**Megahertz (MHz)** – A measure of frequency equivalent to one million cycles per second.

**Middleware** – An intermediate software component located on a wired network between a wireless appliance and the application or data residing on the wired network. Middleware provides appropriate interfaces between the appliance and the host application or server database.

**Mimicking** – Synonymous with spoofing.

**Mission** – The assigned duties to be performed by a resource.

**Misuse** – Use of software that deviates from what is expected based on the software's specifications. If the misuse is malicious in nature, it is typically referred to as abuse.

**Mobile code** – Software obtained from remote computer systems across a network, and downloaded and executed on a local computer system without explicit installation or execution by the recipient. Examples include JavaScript, Java applets, ActiveX controls, and Flash animations. Mobile code can be used for valid applications or can contain a virus that could do harm to a computer system.

**Modes of operation** – A description of the conditions under which an automated information system (AIS) functions, based on the sensitivity of the data processed and the clearance levels and authorizations of the users. Four modes of operation are authorized:

- **Dedicated mode** – An AIS is operating in the dedicated mode when each user who has direct or indirect individual access to the AIS, its peripherals, remote terminals, or remote hosts has all the following:
  - A valid personnel clearance for all information on the system
  - Formal access approval; furthermore, the user has signed nondisclosure agreements for all the information stored and/or processed (including all compartments, subcompartments, and/or special access programs)
  - A valid need-to-know for all information contained within the system
- **System-high mode** – An AIS is operating in the system-high mode when each user who has direct or indirect access to the AIS, its peripherals, remote terminals, or remote hosts has all the following:
  - A valid personnel clearance for all information on the AIS
  - Formal access approval and signed nondisclosure agreements for all the information stored and/or processed (including all compartments, subcompartments, and/or special access programs)
  - A valid need-to-know for some of the information contained within the AIS
- **Compartmented mode** – An AIS is operating in the compartmented mode when each user who has direct or indirect access to the AIS, its peripherals, remote terminals, or remote hosts has all the following:
  - A valid personnel clearance for the most restricted information processed in the AIS
  - Formal access approval and signed nondisclosure agreements for the information which he or she will be able to access
  - A valid need-to-know for the information that he or she will be able to access
- **Multilevel mode** – An AIS is operating in the multilevel mode when all the following statements are satisfied concerning the users who have

direct or indirect access to the AIS, its peripherals, remote terminals, or remote hosts:

- Some do not have a valid personnel clearance for all the information processed in the AIS
- All have the proper clearance and the appropriate formal access approval for the information to which they are to have access
- All have a valid need-to-know for that information to which they are to have access

**Multilevel device** – A device that is used in a manner that permits it to simultaneously process data of two or more security levels without risk of compromise. To accomplish this, sensitivity labels are normally stored on the same physical medium and in the same form (e.g., machine-readable or human-readable) as the data being processed.

**Multilevel secure** – A class of system containing information with different sensitivities that simultaneously permits access by users with different security clearances and needs-to-know but that prevents users from obtaining access to information for which they lack authorization.

**Multilevel security mode** – See *modes of operation*.

**Multiple inheritance** – In object-oriented programming, a situation where a subclass inherits the behavior of multiple superclasses.

**Multiuser mode of operation** – A mode of operation designed for systems that process sensitive, unclassified information in which users might not have a need-to-know for all information processed in the system. This mode is also used for microcomputers processing sensitive unclassified information that cannot meet the requirements of the stand-alone mode of operation.

**Mutually suspicious** – A state that exists between interacting processes (subsystems or programs) in which neither process can expect the other process to function securely with respect to some property.

**National Computer Security Center (NCSC)** – Originally named the DoD Computer Security Center, the NCSC is responsible for encouraging the widespread availability of trusted computer systems throughout the federal government. It is a branch of the National Security Agency (NSA) that also initiates research and develops and publishes standards and criteria for trusted information systems.

**National Information Assurance Certification and Accreditation Process (NIACAP)** – Provides a standard set of activities, general tasks, and a management structure to certify and accredit systems that will maintain the information assurance and security posture of a system or site. The NIACAP is designed to certify that the information system meets documented accreditation requirements and continues to maintain the accredited security posture throughout the system life cycle. This document is issued and maintained by the National Security Telecommunications and Information Systems Security Committee (NSTISSC).

**National Institute of Standards and Technology (NIST)** – NIST is a nonregulatory federal agency within the U.S. Department of Commerce. NIST's mission is to promote U.S. innovation and industrial competitiveness by advancing measurement science, standards, and technology in ways that enhance economic security and improve our quality of life.

**National Security Agency (NSA)** – NSA is a cryptologic intelligence agency of the U.S. Department of Defense responsible for the collection and analysis of foreign communications and foreign signals intelligence, as well as protecting U.S. government communications and information systems.

**NCSC** – See *National Computer Security Center.*

**Need-to-know** – The necessity for access to, knowledge of, or possession of specific information that is required to carry out assigned duties in industrial automation and control systems as well as systems critical to national security.

**Network basic input/output system (NetBIOS)** – A standard interface between networks and PCs that enables applications on different computers to communicate within a LAN. NetBIOS was created by IBM for its early PC network, was adopted by Microsoft, and has since become a de facto industry standard. It is not routable across a wide area network (WAN).

**Network interface card (NIC)** – A network adapter inserted into a computer that enables the computer to be connected to a network.

**Network monitoring** – A form of operational support enabling network management to view the network's inner workings. Most network-monitoring equipment is nonobtrusive and can be used to determine the network's utilization and to locate faults.

**NIACAP** – See *National Information Assurance Certification and Accreditation Process.*

**NIAP** – See *National Information Assurance Partnership.*

**NIST** – See *National Institute of Standards and Technology.*

**Node** – Any network-addressable device on a network, such as a router or network interface card. Any network station.

**Nonrepudiation** – Nonrepudiation in digital systems refers to ensuring that the sender of a message or contract cannot later deny sending the message or contract and that, conversely, a receiver to whom the message was sent cannot deny receiving the message. Digital signatures are used to implement nonrepudiation.

**NSA** – See *National Security Agency.*

**Object** – A passive entity that contains or receives information. Access to an object implies access to the information that it contains. Examples of objects include records, blocks, pages, segments, files, directories, directory trees, and programs, as well as bits, bytes, words, fields, processors, video displays, keyboards, clocks, printers, and network nodes.

**Object request broker (ORB)** – The fundamental building block of the object request architecture (ORA), which manages the communications among the ORA entities. The purpose of the ORB is to support the interaction of objects in heterogeneous, distributed environments. The objects may be on different types of computing platforms.

**Object reuse –** The reassignment and reuse of a storage medium (e.g., page frame, disk sector, and magnetic tape) that once contained one or more objects. To be securely reused and assigned to a new subject, storage media must contain no residual data (data remanence) from the object(s) that were previously contained in the media.

**Object services –** Services that support the ORB in creating and tracking objects as well as performing access control functions.

**OMB –** Office of Management and Budget. As part of the executive branch of the U.S. government, the OMB serves the president of the United States in implementing his vision across the executive branch. OMB is the largest component of the executive office of the president. It reports directly to the president and helps a wide range of executive departments and agencies across the federal government to implement the commitments and priorities of the president.

**OPC –** A set of specifications for the exchange of information in a process control environment.

**NOTE** The abbreviation *OPC* originally came from *OLE for process control,* where *OLE* was short for *object linking and embedding*.

**Open security environment –** An environment that includes those systems in which at least one of the following conditions holds true – (1) application developers (including maintainers) do not have sufficient clearance or authorization to provide an acceptable presumption that they have not introduced malicious logic, and (2) configuration control does not provide sufficient assurance that applications are protected against the introduction of malicious logic prior to and during the operation of system applications.

**Open source software (OSS) –** Software that is publicly available and is provided under a license that is less restrictive than a typical commercial license. A typical license may permit users to change the software and redistribute the software in modified or unmodified form.

**Open System Interconnection (OSI) –** An ISO standard specifying an open system capable of enabling communications between diverse systems. OSI has the following seven layers of distinction – physical, data link, network, transport, session, presentation, and application. These layers provide the functions that enable standardized communications between two application processes.

**Operating system (OS)** – The collection of software that directs a computer's operations, controlling and scheduling the execution of other programs, and managing storage, input/output, and communication resources.

**Operations security** – Controls over hardware, media, and operators who have access to computer systems. Operations security controls safeguard information assets while the data is resident in the computer or otherwise directly associated with the computing environment. The controls address both software and hardware as well as such processes as change control and problem management.

**Operator** – An individual who supports system operations from the operator's console, monitors execution of the system, and controls the flow of jobs.

**Orange Book** – Alternate name for the DoD Trusted Computer System Evaluation Criteria.

**Original equipment manufacturer (OEM)** – A manufacturer of products for integration into other products or systems.

**ORA** – See *object request broker.*

**ORB** – See *object request broker.*

**OS** – See *operating system.*

**OSI** – See *open system interconnection.*

**OSS** – See *open source software.*

**Overt channel** – A path within a computer system or network that is designed for the authorized transfer of data. Compare with covert channel.

**Overwrite procedure** – A software process that replaces data previously stored on storage media with a predetermined set of meaningless data or random patterns. See *magnetic remanence.*

**Packet** – A basic message unit for communication across a network. A packet usually includes routing information, data, and (sometimes) error-detection information.

**Packet-switched** – (1) A type of network that routes data packets based on an address contained in the data packet. Multiple data packets can share the same network resources. (2) A type of communications network that uses shared facilities to route data packets from and to different users. Unlike a circuit-switched network, a packet-switched network does not set up dedicated circuits for each session.

**Passivation layer** – A protective layer that covers an integrated circuit.

**Password** – A protected/private character string that is used to authenticate an identity.

**Penetration** – The act of successfully bypassing a system's security mechanisms.

**Penetration testing** – The portion of security testing in which the evaluators attempt to circumvent the security features of a system. One may assume that the evaluators will use all system design and implementation documentation, which can include listings of system source code, manuals, and circuit diagrams. The evaluators work under the same constraints that are applied to ordinary users.

**Performance modeling** – The use of simulation software to predict network behavior, allowing developers to perform capacity planning. Simulation makes it possible to model the network and impose varying levels of utilization to observe the effects.

**Permissions** – A description of the type of authorized interactions that a subject can have with an object. Examples of permission types include read, write, execute, add, modify, and delete.

**Personnel security** – (1) Procedures established to ensure that all personnel who have access to sensitive information possess the required authority as well as appropriate clearances. (2) Procedures established to evaluate a person's background and thereby provide assurance of necessary trustworthiness.

**PHI** – See *protected health information.*

**Physical security** – The application of physical barriers and control procedures as preventive measures or countermeasures against threats to resources and sensitive information.

**Piggyback** – Gaining unauthorized access to a system via another user's legitimate connection. See *between-the-lines entry.*

**PKI** – See *public key infrastructure.*

**Plaintext** – Message text in clear, human-readable form.

**Platform for Privacy Preferences (P3P)** – Proposed standards developed by the World Wide Web Consortium (W3C) to implement privacy practices on websites.

**Portability** – Defines network connectivity that can be easily established, used, and then dismantled.

**Port scanning** – Connecting to Transmission Control Protocol (TCP) and User Datagram Protocol (UDP) ports in order to determine the services and applications running on the target host.

**Presentation layer** – The layer of the OSI model that negotiates data transfer syntax for the application layer and performs translations between different data types, if necessary.

**Private key encryption** – See *symmetric (private) key encryption.*

**Privileged instructions** – A set of instructions (e.g., interrupt handling or special computer instructions) used to control features such as storage protection features that are generally executable only when the automated system is operating in the executive state.

**PRNG** – Pseudorandom number generator. An algorithm that produces a sequence of bits that are uniquely determined from an initial value called a *seed*. The output of the PRNG "appears" to be random, that is, the output is statistically indistinguishable from random values. A cryptographic PRNG has the additional property that the output is unpredictable, given that the seed is not known.

**Procedural language** – Comprises sequential execution of instructions based on the von Neumann architecture of a CPU, memory, and an input/output device. Variables are part of the sets of instructions used to solve a particular problem, and therefore the data is not separate from the instruction statements.

**Process** – (1) A program in execution. See *domain* and *subject*. (2) A series of operations performed in the making, treatment or transportation of a product or material.

**Program** – (1) A plan of action aimed at accomplishing a clear technical, production, or business objective, with details on what work is to be done, by whom, when, and what means or resources will be used. (2) In computing, a set of coded instructions that a computer executes to solve a problem or produce a desired result.

**Program manager** – The person ultimately responsible for the overall procurement, development, integration, modification, operation, and maintenance of an IT system.

**Protected health information (PHI)** – Individually identifiable health information that is:

- Transmitted by electronic media
- Maintained in any medium described in the definition of electronic media (under HIPAA)
- Transmitted or maintained in any other form or medium

**Protection profile (PP)** – In the Common Criteria (CC), an implementation-independent specification of the security requirements and protections of a product that could be built.

**Protection-critical portions of the TCB** – Those portions of the TCB whose normal function is to deal with access control between subjects and objects. Their correct operation is essential to the protection of the data on the system.

**Protocols** – A set of rules and formats used to specify interactions between communicating entities.

**Prototyping** – A method of determining or verifying requirements and design specifications. The prototype normally consists of network hardware and software that support a proposed solution. The approach to prototyping is typically a trial-and-error experimental process.

**Public key cryptography** – See *asymmetric key encryption*.

**Public key infrastructure (PKI)** – A PKI binds public keys to entities, enables other entities to verify public key bindings, and provides the services needed for ongoing management of keys in a distributed system. The goal of the PKI security architecture is to protect and distribute information that is needed in a widely distributed environment where the users, resources, and stakeholders are in different places at different times. A PKI allows business to be conducted with the justifiable confidence that:

- The person or process sending the transaction is the actual originator.
- The person or process receiving the transaction is the intended recipient.
- Data integrity has not been compromised.

**Purge** – The removal of sensitive data from an automation and control system, IT system, storage device, or peripheral device with storage capacity at the end of a processing period. This action is performed in such a way that there is assurance proportional to the sensitivity of the data that the data cannot be reconstructed.

**Quality** – The degree to which software or hardware meets its specifications and satisfies its intended purpose.

**RADIUS** – See *remote authentication dial-in user service*.

**RC4** – Rivest, Shamir, and Adleman (RSA) cipher algorithm 4.

**Read** – A fundamental operation that results only in the flow of information from an object to a subject.

**Read access** – Permission to read information.

**Read-only memory (ROM)** – Computer memory on which data has been written and stored. Once data has been written onto a ROM chip, it cannot be removed and can only be read.

**Recovery planning** – The advance planning and preparations that are necessary to minimize loss and to ensure the availability of the critical information systems of an organization in the event of a system failure or outage/disruption.

**Recovery procedures** – The actions that are necessary to restore a system's computational capability and data files after a system failure or outage/disruption.

**Reference-monitor concept** – An access-control concept that refers to the use of an abstract machine (mechanism) that mediates all access to objects by subjects.

**Reference-validation mechanism** – An implementation of the reference monitor concept. A security kernel is a type of reference-validation mechanism.

**Regression testing** – Atesting process used to ensure that existing software functions of the product have not been accidentally damaged as an unintended by-product of adding additional software features.

**Relevancies** – Attributes of a software component that are applicable to its attack surface. A system's attack surface is the set of ways in which an adversary can enter the system and potentially cause damage.

**Reliability** – The probability of a given system performing its mission adequately for a specified period of time under expected operating conditions.

**Remote authentication dial-in user service (RADIUS) –** A server for remote user authentication and accounting. Its primary use is for Internet service providers, though it may be used on any network that needs a centralized authentication and/or accounting service for its workstations.

**Requirement –** A description of a functional or nonfunctional (security) behavior that software or hardware must meet.

**Residual risk –** The portion of risk that remains after security measures have been applied.

**Resilience –** The ability to quickly adapt and recover from any known or unknown changes to the environment through holistic implementation of risk management, contingency, and continuity planning. In software, a measure of how effectively software can recover after being compromised. If highly resilient software is compromised, damage to the software will be minimized and it will recover quickly to an acceptable level of service.

**RFC –** Request for comment. Documents that are the working notes of the Internet research and development community published by the Internet Engineering Task Force (IETF). An RFC document may be on essentially any topic related to computer communication and can describe methods, protocols, behaviors, and so on pertaining to the Internet.

**RFP –** Request for proposal. A document issued at an early stage in a procurement process where an organization is invited to submit a proposal to provide goods or services.

**RIP –** See *routing information protocol.*

**Risk –** A measure of the extent to which an entity is threatened by a potential circumstance or event, and typically a function of (1) the adverse impacts that would arise if the circumstance or event occurs; and (2) the likelihood of occurrence.

**Risk analysis –** The process of identifying security risks, determining their magnitude, and identifying areas needing safeguards. Risk analysis is a part of risk management.

**Risk assessment** – The process that systematically identifies potential vulnerabilities to valuable system resources and threats to those resources, quantifies loss exposures and consequences based on probability of occurrence, and (optionally) recommends how to allocate resources to countermeasures to minimize total exposure.

**Risk management** – Process of identifying and applying countermeasures commensurate with the value of the assets protected based on a risk assessment.

**Robustness** – The degree to which a software component or system can function correctly in the presence of invalid or unexpected inputs and unexpected or stressful environmental conditions, including input and conditions that are intentional and malicious. Robust systems are designed to operate under a variety of conditions within a certain range and will fail gracefully outside of that range. Resilient systems have the ability to recover and adapt to any known environmental changes.

**Role** – In role-based access control, permitted actions on resources are identified with roles rather than with individual subject identities.

**ROM** – See *read-only memory.*

**Router** – A network component that provides internetworking at the network layer of a network's architecture by allowing individual networks to become part of a wide area network (WAN). A router works by using logical and physical addresses to connect two or more separate networks. It determines the best path by which to send a packet of information.

**Routing information protocol (RIP)** – A common type of routing protocol. RIP bases its routing path on the distance (number of hops) to the destination. RIP maintains optimum routing paths by sending out routing update messages if the network topology changes.

**RSA** – Rivest, Shamir, and Adleman.

**Safeguards** – See *security safeguards.*

**Safety** – Software or hardware for industrial automation and control systems and IT systems that are life critical should exhibit the property of safety, that is, they should behave as necessary and required even if components of the system fail.

**Safety instrumented system (SIS)** – A system used to implement one or more safety instrumented functions.

**NOTE** A safety instrumented system is composed of any combination of sensors, logic solvers, and actuators.

**Safety integrity level** – Discrete level (one of four) for specifying the safety integrity requirements of the safety instrumented functions to be allocated to a safety instrumented system.

**NOTE** Safety integrity level 4 has the highest level of safety integrity; safety integrity level 1 has the lowest.

**Safety network** – A network that connects safety instrumented systems for the communication of safety-related information.

**Sandbox** – An access-control-based protection mechanism. It is commonly applied to restrict the access rights of mobile code downloaded from a website as an applet. The code is set up to run in a "sandbox" that blocks its access to the local workstation's hard disk, thus preventing malicious activity by the code. The sandbox is usually interpreted by a virtual machine such as the Java Virtual Machine (JVM).

**SCADA systems** – Highly distributed systems used to control geographically dispersed assets, often scattered over thousands of square kilometers, where centralized data acquisition and control are critical to system operation. They are used in distribution systems, such as water distribution and wastewater collection systems, oil and natural gas pipelines, electrical power grids, and railway transportation systems.

**Scalar** – A computer language variable or field that can hold only one value at a time.

**Scanning** – Sending packets or requests to another system to gain information to be used in a subsequent attack.

**SDLC** – See *system development life cycle.*

**Secure coding** – The use of software programming practices that reduce or eliminate software defects or programming errors so that software can be built with a higher level of security and quality assurance.

**Secure configuration management** – The set of procedures that are appropriate for controlling changes to a system's hardware and software structure for the purpose of ensuring that changes will not lead to violations of the system's security policy.

**Secure sockets layer (SSL)** – A protocol used for protecting private information during transmission via the Internet.

**Secure state** – A condition in which no subject can access any object in an unauthorized manner.

**Security** – Measures and controls that ensure the confidentiality, integrity, availability, and accountability of the information processed, transmitted, and stored by industrial automation and control systems or IT systems.

**Security evaluation** – An evaluation that is performed to assess the degree of trust that can be placed in a system for the secure handling of sensitive information. One type, a *product evaluation,* is an evaluation performed on the hardware and software features and assurances of a computer product from a perspective that excludes the application environment. The other type, a *system evaluation,* is made for the purpose of assessing a system's security safeguards with respect to a specific operational mission; it is a major step in the certification and accreditation process.

**Security failure** – An event that is a violation of a particular system's explicit or implicit security policy.

**Security fault analysis** – A security analysis, usually performed on hardware at the gate level, to determine the security properties of a device when a hardware fault is encountered.

**Security features** – The security-relevant functions, mechanisms, and characteristics of system hardware and software. Security features are a subset of system security safeguards.

**Security flaw** – An error of commission or omission in a system that might enable protection mechanisms to be bypassed.

**Security flow analysis** – A security analysis performed on a formal system specification that locates the potential flows of information within the system.

**Security functional requirements** – Requirements, preferably from the Common Criteria, Part 2, that when taken together specify the security behavior of an industrial automation and control system or IT product or system.

**Security kernel** – The hardware, firmware, and software elements of a trusted computing base (TCB) that implement the reference monitor concept. The security kernel must mediate all accesses, must be protected from modification, and must be verifiable as correct.

**Security level** – The combination of a hierarchical classification and a set of nonhierarchical categories that represents the sensitivity of information.

**Security measures** – Elements of software, firmware, hardware, or procedures that are included in a system for the satisfaction of security specifications.

**Security objective** – A statement of intent to counter specified threats and/or satisfy specified organizational security policies and assumptions.

**Security perimeter** – The boundary where security controls are in effect to protect assets.

**Security policy** – The set of laws, rules, and practices that regulates how an organization manages, protects, and distributes sensitive information.

**Security policy model** – A formal presentation of the security policy enforced by the system. It must identify the set of rules and practices that regulate how a system manages, protects, and distributes sensitive information. See *Bell–LaPadula model* and *formal security policy model*.

**Security process** – The series of activities that monitor, evaluate, test, certify, accredit, and maintain the system accreditation throughout the system life cycle.

**Security requirements** – The types and levels of protection that are necessary for equipment, data, information, applications, and facilities to meet the security policy.

**Security requirements baseline** – A description of the minimum requirements necessary for a system to maintain an acceptable level of security.

**Security safeguards (safeguards)** – The proactive protective measures and controls that are prescribed to meet the security requirements specified for a system. Those safeguards can include (but are not necessarily limited to) hardware and software security features, operating procedures, accountability procedures, access and distribution controls, management constraints, personnel security, and physical structures, areas, and devices.

**Security specifications** – A detailed description of the safeguards required to protect a system.

**Security target (ST)** – (1) In the Common Criteria, a listing of the security claims for a particular IT security product. (2) A set of security functional and assurance requirements and specifications to be used as the basis for evaluating an identified product or system.

**Security test and evaluation (ST&E)** – Examination and analysis of the safeguards required to protect an IT system, as they have been applied in an operational environment, to determine the security posture of that system. Security posture is the security status of an industrial control and automation system or an enterprise's networks, information, and systems based on information assurance resources (e.g., people, hardware, software, and policies) and capabilities in place to manage the defense of the automation system or enterprise and to react as the situation changes.

**Security testing** – A process that is used to determine that the security features of a system are implemented as designed. This process includes hands-on functional testing, penetration testing, and verification.

**Sensitive information** – Information that if lost, misused, modified, or accessed by unauthorized individuals could affect the national interest, industrial automation and control system safety and operations, or IT systems.

**Sensitivity label** – A piece of information that represents the security level of an object. Sensitivity labels are used by the TCB as the basis for mandatory access control decisions.

**Sensors and actuators** – Measuring elements (sensors) and actuating elements (actuators) connected to process equipment and to the control system.

**Service roles** – The context in which a software component operates, consisting of type of service, authentication mechanism, and identities.

**Session layer** – One of the seven OSI model layers; it establishes, manages, and terminates sessions between applications.

**Shared key authentication** – A type of authentication that assumes each station has received a secret shared key through a secure channel, independent from an 802.11 network. Stations authenticate through shared knowledge of the secret key. Use of shared key authentication requires implementation of the 802.11 Wired Equivalent Privacy (WEP) algorithm.

**Simple Mail Transfer Protocol (SMTP)** – The Internet email protocol.

**Simple Network Management Protocol (SNMP)** – Internet standard protocol for managing devices on IP computer networks.

**Simple security condition** – See *simple security property.*

**Simple security property** – A Bell–LaPadula security model rule enabling a subject read access to an object only if the security level of the subject dominates the security level of the object. Synonymous with simple security condition.

**SIS** – See *safety instrumented system.*

**SLC** – See *software life cycle.*

**Social engineering –** Attacks targeting an organization's employees through the use of social skills to obtain sensitive information.

**Software assurance –** The level of confidence that software is free from vulnerabilities, either intentionally designed into the software or accidentally inserted at any time during its life cycle, and that the software functions in the intended manner.

**Software development life cycle process –** The process of using a model to translate user needs into a software product.

**Software engineering –** The science and art of specifying, designing, implementing, and evolving programs, documentation, and operating procedures whereby computers can be made useful to humans.

**Software life cycle (SLC) –** Comprises software concept, requirements analysis, architectural design, detailed design, code construction and unit testing, system testing, release, and code distribution and maintenance.

**Software process –** A set of activities, methods, and practices that are used to develop and maintain software and associated products.

**Software process capability –** The range of expected results that can be achieved by following a software process.

**Software process maturity –** The extent to which a software process is defined, managed, measured, controlled, and effective.

**Software process performance –** The result achieved by following a software process.

**Software security –** General-purpose executive, utility, or software development tools and applications programs or routines used to protect data that are handled by a system.

**Software system test and evaluation process –** A process that plans, develops, and documents the quantitative demonstration of the fulfillment of all baseline functional performance, operational, and interface requirements of a software system.

**Software weakness** – A portion of code that may lead to a vulnerability.

**Software-intensive system** – A system in which the majority of components and functionalities are implemented in software.

**SONET** – See *synchronous optical networking*.

**Source code** – A series of statements written in a human-readable computer programming language.

**Spoofing** – Attempting to gain access to a system by posing as an authorized user. Synonymous with impersonating, masquerading, or mimicking.

**SQL** – See *Structured Query Language*.

**SQL injection** – The process of an attacker inserting SQL statements into a query by exploiting vulnerability for the purpose of sending commands to a web server database.

**SSL** – See *Secure Sockets Layer*.

**SSO** – See *system security officer*.

**ST** – See *security target*.

**ST&E** – See *security test and evaluation*.

**Standalone (shared system)** – A system that is physically and electrically isolated from all other systems and intended to be used by more than one person, either simultaneously (e.g., a system that has multiple terminals) or serially, with data belonging to one user remaining available to the system while another user uses the system (e.g., a personal computer that has nonremovable storage media, such as a hard disk).

**Standalone (single-user system)** – A system that is physically and electrically isolated from all other systems and is intended to be used by one person at a time, with no data belonging to other users remaining in the system (e.g., a personal computer that has removable storage media, such as a flash drive).

**Star property (* property)** – A Bell–LaPadula security model rule giving a subject write access to an object only if the security level of the object dominates the security level of the subject. This is also called the *confinement property.*

**State variable** – A variable that represents either the state of the system or the state of some system resource.

**Structured Query Language (SQL)** – An international standard for defining and accessing relational databases.

**Subject** – An active entity, generally in the form of a person, process, or device, that causes information to flow among objects or that changes the system state.

**Subversion** – An intentional violation of software's integrity or its security controls in order to compromise the software or the using system.

**Supervisor state** – Synonymous with executive state.

**Survivability** – The capability of software or hardware to fulfill its objectives in the presence of attacks, failures, or accidents.

**Symmetric (private) key encryption** – Cryptographic system in which the sender and receiver both know a secret key that is used to encrypt and decrypt a message.

**Synchronous optical networking (SONET)** – A fiber-optic transmission system for high-speed digital traffic. SONET is part of the Broadband Integrated Services Digital Network (BISDN) standard.

**Synchronous transmission** – A type of communications data synchronization whereby frames are sent within defined time periods. It uses a clock to control the timing of bits being sent in a frame. See *asynchronous transmission*.

**System** – Any organized assembly of resources and procedures united and regulated by interaction or interdependence to accomplish a set of specific functions.

**System development life cycle (SDLC)** – The scope of activities associated with a system that encompasses the system's initiation, development and acquisition; implementation, operation and maintenance; and ultimately its disposal that instigates another system initiation.

**System development methodologies** – Methodologies developed through software engineering to manage the complexity of system development. Development methodologies include software engineering aids and high-level design analysis tools.

**System entity** – A system subject (user or process) or object.

**System integrity** – A characteristic of a system that performs its intended function in an unimpaired manner, free from deliberate or inadvertent unauthorized manipulation of the system. Also see *integrity*.

**Security mode** – The lowest security level supported by a system at a particular time or in a particular environment.

**System security officer (SSO)** – An individual having the responsibility to ensure the security of an industrial automation and control system or IT system, excluding, for example, guards, physical security personnel, law enforcement officials, and disaster recovery officials.

**System testing** – A type of testing that verifies the installation of the entire network. Testers normally complete system testing in a simulated production environment, simulating actual users in order to ensure that the network meets all stated requirements.

**Tainted input** – Input data that has not been examined or sanitized prior to use by an application.

**Tampering** – Any unauthorized modification that alters the proper functioning of equipment or a system in a manner that degrades the security or functionality that it provides.

**Target of evaluation (TOE)** – In the Common Criteria, TOE refers to the product to be tested.

**TCB** – See *trusted computing base.*

**TCP** – See *Transmission Control Protocol.*

**Technical attack** – An attack that can be perpetrated by circumventing or nullifying hardware and software protection mechanisms, rather than by subverting system personnel or other users through social engineering.

**Technical vulnerability** – A hardware, firmware, communication, or software flaw that leaves a computer processing system open for potential exploitation, either externally or internally, thereby resulting in a risk to the owner, user, or manager of the system.

**TELNET (not an acronym)** – A virtual terminal protocol used in the Internet, enabling users to log in to a remote host. TELNET is defined as part of the TCP/IP protocol suite.

**TEMPEST** – See *Transient ElectroMagnetic Pulse Emanations Standard.*

**Test case** – An executable test with a specific set of input values and a corresponding expected result.

**Threat agent** – A means of exploiting a vulnerability in a system, operation, or facility.

**Threat analysis** – The examination of all actions and events that might adversely affect a system or operation.

**Threat** – Any circumstance or event with the potential to cause harm to an industrial automation and control system or IT system in the form of destruction, disclosure, adverse modification of data, and/or denial of service.

**Threat factor** – The characterization of a threat based on a violation of confidentiality, integrity, or availability (CIA); the asset being targeted; the role of the user; and the type of accessibility. It helps refine the application context to aid in analyzing the threat.

**Threat monitoring** – The analysis, assessment, and review of audit trails and other data that are collected for the purpose of searching for system events that might constitute violations or attempted violations of system security.

**TLS** – See *Transport Layer Security.*

**TOE** – See *target of evaluation.*

**Top-level specification** – A nonprocedural description of system behavior at the most abstract level; typically, a functional specification that omits all implementation details.

**Topology** – A description of a network's geographical layout of nodes and links.

**Traceroute** – Software utility used to determine the path to a target computer.

**Transient ElectroMagnetic Pulse Emanations Standard (TEMPEST)** – The U.S. government standard for control of spurious compromising emanations emitted by electrical equipment; also used to refer to the investigation, study, and control of such emanations.

**Transmission Control Protocol (TCP)** – A commonly used protocol for establishing and maintaining communications between applications on different computers. TCP provides full-duplex, acknowledged, and flow-controlled service to upper-layer protocols and applications.

**Transmission Control Protocol/Internet Protocol (TCP/IP)** – A de facto industry-standard protocol for interconnecting disparate networks. TCP and IP are standard protocols that define both the reliable full-duplex transport level and the connectionless, best-effort unit of information passed across a network such as the Internet.

**Transport layer** – OSI model layer that provides mechanisms for the establishment, maintenance, and orderly termination of virtual circuits while shielding the higher layers from the network implementation details.

**Transport L ayer Security (TLS)** – An authentication and security protocol widely implemented in browsers and web servers.

**Trap door** – A hidden software or hardware mechanism that can be triggered to permit system protection mechanisms to be circumvented. It is activated in a manner that appears innocent—for example, a special "random" key sequence at a terminal. Software developers often introduce trap doors in their code to enable them to re-enter the system and perform certain functions.

**Trojan horse** – A computer program that has an apparent or actually useful function but contains additional (hidden) functions that surreptitiously exploit the legitimate authorizations of the invoking process to the detriment of security or integrity.

**True positive** – When a security tool reports a weakness that is actually present. See *false positive, false negative.*

**Trust** – In a relationship between entities, the confidence that each entity will behave as expected—dependably, securely, and reliably— within a specified context.

**Trusted computer system** – A system that employs sufficient hardware and software security assurance measures to enable its use for the simultaneous processing of a range of sensitive or classified information.

**Trusted computing base (TCB)** – The totality of protection mechanisms within a computer system, including hardware, firmware, and software, the combination of which is responsible for enforcing a security policy. A TCB consists of one or more components that together enforce a unified security policy over a product or system. The ability of a TCB to correctly enforce a unified security policy depends solely on the mechanisms within the TCB and on the correct input of parameters by system administrative personnel (e.g., a user's clearance level) related to the security policy.

**Trusted distribution** – A method for distributing the TCB hardware, software, and firmware components (both originals and updates), which provides the means for protecting the TCB from modification during distribution and for detecting any changes to the TCB that might occur.

**Trusted path** – A mechanism by which a person at a terminal can communicate directly with the TCB. This mechanism can be activated only by the person or by the TCB and cannot be imitated by untrusted software.

**Trusted process** – A process whose incorrect or malicious execution is capable of violating system security policy.

**Trusted software** – The software portion of the TCB.

**Trustworthiness** – The attribute of a person or organization that provides confidence to others of the qualifications, capabilities, and reliability of that entity to perform specific tasks and fulfill assigned responsibilities.

**UDP** – See *User Datagram Protocol.*

**Unit** – See *lines, units, cells.*

**Untrusted process** – A process that has not been evaluated or examined for adherence to the security policy. It might include incorrect or malicious code that attempts to circumvent the security mechanisms.

**Use case** – A technique for capturing potential functional requirements that employs the use of one or more scenarios that convey how the system should interact with the end user or another system to achieve a specific goal.

**NOTE** Use cases typically treat the system as a black box, and the interactions with the system, including system responses, are as perceived from outside the system. Use cases are popular because they simplify the description of requirements and avoid the problem of making assumptions about how this functionality will be accomplished.

**User** – (1) A person or process that is accessing an industrial automation and control system or IT system either by direct connection (e.g., via terminals) or by indirect connection (in other words, preparing input data or receiving output that is not reviewed for content or classification by a responsible individual. (2) A person or process authorized to access an industrial automation and control system or IT system.

**User Datagram Protocol (UDP)** – UDP uses the underlying Internet Protocol (IP) to transport a message. This is an unreliable, connectionless delivery scheme. It does not use acknowledgments to ensure that messages arrive and does not provide feedback to control the rate of information flow. UDP messages can be lost, duplicated, or arrive out of order. It is useful in sending high volumes of data such as streaming video.

**User ID –** A unique symbol or character string that is used by a system to identify a specific user.

**Validation –** A process for evaluating whether software will satisfy the specified requirements (functional and nonfunctional) for its intended purpose.

**Verification –** A process for evaluating whether software conforms to the specifications, regulations, or other conditions imposed on it while it was being developed.

**Verification and validation (V&V) –** All of the verification and validation activities that are undertaken to ensure that software will function according to its specification.

**Virus –** A self-replicating or self-reproducing program that spreads by inserting copies of itself into other executable code or documents.

**Vulnerability –** A weakness in an information system, system security procedures, internal controls, or implementation that could be exploited or triggered by a threat source.

**Vulnerability analysis –** A method of determining vulnerability that includes the susceptibility of a particular system to a specific attack and the opportunities that are available to a threat agent to mount that attack.

**Vulnerability assessment –** Systematic examination of an industrial automation and control system or IT system or product to determine the adequacy of security measures, identify security deficiencies, provide data from which to predict the effectiveness of proposed security measures, and confirm the adequacy of such measures after implementation.

**Weakness –** A bug found in software that has the potential of being exploited as a vulnerability when the software is operational.

**White box test –** A test in which the ethical hacking team has full knowledge of the target information system.

**White hat hacker** – An individual who conducts ethical hacking to help secure and protect an organization's information systems.

**Wide area network (WAN)** – A network that interconnects users over a wide area, usually encompassing different metropolitan areas.

**Wired equivalency privacy (WEP)** – The algorithm of the IEEE 802.11 wireless LAN standard that is used to protect transmitted information from disclosure. WEP is designed to prevent the violation of the confidentiality of data transmitted over a wireless LAN. WEP generates secret shared encryption keys that both source and destination stations use to alter frame bits to avoid disclosure to eavesdroppers. WEP is no longer considered a viable encryption mechanism due to known weaknesses.

**Wired equivalency privacy (WEP) encryption** – An encryption method specified in the IEEE 802.11 standard for wireless security. See *Wired Equivalency Privacy (WEP)*.

**Wireless** – A property of a computing device whereby it can access a network without a wired connection.

**WLAN** – Wireless local area network.

**Work factor** – An estimate of the effort or time needed by a potential intruder who has the specified expertise and resources to overcome a protective measure.

**Work function** – The difficulty of recovering plaintext from ciphertext, as measured by cost and/or time. The security of the system is directly proportional to the value of the work function. The work function need only be large enough to suffice for the intended application. If the message to be protected loses its value after a short period of time, the work function need only be large enough to ensure that the decryption would be highly infeasible in that period of time.

**Write** – A fundamental operation that results only in the flow of information from a subject to an object.

**Write access** – Permission to write to an object.

**Zeroizing** – The act of erasing sensitive parameters from a cryptographic module.

**Zeroization** – The practice of erasing sensitive parameters (especially keys) from a cryptographic module to prevent their disclosure. The keys are typically zeroized (erased) in response to an attempt to tamper with the module.

# Bibliography

ANSI/ISA-51.1-1979 (R1993). *Process Instrumentation Terminology*. Research Triangle Park, NC: ISA (International Society of Automation).

ANSI/ISA-75.05.01-2000 (R2005). *Control Valve Terminology*. Research Triangle Park, NC: ISA (International Society of Automation).

ANSI/ISA-84.00.01-2004 (IEC 61511 Mod). *Functional Safety: Safety Instrumented Systems for the Process Industry Sector*. Research Triangle Park, NC: ISA (International Society of Automation).

ANSI/ISA-88.01-2010 (IEC 62264-1 Mod). *Batch Control – Part 1: Models and Terminology*. Research Triangle Park, NC: ISA (International Society of Automation).

ANSI/ISA-95.00.01-2010. *Enterprise-Control System Integration – Part 1: Models and Terminology*. Research Triangle Park, NC: ISA (International Society of Automation).

ANSI/ISA-99.00.01-2007. *Security for Industrial Automation and Control Systems – Part 1: Terminology, Concepts and Models*. Research Triangle Park, NC: ISA (International Society of Automation).

ANSI/ISA-99.02.01-2009. *Security for Industrial Automation and Control Systems: Establishing an Industrial Automation and Control Systems Security Program*. Research Triangle Park, NC: ISA (International Society of Automation).

ANSI/ISA-TR99.00.01-2007. *Security for Industrial Automation and Control Systems, Technical Report 1: Security Technologies for Industrial Automation and Control Systems*. Research Triangle Park, NC: ISA (International Society of Automation).

ANSI/ISA-TR99.00.02-2004. *Integrating Electronic Security into the Manufacturing and Control Systems Environment*. Research Triangle Park, NC: ISA (International Society of Automation).

Bailey, D., and E. Wright. *Practical SCADA for Industry.* Milpitas, CA: IDC Technologies, 2003.

Boyer, S. *SCADA Supervisory Control and Data Acquisition, 2nd ed.* Research Triangle Park, NC: ISA (International Society of Automation), 1999.

Carnegie Mellon Software Engineering Institute. *Capability Maturity Model Integration (CMMI) for Software Engineering, v1.1.* Pittsburgh: Carnegie Mellon University, 2002.

Committee on National Security Systems (CNSS). CNSS Instruction No. 4009, "National Information Assurance (IA) Glossary." Washington, DC: CNSS, April 2010.

Erickson, K., and John L. Hedrick. *Plant Wide Process Control*. Hoboken: John Wiley & Sons, 1999.

Falco, Joe, et al. "IT Security for Industrial Control Systems." *NISTIR 6859* (NIST Interagency Report), 2003. http://www.nist.gov/customcf/get_pdf.cfm?pub_id=821684.

Federal Information Processing Standards (FIPS) PUB 140-2. "Section 2, Glossary of Terms and Acronyms." *Security Requirements for Cryptographic Modules.* Washington, DC: NIST (National Institute of Standards and Technology), 2001.

Federal Information Processing Standards (FIPS) PUB 199. *Standards for Security Categorization of Federal Information and Information Systems.* Washington, DC: NIST (National Institute of Standards and Technology), 2004.

Federal Information Processing Standards (FIPS) PUB 200. *Minimum Security Requirements for Federal Information and Information Systems.* Washington, DC: NIST (National Institute of Standards and Technology), 2006.

IEC/PAS 62409: 2005 Ed. 1.0. *Real-time Ethernet for Plant Automation.* Geneva 20 – Switzerland: IEC (International Electrotechnical Commission).

IEC/PAS 62410: 2005 Ed. 1.0. *Real-time Ethernet SERCOS III.* Geneva 20 – Switzerland: IEC (International Electrotechnical Commission). *Glossary* at http://std.iec.ch/glossary.

ISA (International Society of Automation). *The Automation, Systems, and Instrumentation Dictionary, Fourth Edition.* Research Triangle Park, NC: ISA, 2003.

ISA-dTR62443-1-2. *Security for Industrial Automation and Control Systems - Master Glossary of Terms and Abbreviations.* Research Triangle Park, NC: ISA (International Society of Automation), 2011.

ISA/IEC-TR62442-2-2. *Implementation Guidance for an IACS Security Management System.* Research Triangle Park, NC: ISA (International Society of Automation)

ANSI/ISA-62443-3-3 (99.03.03)-2013. *Security for Industrial Automation and Control Systems – Part 3-3: System Security Requirements and Security Levels.* Research Triangle Park, NC: ISA (International Society of Automation).

ISO/IEC 10746-1:1998. *Information technology – Open Distributed Processing – Reference model: Overview.* Geneva 20 – Switzerland: IEC (International Electrotechnical Commission) and ISO (International Organizations for Standardization).

ISO/IEC 15408-1:2009. *Information technology – Security techniques – Evaluation criteria for IT security – Part 1: Introduction and general model*. Geneva 20 – Switzerland: IEC (International Electrotechnical Commission) and ISO (International Organizations for Standardization).

ISO/IEC 15408-2:2008. *Information technology – Security techniques – Evaluation criteria for IT security – Part 2: Security functional requirements.* Geneva 20 – Switzerland: IEC (International Electrotechnical Commission) and ISO (International Organizations for Standardization).

ISO/IEC 15408-3:2008. *Information technology – Security techniques – Evaluation criteria for IT security – Part 3: Security assurance requirements.* Geneva 20 – Switzerland: IEC (International Electrotechnical Commission) and ISO (International Organizations for Standardization).

ISO/IEC 17799:2005. *Information technology – Security techniques – Code of practice for information security management.* Geneva 20 – Switzerland: IEC (International Electrotechnical Commission) and ISO (International Organizations for Standardization).

ISO/IEC 27001:2005. *Information technology – Security techniques – Information security management systems – Requirements.* Geneva 20 – Switzerland: IEC (International Electrotechnical Commission) and ISO (International Organizations for Standardization).

ISO/IEC 7498-2:1989. *Information processing systems – Open System Interconnection – Basic reference Model – Part 2: Security Architecture.* Geneva 20 – Switzerland: IEC (International Electrotechnical Commission) and ISO (International Organizations for Standardization).

NIST Process Control Security Requirements Forum (PCSRF). *Industrial Control System-System Protection Profile (ICS-SPP).* Version 1.0. Washington, DC: NIST (National Institute of Standards and Technology), 2004.

NIST SP 800-18. *Guide for Developing Security Plans for Federal Information Systems.* Revision 1. Washington, DC: NIST (National Institute of Standards and Technology), 2006.

NIST SP 800-28. *Guidelines on Active Content and Mobile Code.* Version 2. Washington, DC: NIST (National Institute of Standards and Technology), 2008.

NIST SP 800-30. *Risk Management Guide for Information Technology Systems.* Washington, DC: NIST (National Institute of Standards and Technology), 2002.

NIST SP 800-30. *Guide for Conducting Risk Assessments.* Revision 1. Washington, DC: NIST (National Institute of Standards and Technology), 2012.

NIST SP 800-34. *Contingency Planning Guide for Information Technology Systems.* Revision 1. Washington, DC: (National Institute of Standards and Technology), 2010.

NIST SP 800-37. *Guide for Applying the Risk Management Framework to Federal Information Systems.* Revision 1. Washington, DC: NIST (National Institute of Standards and Technology), 2010.

NIST SP 800-39. *Managing Information Security Risk: Organization, Mission, and Information System View*. Washington, DC: NIST (National Institute of Standards and Technology), 2011.

NIST SP 800-47. *Security Guide for Interconnecting Information Technology Systems*. Washington, DC: NIST (National Institute of Standards and Technology), 2002.

NIST SP 800-53. *Recommended Security Controls for Federal Information Systems*. Revision 3. Washington, DC: NIST (National Institute of Standards and Technology), 2010.

NIST SP 800-53. *Security and Privacy Controls for Federal Information Systems and Organizations*. Revision 4. Washington, DC: NIST (National Institute of Standards and Technology), 2013.

NIST SP 800-61. *Computer Security Incident Handling Guide.* Revision 2 (draft). Washington, DC: NIST (National Institute of Standards and Technology), 2012.

NIST SP 800-82. *Guide to Industrial Control Systems (ICS) Security*. Revision 2. Washington, DC: NIST (National Institute of Standards and Technology), 2015.

NIST SP 800-137. *Information Security Continuous Monitoring (ISCM) for Federal Information Systems and Organizations.* Washington, DC: NIST (National Institute of Standards and Technology), 2011.

# Index